# HUMAN SEX AND SEXUALITY

Second Revised and Enlarged Edition

by
EDWIN B. STEEN

Professor Emeritus of Biology
Western Michigan University, Kalamazoo

and

JAMES H. PRICE

Professor of Health Education
The University of Toledo, Ohio

DOVER PUBLICATIONS, INC. · New York

Published in Canada by General Publishing Company, Ltd., 30 Lesmill Road, Don Mills, Toronto, Ontario.

Published in the United Kingdom by Constable and Company, Ltd., 10 Orange Street, London WC2H 7EG.

This Dover edition, first published in 1988, is a revised and enlarged republication of the work first published by John Wiley & Sons, New York, 1977. For the Dover edition the authors have revised several portions of the original text; rewritten the Appendixes, including a greatly expanded Dictionary of Sexual Terms; and prepared a new Preface, to which they have attached a new chapter further updating portions of the first edition.

Manufactured in the United States of America
Dover Publications, Inc., 31 East 2nd Street, Mineola, N.Y. 11501

*Library of Congress Cataloging-in-Publication Data*

Steen, Edwin Benzel, 1901–
    Human sex and sexuality.

    Bibliography: p.
    Includes index.
    1. Sex. 2. Family. 3. Hygiene, Sexual. I. Price, James H. (James Harold), 1943–    . II. Title. [DNLM: 1. Reproduction. 2. Sex. 3. Sex Behavior. 4. Sex Education. HQ 34 S814h]
HQ31.S787 1987          613.9                87-27526
ISBN 0-486-25544-1 (pbk.)

# PREFACE TO THE DOVER EDITION

This book is a revised edition of our original work. Since its publication in 1977 a wide variety of events in the field of human sexuality have occurred and been reported in the scientific literature. A comprehensive rewriting of this book, however, was deemed unnecessary, since most of our original work is still up-to-date and was written as an overview of the primary topics of human sexuality, not as an all-inclusive tome. Thus, our major goal in this revised edition is to update our original work as necessary without rewriting it altogether.

In numerous areas throughout the book, statistics have been updated and small sections have been rewritten to reflect current thinking. At times, however, new material could not be easily added directly to the original text. Thus, we have attached to this preface a new chapter updating the original edition. Included are nine new sections on the following topics: toxic shock syndrome, the contraceptive sponge, the contraceptive pill for men, sexual variations (urophilia, coprophilia, urethralism, klismaphilia, and apotemnophilia), AIDS, chlamydial infections, viral hepatitis, Molluscum contagiosum, and, finally, sexual harassment. The reader may either read these new sections at once or else skip them for now and return to them after reading the sections in the original text to which they are related.

In closing we would like to express our thanks to Mr. Hayward Cirker, the president of Dover Publications, Inc., for having the insight to reprint our book in an updated version. We hope that this book, with its greatly expanded dictionary, will be a useful aid in assisting those who want to learn more about the field of human sexuality.

**EDWIN B. STEEN**
**JAMES H. PRICE**

# HUMAN SEX AND SEXUALITY—NEW DEVELOPMENTS

## TOXIC SHOCK SYNDROME (TSS)

Toxic shock syndrome is a condition characterized by high fever, sore throat, headache, vomiting, diarrhea, and skin rash, followed by peeling of the skin. It was first reported in 1978 in a group of children, boys and girls aged 8 to 17 years. Untreated, it may rapidly progress to severe and intractable shock resulting in cardiovascular collapse and death. TSS is caused by a toxin (poison) produced by a species of bacteria, *Staphylococcus aureus,* normally found on or in the body (skin, mouth, nose). It is found in the vagina of 2 to 15 percent of all women. In women affected with toxic shock syndrome, it is present in the vagina in almost 100 percent of the cases. This indicates that it is most likely to be the causative agent.

In almost all cases in menstruating women, it has been found that they were using vaginal tampons. It is thought that tampons cause TSS in three ways: (1) they create a favorable environment for the growth of bacteria in the vaginal canal; (2) they may absorb selected micronutrients from the vaginal canal, causing a disturbance in the natural defense mechanisms of the vagina; (3) they may rub against the vaginal wall and cause microlacerations through which bacteria can gain entrance into the bloodstream. The bacteria might also gain entry to the peritoneal cavity via the uterus.

A typical victim of TSS is a young woman of reproductive age (usually under 30) who uses tampons to absorb the menses. Ninety-five percent of TSS cases are women. The death rate for TSS is approximately 8 to 10 percent. Not all tampons have the same incidence of association with TSS. The Rely brand of tampons (both regular and super) had a substantially greater rate of TSS than other brands. Rely tampons were withdrawn from the market by the company. Prevention of TSS would entail avoiding the use of tampons throughout the menstrual period, and switching to the use of napkins or minipads. Tampons promoted for high absorbency are especially to be avoided. The effectiveness of frequently changing tampons on the rate of TSS is unknown.

## THE CONTRACEPTIVE SPONGE

The newest contraceptive device is the contraceptive sponge. The sponge is made of polyurethane, which is impregnated with a spermicidal substance. The sponge is sold as an over-the-counter product; thus it is not necessary to see a physician for fitting or for a prescription.

The contraceptive sponge is inserted into the vagina before intercourse. Before insertion, the sponge must be moistened with water to activate the spermicide. The sponge acts as a spermicidal agent, as a mechanical barrier, and as an absorbing device to trap sperm. The effectiveness of the sponge is thought to be the same as that of the diaphragm.

Advantages of the sponge include that it can be inserted up to 18 hours before intercourse and can be left in place for 24 hours. A second round of intercourse does not require a replacement of the sponge. The sponge is inexpensive and disposable. A disadvantage of the sponge is that about 5 percent of its users claim a mild irritation with regular use. A more serious, yet rare, side effect is toxic shock syndrome. Women who might be susceptible to toxic shock syndrome should not use the contraceptive sponge.

## THE CONTRACEPTIVE PILL FOR MEN

A birth-control pill for men has recently been developed in China. Its active ingredient is *gossypol,* a derivative of cottonseed oil. It acts by blocking or greatly reducing sperm production and it does not apparently affect hormone production or diminish sexual desire. The Republic of China is using gossypol as a male contraceptive pill. The pill is taken daily for three months followed by a maintenance pill twice a week thereafter. Side effects, noted in 3 to 10 percent of men, include fatigue, reduced appetite, dizziness, digestive disorders, decreased libido and potency. These are reduced on maintenance dosage. The pill appears to be 99-percent effective and its effects are reversible. Studies are now being conducted in the United States to test its effectiveness and long-term safety.

## SEXUAL VARIATIONS

*Urophilia.* Urophilia, also known as urolagnia, is sexual arousal and interest in urine and urination. The interest may take a variety of forms, including watching others urinate, urinating on others, being urinated on (known as "golden showers"), or even drinking urine. This form of sexual variation is frequently associated with sadomasochistic activity and represents a form of domination.

*Coprophilia.* Coprophilia is sexual arousal through interest in feces and bowel movements. Coprophilic activity may include watching others have bowel movements, having bowel movements on others, having others defecate on one's self, or simply coming in contact with feces. A special form of this behavior is the eating of feces, known as *coprophagia.*

*Urethralism.* Urethralism is sexual arousal from the insertion of objects into one's urethra. This behavior is common in children, but with them it is not done for sexual stimulation but as a form of body experimentation. Rarely the object may be forced up into the bladder. Physicians have removed a wide variety of objects from urinary bladders as a result of urethralism.

*Klismaphilia.* Klismaphilia is sexual arousal through interest in enemas. The interest may take the form of administering an enema to another or having an enema administered to one's self. Sexual gratification is frequently obtained by hiring prostitutes to engage in the behavior (known as "water sports").

*Apotemnophilia.* Apotemnophilia, also known as *amputism,* is sexual arousal through interest in amputations. The interest may take one of two forms. The first form involves a strong sexual desire to have sexual activity with a person with an amputation. The second form is self-amputism, in which the individual may actually fake illness in order to have a finger or limb amputated. More often, the self-amputism takes the form of frequent sexual fantasies involving having an amputation.


## ACQUIRED IMMUNE DEFICIENCY SYNDROME (AIDS)

The newest sexually transmitted disease, and one of the most frightening, is a viral infection known as *acquired immune deficiency syndrome* (also known by the acronym AIDS). AIDS is thought to be an infection caused by a virus known as HTLV-III (human T-cell lymphotropic virus type III). The virus is thought to cause a breakdown in the immune system that normally helps to protect the body against infections. AIDS victims, because of their impaired immune system, are likely to develop a variety of opportunistic infections (usually pneumonia) that are typically found only in patients being treated for cancer or with transplants. The drugs used to treat these patients lower their resistance to infections by impairing their immune systems. Massive infections in AIDS patients or the development of a rare form of cancer called *Kaposi's sarcoma* are the usual causes of death. So far 50 percent of those having AIDS have died. Of those who have had the disease for 2 years, 75 percent have died, while AIDS has proved fatal in all patients who have had it for 5 years.

The disease was first reported in the summer of 1981, 5 cases in Los Angeles. Since that initial year there has been a doubling of the cases every 6 months so that by the end of 1984 there were almost 10,000 confirmed cases of AIDS. By the beginning of 1987 there were 30,000 cases confirmed, and if the current rate continues there may be several million cases by 1990. Ninety percent of the adult patients are 20 to 49 years old. Sixty percent of the AIDS patients are white; 25 percent are black; and 14 percent are Hispanic. Ninety-four percent of the patients are men. Of the initial 1,000 cases, 73 percent were homosexual or bisexual men, 16 percent were intravenous drug abusers, 5 percent Haitian men, and 1 percent hemophiliacs. Since the initial reporting, Haitians are no longer considered to be a high-risk group because many of these men were found to have homosexual contacts or to be intravenous drug users. However, it has recently been shown that it is possible to acquire AIDS by heterosexual activity (usually with prostitutes). Thus, it has been suggested that heterosexuals who have a variety of different partners (i.e., 10–12 different partners per year) may be a high-risk group. Over 100 children have acquired AIDS. Seventy-two percent of these children came from families in which one or both parents had AIDS or were in the high-risk groups, another 18 percent had received blood transfusions or were hemophiliacs who needed blood components. This further confirms that the heterosexual transmission may not necessarily be through intercourse, but may be through contact with other body fluids (possibly saliva).

The incubation period for AIDS has been variously estimated to be from 6 months to 5 years. Symptoms of the disease are variable and may include progressive unexplained weight loss, a persistent fever, swollen lymph nodes, repeated infections, lack of energy, and possibly reddish-purple spots on the skin about the size of a coin (common in Kaposi's sarcoma). As of 1987 there has been no successful treatment developed for AIDS. So far the cost of AIDS to the nation has been billions of dollars in medical care and loss of income.

## CHLAMYDIAL INFECTIONS

These infections are caused by a species of bacteria, *Chlamydia trachomatis,* an intracellular parasite acquired by humans through close intimate physical contact. Public health officials have been so concerned about the threat of AIDS that they have virtually ignored the nation's growing epidemic of chlamydia infections. It is believed that chlamydia infections are the most common of all STD's, estimated at from 4 to 10 million new cases each year. The actual number of infections is unknown but chlamydia is believed to be 10 times as common as herpes and twice as common as gonorrhea.

*Chlamydia trachomatis* is responsible for from one-third to two-thirds of pelvic inflammatory diseases (PID) in women and if untreated will cause sterility. The infection may also cause sterility in men. Unfortunately, three-fourths of the women and 10 percent of the men with chlamydia will have no symptoms with their infections. It has been found that the highest rates of chlamydial infections occur in 20- to 24-year-olds, in blacks, and in those with multiple sexual partners.

Chlamydia infections can be easily treated with antibiotics such as erythromycin and tetracycline. However, because of the high rate of asymptomatic infections it would be better to attempt to prevent acquiring the infection. Prevention of chlamydia infections would include limiting the number of sexual partners, the use of barrier contraceptives, such as condoms and diaphragms, and the use of spermicides along with other contraceptives to help reduce transmission.

## VIRAL HEPATITIS

Viral hepatitis is a viral infection of the liver that can vary radically in severity. It may be entirely symptomless, cause acute debilitating jaundice, or even result in death (in about 1 percent of those hospitalized). While there are three forms of viral hepatitis, only two forms are thought to be sexually transmitted: hepatitis A and hepatitis B.

Hepatitis A, formerly called infectious hepatitis, has an incubation period of approximately 2 to 7 weeks. It is spread by either direct or indirect contact with feces (e.g., through oral-anal contact or oral-penile contact after anal intercourse). Thus, homosexual males have a higher rate of hepatitis A infections than heterosexual males.

Hepatitis B, formerly known as serum hepatitis, can be spread by blood or other biological fluids such as saliva, seminal fluids, and vaginal secretions. Hepatitis B has an incubation period of approximately 4 to 16 weeks. It is thought that as much as 40 percent of the 200,000 annual cases are sexually transmitted. Homosexual men have the highest rates of occurrence of any subsample of the population. Almost 10 percent of the infected patients become chronic carriers of the virus, which may last months, years, or a lifetime. Carriers typically show no symptoms but serve as sources of new infections.

Hepatitis is diagnosed utilizing blood tests. Treatment is usually given according to symptoms and hospitalization is required only in severe cases. Hepatitis B patients are at increased risk of developing cancer of the liver. A safe and effective vaccine is available to immunize individuals against viral hepatitis B.

## MOLLUSCUM CONTAGIOSUM

Molluscum contagiosum is a chronic disease of the skin characterized by the formation of smooth, waxy papules 2 to 10 mm in diameter. They may be sessile or pedunculated, and occur most generally in the genital or pubic area. The papules contain a white, semisolid material with large cells that contain inclusion bodies. The condition is caused by a virus that is transmitted by direct contact.

## SEXUAL HARASSMENT

Sexual harassment is defined as unwanted sexual advances and requests for sexual favors that create discomfort and/or interfere with normal activities. Sexual harassment can occur anywhere, even on the street, but is most likely to occur in the workplace and educational settings. Whether a condition can be considered sexual harassment depends to a great extent upon the perceptions of the individuals involved. Sexual harassment exists on a continuum, ranging from catcalls and whistles, to leers and staring at a woman's chest, to outright demands for physical intimacy in exchange for promotion, retention of a job, or higher grades in a class.

Recent studies have found sexual harassment to be very widespread. Women are more likely than men to be victims of sexual harassment. A survey of federal employees found that 42 percent of the women and 15 percent of the men surveyed had been sexually harassed. A national survey of psychologists found that almost 17 percent of the women who responded and 3 percent of the men had experienced sexual contact with their teachers when they were graduate students. While it is possible for faculty-student affairs to be mutually agreed upon, it must be recognized that the professor has subtle (or overt) power over the student's grade and possibly future academic career.

The effects of sexual harassment on the victim may include psychological, economic, and physical consequences. The victim of harassment may feel anger, humiliation, embarrassment, degradation, and helplessness. The feelings of helplessness and isolation may cause the victim to comply with the sexual demands. Sexual demands are sometimes complied with, not because the victim feels flattered, but because he or she feels threatened. Since jobs are not always easy to find, many women, especially those who are single parents, are more likely to put up with the harassment. Others may fear being fired and suffering the economic consequences; thus they too may give in to the advances. Finally, physical problems may develop from the stress of chronic sexual harassment, or such sexual problems as sexually transmitted diseases could result from these unwanted encounters.

In 1980 the Equal Employment Opportunity Commission issued guidelines emphasizing that both verbal and physical harassment are illegal. A company that does not take immediate action against a supervisor who is involved in sexual harassment is legally liable. A company is equally liable if it coerces a worker into providing sexual services to clients. A person sexually harassed can file an official complaint with the city or state Human Rights Commission or the local office of the federal Equal Employment Opportunity Commission.

# PREFACE TO THE FIRST EDITION

During the past 10 or 15 years, the period of the so-called "Sexual Revolution," profound changes have taken place in attitudes toward sex and sexual behavior among both the younger and older generations. Standards in sexual behavior have been changing and there is an increased interest in the general subject of sex and sexuality. Programs in sex education are now being initiated in schools at nearly all levels and there has been a flood of books dealing with the subject. These books seek to satisfy the almost insatiable demand for information on or about all aspects of sex.

In this book, we present the basic facts pertaining to the development of sex and sexuality, hoping that a better understanding of its various manifestations will be acquired. We summarize the views on how sex is determined and how the sexuality of an individual develops and we explain the various types of sexual activities including both what is considered "normal" and variant behavior. Since sex and reproduction are inseparable, we give a detailed description of the male and female reproductive systems and explain the processes of pregnancy, childbirth, and associated phenomena. The sexual act, the responses of the male and female, methods of contraception, abortion, venereal disease, and other related topics are covered. On all subjects, we tried to present the latest findings and developments.

We give special attention to topics that seem to especially concern young people today: sexual dysfunction, sperm banks, teenage pregnancy, rape, drugs and sex, nudism and nudity, obscenity and pornography, sex and the law, sex and aging, and other related topics. Since much misinformation prevails, a section dealing with fallacies and misconceptions concerning sex has been included, that hopefully, will allay needless fears and worries pertaining to various matters of concern, especially to the young or ill informed.

Since an understanding of any subject requires a thorough familiarity with the terminology employed, a complete dictionary of terms pertaining to sex, reproduction, and development is included. In it are listed slang terms and terms used in everyday language. Finally, we provide a comprehensive list of references.

EDWIN B. STEEN
JAMES H. PRICE

# ACKNOWLEDGMENTS

We express thanks and appreciation to the following people for their contributions in the preparation of this text. First, to the many authors and researchers whose discoveries and writings provided the basic information from which this work was prepared. Since the sources of information were numerous and varied, individual acknowledgment is impractical. Second, to Dr. Gordon Duncan and Dr. R. J. Ericsson, former Adjunct Professors of Biology, Western Michigan University for their stimulating course in reproductive physiology, which incorporated new ideas and concepts resulting from their research with the Upjohn Company, Kalamazoo, Michigan, and Dr. Frederick P. Gault, Professor of Psychology, Western Michigan University, for helpful suggestions and comments.

Finally we are grateful to the staff of John Wiley & Sons, especially Andrew Ford and Malcolm Easterlin, for their overall supervision in preparation of the manuscript; Susan Giniger, for her careful editing; and Terrie Harmsen, for directing the work through production stages. We also appreciate the work of reviewers, especially Dr. J. J. Schifferes and Professor Robert Synovitz, for their comments and suggestions. To all members of the John Wiley staff who had a part in the production of this book, we express our appreciation.

E. B. S.
J. H. P.

# CONTENTS

# ONE
# THE DEVELOPMENT
# OF HUMAN SEXUALITY

## SEX AND SEXUALITY

The word "sex" is commonly used to refer to the genital organs and the activities in which they are involved, especially the production of offspring. However, "sex" involves much more than procreation. *Sex,* or the broad term *sexuality,* involves an individual's entire personality. It involves the identification of a person with a gender (males as distinguished from females), and with that identification, a person develops feelings, attitudes, and behaviors that are appropriate for that sex. He or she will learn to think, feel, and act as a male or female and, as such, will influence and be influenced by everyone with whom he or she comes into contact socially. Every aspect of behavior will in some respect be involved. It will determine one's life-style or manner of living; one's feelings and reactions to others of his own or the opposite sex. In other words, sexuality involves the entire person.

Recent studies in the anatomy and physiology of reproduction and in the psychology of sexual behavior have brought out the fact that a person's sexuality results from many factors and conditions that act at different times in one's lifetime. The nature of each individual is determined by two types of influences, *genetic* and *environmental.*

Genetic or hereditary factors are inherent within the germ cells, the egg and sperm. These determine the basic structure, mode of development, and the functioning of the organ systems. Superimposed on these are environmental factors. These include all the influences, physical and chemical, which act on a developing embryo or fetus while it is growing within the mother and those conditions that act on a newborn child from birth to maturity and throughout one's entire life. These two forces will determine the way an individual develops and his behavior within our society.

Basic sex, whether male or female (referred to as *genetic* or *chromosomal sex*), depends on the chromosomal content of an individual. Chromosomes are self-duplicating bodies present in every cell. They are of primary importance in that they contain the hereditary determiners or genes and they are responsible for the determination of sex. Every species of animal or plant possesses a characteristic number of chromosomes, the number in humans being 46 (23 pairs). A single chromosome of each pair is received from each parent at conception. The pairs of chromosomes are essentially alike in both sexes except for one pair, the *sex chromosomes*. In the female, this pair is designated XX, in the male XY.

Genetic sex is determined at the time of conception when the egg is fertilized (Fig. 1). Eggs and sperm possess only one half of the normal chromosome complement. All eggs contain one X chromosome. Sperm are of two types: one half contain an X chromosome and one half a Y chromosome. If an X-bearing sperm fertilizes an X-bearing egg, the resulting zygote will contain two X chromosomes and will develop into a female (XX). If a Y-bearing sperm fertilizes an X-bearing egg, the resulting zygote will develop into a male (XY). Anomalies sometimes occur in chromosome numbers. An individual may possess too many or too few, or chromosomes

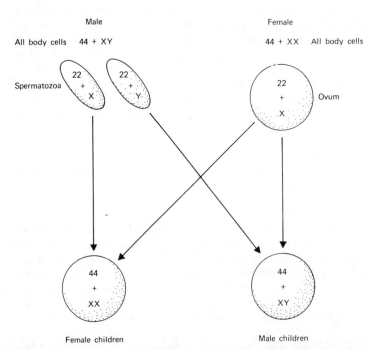

**Figure 1    Sex determination in humans. All eggs contain an X chromosome. One half of the sperm contain an X chromosome, one half contain a Y chromosome. Sex of child depends on which type of sperm fertilizes the egg.**

may be defective. Chromosomes may be altered by environmental factors, as irradiation. When this occurs, developmental processes are usually affected with serious effects on the developing individual.

Gonadal or biologic sex is that which is determined by the effect of hormones on the body. The gonads not only produce reproductive cells but they also serve as endocrine organs elaborating hormones. Hormones are chemical substances produced by glands lacking ducts. These substances are released into and distributed by the blood stream to all parts of the body where they exert their effects sometimes remote from the place of origin. Sex hormones include androgens or male hormones and estrogens and progesterone, female hormones. These are important in the development of accessory sex organs and are responsible for the development of secondary sex characteristics such as body form and appearance—overt characteristics that are generally used in identifying a person as a male or female. Disorders in the production of hormones may result in the virilization of women or the feminization of men.

## EARLY SEXUAL DEVELOPMENT

Sexual identity or gender identity (core gender) is the development of a sense of masculinity or femininity, that is, acquiring a male or female identity. This depends principally on sex assignment (being designated male or female) and rearing. At birth, a child is potentially capable of acquiring either a male or female identity regardless of his or her gonadal sex: Sexual identity is developed gradually but becomes well established during the first three years of childhood.

Sometimes mistakes are made in the assignment of sex to a newborn baby. Structural abnormalities may occur due to the nature of the sex chromosomes or from the effects of hormones acting upon various structures during intrauterine development. As a result, the reproductive organs may not be readily distinguished as entirely male or female. In the female, a large clitoris may be mistaken for a small penis and in the male, a small opening on the under side of the penis (hypospadias) may be mistaken for a vaginal opening. As a consequence a mistake may be made in the diagnosis of sex. When such occurs, the sex of assignment and rearing usually prevails since the child is capable of acquiring either a male or female identity. When changes in gender assignment are made, if such occurs by the age of three, few difficulties of a psychological nature are noticed, but after the age of three, the disturbances are pronounced.

By the age of three, and usually earlier, most children can correctly identify themselves as being a boy or girl and within the next year or two, almost all can apply the appropriate label to others. By the age of five, most children perceive the major physical and behavioral differences between parents and have learned or acquired the basic sex role stereotypes of their society.

How do children learn these sex roles? *Sex role* is the acquisition of behavioral characteristics that are consistent with a person's biological makeup, that is, masculine behavior in a male and feminine behavior in a female. What makes a child prefer one sex to another? What motivates a young boy to want to grow up to be a "daddy" and a little girl to want to be a "mommy"? Psychologists are trying to answer these questions. There have been and still remain differences in opinion as to the determining factors in the development of masculinity or femininity in a child. The three principal views as to the development of gender identity are the psychoanalytic or Freudian theory and the social-learning and cognitive-developmental theories held by developmental psychologists.

The *psychoanalytic view* of gender identity is based on the work of Sigmund Freud (1856–1939), well known for his "libido theory" of sexuality and the part played by the libido in psychoneuroses. He believed in the existence of a sexual instinct that manifested itself both bodily and mentally. Libido was regarded as the force by which the sexual instinct was expressed. He regarded all individuals as being basically bisexual with the development of the male and female components of each sex dependent on basic instinctual drives and subsequent relationships with parents.

Freud believed that sexuality developed gradually from birth. He considered an infant as capable of experiencing erotic sensations and that the sexual instinct developed gradually during the first five or six years with the most critical phases of psychosexual development and differentiation occurring during this period. Many nonsexual bodily functions as feeding and defecation were regarded as manifestations of sexual feelings (auto-eroticism), the oral and anal regions thus constituting erotogenic zones. He designated the period from birth through the first three years as constituting the *oral* and *anal stages*.

The *phallic stage* follows the oral and anal stages characterized by sexual curiosity during which the child becomes aware of the anatomical differences between boys and girls. The boy becomes aware that he possesses a penis and girls that they lack such an organ. Since males at this age may engage in masturbatory activity, they may develop fear of losing this organ that gives them so much pleasure, causing what Freud called a "castration complex." In the female, lack of this organ may cause "penis envy." During this period close attachments to parents may develop, the boy tending to love his mother and reject his father (the Oedipus complex) and the girl may devote her love to her father (Electra feelings). A common idea held by children during these years is that birth occurs by way of the bowel. Also sexual relations sometimes observed between parents may be interpreted as being cruel and sadistic activities on the part of the father.

Following the phallic stage from about the age of six to puberty is a *latent period* during which sexual development is relatively stationary. The erotic feelings that a child might have for a particular person tend to die down and do not return until puberty. Sexual activity is not entirely dormant since some sexual experimentation may take place. With the advent of *puberty,* the final stages of genital development are attained with a variety of sexual activities and habits being manifested. These activities, both those considered normal and those classified as deviations are considered to be the result of one's inborn sexual constitution and experiences encountered, especially those of early life. Table 1 lists the stages in sexual development according to the Freudian theory with some characteristics of each.

Freud believed that the parent was more than merely a model for the initiation of sex-appropriate behavior. Identification rather than imitation was thought to occur. Identification involves the acquisition of many aspects of the model's personality and eventual development into a superego or conscience. He theorized that excessive frustration of sexual needs at any of the three stages could result in fixation at that stage or regression into an earlier psychological stage and that this accounted for most adult neuroses. He also regarded most perversions or deviations in adult

**Table 1   Stages in sexual development according to Freudian theory**

| STAGE | AGE | CHARACTERISTIC ACTIVITIES |
|---|---|---|
| Oral stage | Birth to one year | Pleasurable activities involve the mouth (sucking, eating, chewing, biting, spitting). |
| Anal stage | One year to three years | Pleasure centers about retention and expulsion of feces. Proper toilet training of importance; should neither be too permissive nor too strict. |
| Phallic stage | Three years to six years | Pleasure centers about genital organs. Oedipus complex and castration anxiety develop in males; Electra complex and penis envy in females. |
| Latency period | Six years to puberty | Oedipal and Electral complexes are resolved; child identifies with parent of the same sex. Erotic impulses toward opposite sex repressed. |
| Genital stage | Adolescence | Heterosexual love and sexuality develop replacing egocentric and incestous love. Sex drives are channeled into group activities and preparation for work and marriage. |

(Table modified from *Life and Health,* 1972. CRM, Inc. By permission of Random House, Inc.

behavior as reversions to an infantile type of activity, deviations falling into two categories, deviations in the *object* of sexual instinct and deviations in the *aim*.

Today, few psychologists accept all of Freud's views and opinions; however, many have accepted the general concept of identification but have suggested that this may be motivated by envy of the model's authority and competence.

The *social-learning concept* of the development of gender identity emphasizes the importance of learning. It is the view most commonly held to explain sex role development. It holds that children come to behave in sex-appropriate ways because they are actively taught to do so by their parents, siblings, peers, teachers, and the mass media.

In our society, boys are given boyish names, dressed in boyish clothes and encouraged to act as "boys." Activities in sports are encouraged and aggressive behavior observed. He is taught not to cry when hurt and to avoid any activity that would put him in the class of being a "sissy." He is taught to act like "daddy."

From the social-learning perspective, sex roles, like most behavior, are controlled by their consequences. Actions that are approved and bring rewards tend to recur; those that bring on punishment or disapproval tend to be avoided and disappear. Imitation plays an important role in the learning process. A child learns sex-appropriate behavior by observing models and, in this context, the parent of the same sex as the child and with whom the child can identify, assumes a role of primary importance.

In the *cognitive-developmental approach,* intellectual development is considered the key to gender identity. This concept has been developed from observations made on the relation of mental maturity to sex role development. By the ages of five or six, children consistently attribute power and prestige (aggression and exposure to danger) to males and homemaking and child care to females. Such stereotyping develops early and is more pronounced between the ages of five and eight than later. This suggests that the stereotypes are not acquired through direct learning but rather are created mentally by the child for himself on the basis of male-female differences in body build and in extrafamilial society roles.

In other words, by the age of two or three, children, because of increasing mental maturity, can apply the proper sex label, male or female, to themselves. They then slowly learn to apply the labels to others usually on the basis of such external clues as clothing. By the age of five or six, their beliefs about their own gender identity have become permanently established. Once children have labeled themselves, they are motivated to behave in ways that are consistent with their labels. They come to value things and activities that are associated with being a boy or girl. They recognize that they are boys or girls and therefore want to do those things that are done by their sex and to get approval for doing them. This, it should be noted, is the reverse of the social-learning concept, which states that a child primarily wants rewards, therefore when the child is rewarded for doing things appropriate for his or her sex, the child then wants to be of that sex.

These, briefly, are the theories of psychosexual development. All hold that gender development occurs over a period of time and does not result from a single or unexpected sexual event. There is evidence that most children engage in some type of sexual activity (masturbation, exploratory games, and in some cases coition or homosexual activity) between the ages of 6 and 12 and few show any adverse psychological effects from such. The single traumatic experience that ruins a child for life is rare, especially during later childhood. The effects of such an experience as rape, incest, or homosexual experiences probably depends on the healthiness or unhealthiness of the child's prior teaching and upbringing.

The importance of the role parents play in the establishment of proper sex identity is now generally recognized. Their attitudes and behavior toward their children and their relationships to each other and to their associates influence profoundly attitudes and behavior of their children. The weak father, or the excessively domineering father; the overprotective or seductive mother are often encountered in cases of homosexuality, transsexualism, and other types of variant behavior. In the field of sex education, the proper presentation of sex information and an understanding of the sexuality of their children contributes much to the development of proper gender identity that is essential for normal and complete sexual relationships as adults.

## STAGES IN SEXUAL DEVELOPMENT
Sexual development in humans encompasses the entire span of life. It comprises four periods: infancy and childhood, puberty and adolescence, adulthood (early and middle), and late adulthood (menopause and aging). A summary of the changes that take place during each of these periods follows.

### Infancy and Childhood
Children are capable of responding sexually at an early age, infants in both sexes experiencing pleasure from stimulation of the genital organs or parts of the body considered to be erogenous zones. Male infants experience erections at an early age and children are known to experience orgasmic reactions throughout their preadolescent period. Young children usually engage in genital play. Most manipulate their own genitalia and sometimes mutual play is engaged in.

In their early years, children pay little attention to the sex of their playmates, gender being largely ignored. At about the age of five or six, they tend to associate principally with members of their own sex, boys with boys and girls with girls. Sometimes overt hostility to the opposite sex is shown. During this time, the gender role is being refined. Boys are engaging in activities commonly associated with masculinity. They are being taught to be boys and little men. At the same time, girls are engaging in feminine activities. They are being taught to be girls or little women.

Children learn about sex in various ways. They learn from other children and they experiment in various ways with both their own bodies and those of other children. Among boys, manipulation of the genital organs is common and when it is discovered by parents, their disapproval accompanied by stern warnings is often the child's first lesson from his parents in sex education. A child learns much from his parents. Their attitudes toward sex are revealed by their reactions and answers to his or her questions, their behavior toward one another and their willingness or their reluctance to discuss matters pertaining to sex in an open manner without embarassment. In general, parental attitudes toward sex are reflected in the behavior of their children.

### Puberty and Adolescence

As children approach puberty, the period during which the reproductive organs become functional, marked physical and emotional changes occur. These changes are indicators of the maturational progress taking place in which the child is developing into an adult. In girls, the most obvious physical changes are as follows. The breasts begin to enlarge, height increases, pubic hair makes its appearance, the menarche or first menstruation occurs, and axillary hair begins to grow. Comparable changes occur in boys. The penis and testes increase in size, pubic hair makes its appearance, the voice begins to change, the first ejaculations occur, axillary hair and the beard begin to grow. In both sexes there is the development of sexual feelings and longings and their interest in individuals of the same sex is replaced by interest in the opposite sex.

Problems often develop because the average child does not understand these changes and usually receives little or inadequate information concerning them. Girls, unless prepared for the onset of the first menstruation, may be unduly disturbed. Excessive or underdevelopment of the breasts are matters of common concern. Boys may be frightened at the first ejaculation or be concerned over the size of the penis (too large or too small). Excessive concern over masturbatory activity by both boys and girls is common. In general, these concerns are transitory and a proper understanding of their significance obtained from informed parents, teachers, or counselors does much toward the development of proper sexual attitudes for the ensuing years.

Puberty, which occurs between the ages of 10 and 16, occurs earlier in girls than in boys. In girls, first signs occur between the ages of 10 and 14 (average 11) and in boys between the ages of 12 and 16 (average 13). The average age of the onset of puberty has been steadily declining during recent decades, a phenomenon that is not understood and its causes are uncertain.

The most significant physiological changes that occur at puberty in the female are the development of the ovaries and the production of ova and sex hormones (estrogens and progesterone). In the male comparable changes are the development of the testes and the production of spermatozoa and sex hormones (androgens, principally testosterone). The production of sex hormones is the primary factor responsible for the development of secondary sex characteristics. Incidentally, some androgens are produced by the ovary and some estrogens by the testes. The adrenal gland is also the source of sex hormones in both males and females.

Although theoretically, girls on reaching puberty are capable of conceiving and giving birth to a baby, they do so relatively infrequently. This is because during the first and succeeding menstrual cycles, although follicles are developed in the ovary, eggs are not liberated. These are called *anovulatory cycles.* This usually continues for a period of two or three years (to about age 13) and constitutes a *sterility period.* Failure of the follicles to release ova is due to an imbalance in pituitary and ovarian hormones. As the cyclic action of the pituitary gland and the ovaries becomes synchronized and well established, a period of *nubility* (suitability for marriage) is attained. Ovulation begins to occur regularly and impregnation becomes possible, as is indicated by the number of teenage pregnancies. However the optimal time for childbearing is not reached until the age of 23 at which time, and up to the age of 29, mothers have the highest percentage of healthy, normal babies. Young mothers below the age of 23 have a greater likelihood of bearing premature or stillborn babies or babies that have not developed properly.

Sterility in early adolescence applies to the male as well as the female. In the first seminal emissions, sperm may be absent or few in number. Hormonal factors comparable to those of the female are of importance in the development of seminiferous tubules in the testes and the production of an adequate number of viable sperm.

*Adolescence* is the period beginning with puberty and extending through the teen years to adulthood. It is the period during which a girl develops into a mature woman and a boy into a man. It is usually a period of stress. Although the adolescent is capable of reproduction, he or she is not fully mature, either physically, mentally, or emotionally. Self-awareness and its expression are characteristics frequently resulting in conflicts between the adolescent and his parents. An adolescent develops a need for self-identity and personal freedom. The parent is usually reluctant to grant such. During adolescence, the child must give up his childish love for his parents and he or she cannot serve as a love object of his parents. To bring about this change requires a great deal of understanding on the part of both.

Masturbation is a common practice among adolescents, recent studies revealing that about 50 percent of adolescents admit to having engaged in it. It is more common among boys than girls, but girls usually begin the practice earlier than boys and engage in it less frequently. It may be practiced for various reasons as frustration or inability to engage in sexual activities or failure to have an orgasm during petting or intercourse. It may be engaged in as a substitute for, a method of avoidance, or to lessen the desire for intimate sexual relations. Following masturbation, most individuals experience a feeling of physical relief and release from sexual tension.

Old fears and superstitions about the possible ill effects of masturbation seem to have largely disappeared, and feelings of anxiety and guilt seem to be minimal, especially among the better-educated groups of our society. It is now considered to be a more or less common activity and a normal response in sexual development that does not result in either physical or mental injury.

During the teen years, most adolescents begin to engage in heterosexual activity. Going steady or falling in love takes place. This is usually accompanied by light petting (embracing, caressing, kissing), or heavy petting (fondling breasts or genital organs). This frequently results in going the limit or engaging in sexual intercourse, which occurs in about 50 percent of all adolescents. The first act of intercourse is of significance in a number of ways, varying with the individual and the age at which it occurs. For boys it is usually sought for with anticipation of pleasure and its consummation acts to build up feelings of accomplishment and self-esteem. Girls usually do not actively seek intercourse but after engaging in it may have feelings of maturity and especially the feelings of being experienced and more knowledgeable in sexual matters. Occasionally feelings of guilt and disappointment result, occurring more commonly in boys than in girls.

Relationships of adolescents with their parents are frequently difficult. Much of the problem arises from the inability of parents to communicate freely with their children about sexual matters. Sex education of their offspring is one of the worst-managed duties of parenthood. The subject is too often avoided entirely or, if it is handled at all, it is on a "too little, too late" basis. Subjects that adolescents want more information about (e.g., masturbation, contraception, and venereal disease) are difficult for parents to discuss in an objective manner and often parents are as ill informed on these subjects as their children. As a result, most adolescents get the major portion of their sex information from their friends or associates. These are usually poor sources as their education has been just as poorly handled and their knowledge of sexual matters is just as limited. As a result, misconceptions and erroneous ideas are picked up, leading to difficulties in sexual adjustment and the development of a satisfactory sexual life.

### Early and Middle Adulthood

Among adults, there is a wide variety in the forms of sexual expression, depending on the various factors that have operated during the development of each individual. Some individuals may reject overt sexual activity and lead an inactive or celibate life; others adjust to heterosexual activities, some within the marital relationship, others outside of it. Some individuals exhibit a bisexuality, engaging in both heterosexual and homosexual activities, while others reject heterosexuality and depend on self-stimulation or homosexual relations as a sexual outlet.

The social structure of our society is centered about heterosexual peer associations eventually leading to marriage and the development of a nuclear family consisting of a husband, wife, and offspring. The role of the family has been changing in recent years and its importance in our modern, complex society has, in some respects, been questioned. However it is still the principal source of companionship for most adults and it serves as the primary outlet of sexual activity. It is the unit in which children are brought into existence and which determines the nature of the future life of each child.

In most marriages, satisfactory relationships, both sexual and nonsexual, may be established and maintained. However, for various reasons, many marriages do not turn out satisfactorily and end in divorce. In recent years, the total number of divorces has been approximately one third the total number of marriages, however for first marriages, approximately 20 to 25 percent end up in divorce, desertion, or annulment. An important factor in marital failure is a lack of basic knowledge of and an understanding of human sexuality. Frigidity and impotence are common problems; extramarital affairs are common. Masturbation and homosexual activities sometimes become the chief sexual outlets.

It should be noted that, in frequency of total sexual outlet in the male, the peak of sexuality, as indicated by the number of orgasms, is reached in the late teens. It continues high through the twenties, then gradually declines throughout the remaining years of his sexual life. For females, however, the frequency is low in the teens, increases steadily during the twenties, reaches a peak at about age 30, then undergoes a steady decline comparable to that in the male. By comparison, the peak decade for sexual activity for males is between the ages of 20 and 30; for females, between the ages of 30 and 40.

At this time of life, a thorough knowledge and understanding of one's sexuality does much toward the attainment and fulfillment of a satisfactory sexual life, especially an understanding of the relationship of sex to love. Love is a strong, emotional attachment that develops between two persons accompanied by a feeling of companionship and compatibility. It involves giving and sharing, trust, and faithfulness. Falling in love is usually a prelude to marriage. Love makes the other person feel important and, in mutual love, one seeks to find in another person those qualities that he or she feels to be necessary for the full development of his or her personality. In our society, for most individuals, love and sex directed toward the same object provide the optimal means for sexual expression and satisfaction.

### Late Adulthood

This is the period that, in women, begins with the menopause and, in men, roughly the age of 50 to 60. It has been commonly thought that this period marks the end of one's sexual life. However, recent studies have shown that such is not the case, that sexual desires and interests in sex generally persist, and that continued sexual activity is not only possible but desirable. Premature cessation of sexual activity tends to accelerate the onset of physiological changes and concomitant psychological changes associated with aging. Consequently continued sexual functioning enables both the male and female to feel that they are needed and that they are capable of giving and receiving love and affection and providing sexual satisfaction at a time when self-assurance is essential for one's physical and mental well-being.

It is now recognized that there is no definite time when sexual desire and sexual activities cease. In women, the menopause marks the end of childbearing but, in general, sexual response may continue at about the same level as that experienced previously. Reduced hormone levels, especially estrogen, due to the cessation of ovarian function, may bring about some retrogressive changes in the reproductive organs such as atrophy of the ovaries and uterus and some physical changes as the tendency to gain weight and an increase in hirsutism. However, these need not necessarily reduce satisfaction obtained from sexual intercourse. In fact, satisfaction may be enhanced by the elimination of the fear of possible pregnancy. Fears that lessened femininity might result from operations involving the removal of the ovaries or uterus or both are usually groundless.

In men, impotence or fear of impotence is a common problem. This condition is not necessarily limited to older men. Erection in the male depends on both physical and psychic stimuli that reflexly bring about circulatory changes that result in engorgement of the penis with blood bringing about an erection. Fear of inadequate sexual functioning, fear of unacceptability, or fear of just getting old may be important precipitating factors in bringing about this condition. Physical conditions that develop in aging may create problems and increase fears of inadequacy. In men, prostatic enlargement, heart disease, and high blood pressure are often considered to signal the end of sexual activity. Such need not be the case. Sexual activity may continue even after the removal of the prostate and the fear that the exertion of the sexual act will induce a heart attack or a stroke is usually unwarranted.

In general, a continuation of sexual activities during the years of advancing age is considered to be desirable. Both men and women may remain sexually active into their seventies and eighties. Regularity in sexual activities plays an important role in the maintenance of a good physical condition, a stimulating mental outlook, and probably the most important of all, an interest in living and a zest for life.

# TWO
# MASCULINITY
# AND FEMININITY

*Maleness* and *femaleness* are characteristics that individuals possess that are associated with their being either a male or female, the *male* producing motile gametes or spermatozoa, and the *female* producing the larger, nonmotile gametes or ova. These characteristics include structural, physiological, and behavioral qualities by which a male is differentiated from a female. *Gender identity* is the psychological belief or awareness that one is a male or female, based on the possession of male or female sex organs and secondary sex characteristics. *Gender role* or *sex role* is the outward expression of gender identity. It includes all of a person's behavior, manner of speech, dress, and activities by which that person indicates to others the degree to which he or she is a male or female. *Masculinity* and *femininity* refer to those behavioral characteristics that are applied to and associated with males and females by which identification with anatomical or genetic sex is established. These characteristics are not innate but are largely the result of past cultural development, the economy and social life, the nature of the home and family life, political indoctrination, and religious and theological beliefs.

In Western culture, until fairly recent times, the concepts of masculinity and femininity were firmly established with definite traits, characteristics, and behavior assigned to each of the sexes. The culture was primarily androcentric, that is, dominated by males with masculine interests being of primary concern. The basis of this was in various religious and philosophical concepts that considered women to be inferior to men, physically and mentally. It was taught that woman was created from man for man, that her role was primarily reproductive with homemaking and care of the young her primary responsibilities.

This resulted in a dichotomy in which the sexes lived in two separate worlds, women being largely excluded from the world of business and public life. There has seldom been a society that has not been more restrictive of females than males. Throughout history, men have generally occupied the positions of power and prestige in most societies. Judeo-

Christian teachings and tradition have worked together to establish male dominance based on the assumption that man was created in the image of God hence possessed divine will and reason. Woman, being created from man, primarily for procreative purposes, was considered inferior in mind and body and consequently relegated to an inferior position. In the religious world, the positions of power—ministers, priests, bishops, rabbis, and popes—have always been occupied by men. Even the ruler of the universe has been ascribed male characteristics and prayers are addressed to "Him" and "He" is referred to as "Father." In the secular world, kings, emperors, prime ministers, most members of ruling assemblies, and most governors of subsidiary divisions have, with strikingly few exceptions, been men.

Consequently, masculinity has usually been characterized by strength and power, courage, endurance, aggressiveness, independence, lack of emotional sensitiveness, and success. Most males attempt to identify with this stereotype of masculinity. The American male is consequently urged and directed to achieve and be successful in the world of business or in whatever activity he may engage in. This may entail a sacrifice in time devoted to their families, a factor of importance in the present-day genera- tion gap. Passive activities (except those of a creative nature), artistic abilities, and cultural interests have, in general, been discouraged and ignored in the training of boys.

Because the life of a woman was primarily considered to be confined to the home where her principal role was that of a wife and mother, her success was largely determined by her success in raising a family. She was expected to be a gentle, meek, submissive, and understanding helpmate; to be modest and, above all, chaste, characteristics commonly identified with the feminine stereotype. The close association of women with the home and her greater involvement with the children has resulted in a greater emotional attachment between mother and child, a relationship that often impeded the development of children into mature, independent adults. Her success depending on her husband's success, a primary factor in social status, often led to excessive pressure being put on the husband and father to be successful, often a disturbing factor in family life.

The female stereotype also emphasized the importance of physical attractiveness and sexuality as indicated by the size of the cosmetic and fashion industries and their efforts through the mass media to exploit these characteristics. Although the public image of femininity is still that of an attractive female with sex appeal, the common stereotype of a "dumb, stacked, broad" is tending to disappear along with its antithesis, the masculine image of a career woman.

Male and female stereotypes are largely acquired through training and the example set by parents. Boys are taught to be boys or little men and girls are taught to be little women. Boys learn, as they grow up, that certain behavior regarded as effeminate is to be avoided. However as girls grow up, if they exhibit masculine behavior, less pressure is put on them to avoid this behavior as the masculine role is generally regarded as superior. It is interesting to note that boys who at some time have been identified as "sissies" and girls who have sometimes been identified as "tomboys" often perform above average in school. This is probably the result of sex stereotyping that even reaches into early schooling in which certain subjects as mathematics and science are considered to be "male" subjects and subjects as reading, history, and art are considered to be "female" subjects. It is generally noted that when boys are superior students, they are generally identified as having warm parents; when girls are superior students, they usually come from homes in which considerable freedom is granted.

An interesting insight into sex stereotypes has been brought to light in various anthropological studies. Dr. Margaret Mead, an outstanding anthropologist, studied three different groups of primitive people in New Guinea. The results of her studies were reported in a book entitled *Sex and Temperament in Three Primitive Societies*. The first group included the warm, gentle, and mild Arapesh. Both males and females of the Arapesh were tender, sensitive, and "maternalistic." Violent and aggressive individuals of either sex were considered deviant. Both sexes were known to initiate sexual relations, neither being aggressively sexed.

The second group of people, the Mundugumor, were aggressive headhunters and cannibals. They had a "masculine" orientation for both sexes—competitive, strong, independent, and aggressive sexuality. Power, violence, and jealousy were a way of life for the Mundugumor. The deviant persons in this society were usually the meek and mild of either sex.

Finally, in the Tchambuli, the sex roles were similar to the sex roles in American society except they were reversed. The women were the breadwinners and the real rulers of the family. The men were emotional, dependent, vain, and spent much of their time in dancing, gossiping, and doing art work. In sexual activity, the woman was the aggressor. The deviants in this society were the passive women and the aggressive men.

To summarize, both sexes of the Arapesh exhibit the sex role behavior of our contemporary concept of femininity; the Mundugumor exhibit our contemporary sex role behavior for masculinity for both of the sexes; among the Tchambuli, there is a reversal of our contemporary sex roles with the females exhibiting "masculine" behavior and the males "feminine" behavior. In other words, the critical factor is not what a man or woman does but whether the individual feels comfortable about his gender role.

To eliminate the sex role bias in America and to change the nature of the male and female stereotypes, there has been, within the past decade, a resurgence of the feminist movement, the movement that won the franchise for women in the 1920s. This activity, known as the Women's Liberation Movement, was sparked by Betty Friedan's book, *The Feminine Mystique* (1963), which exposed the extent to which men determined the structure of social institutions, regulated and controlled economic life, defined sex roles and judged sexual behavior, and, in general, controlled the lives and life-styles of men and women.

The primary theme of the feminist movement is the emancipation of women, the freeing of women from economic, social, and sexual restraints that have prevented them from participating in the freedom experienced by males. This movement has resulted in a demand for, and the development of, new sexual roles to the extent that the changes already brought about have been termed a *sexual revolution*. Among the changes that are in progress are the modification of the double standard of sexual morality, husband-wife relationships in marriage, the development of new life-styles, and greater freedom of sexual expression by females.

The double standard with respect to premarital sexual intercourse is now being challenged. This generally held concept holds that it is permissible for men to engage in sexual intercourse before marriage but wrong for women. Women who abstained and remained virgins were "good" girls; those who succumbed were "bad" girls. This placed the entire responsibility of sexual morality on the female. The male could be aggressive and demanding but it was the responsibility of the girl to set the limit. This double standard carried over into marriage in which the husband was granted the privilege of extramarital affairs but not the wife. Adultery on the part of the wife is grounds for divorce but not usually for the husband. Attitudes toward the double standard are changing, especially with the increasing acceptance of permissiveness with affection, that is, premarital intercourse is permissible when a continuing and warm, emotional relationship exists.

Husband-wife relationships are changing to a marked degree. The old view of possessiveness with the wife being considered in the category of property has largely disappeared. Instead, marriage is developing more into a partnership in which the wife and husband share more or less equally in economic responsibilities for support of the family, care of the home, and supervision of offspring. The traditional stereotypes of home labor being divided into certain work for women and certain work for men are being discarded and in its place, work in essential activities will be accomplished on the basis of aptitude and ability.

The working out of satisfactory husband-wife relationships however is difficult as is indicated by the large number of unhappy marriages and subsequent divorces. As a result there has developed a "Movement for Alternate Life Styles," which is aimed at granting more sexual freedom with a variety of interpersonal relationships. Various types of sexual relationships as trial marriage, serial monogamy, group marriage, communal life, swinging or mate swapping, and open or contractual marriage are proposed as options available. The principle underlying these various life-styles is that women have the same rights as men with respect to sexual activities.

The demand for greater freedom in sexual expression by the female is one of the significant features of the sexual revolution. In the Victorian era, although a period of sexual repression for both males and females, the sexual needs of the male were of primary consideration. The female was considered to be merely the agent through which his sexual demands would be satisfied. Man was the active individual, woman the passive. For women, sex was often considered to be an unpleasant chore performed as a "duty." Its morality was questioned except when engaged in for procreation.

With the advent of Women's Liberation, women are becoming more conscious of their sexuality and are being educated concerning the nature of their sexual responses and the various means of satisfaction. They are initiating the sexual act and are demanding satisfaction. As a result the role of the male as being the sexually dominant individual of the pair is being threatened and furthermore, the male in many cases is not able to meet the demands of his partner. Female frigidity, which was a common neurosis during the Victorian period, is on the decline while male impotence, with its traumatic effect on the male ego, is becoming increasingly common.

What is the future of the Women's Liberation Movement? On all fronts, in business, in the schools, in religion, in sports and recreational activities, and in home and family life, the "sexist" nature of our present society with its male and female stereotyped roles of the past is being questioned and overturned. In nearly every field of human activity, opportunities that were formerly restricted to men are now being made available to women. Sexual freedom, which has been largely a male prerogative of the past is now being granted to women and women are attaining a state of equality with and freedom from domination by the male. Marriage is now becoming a partnership in which economic responsibilities, the raising and care of offspring, the development and fulfillment of intellectual and emotional needs of both participants, and the satisfaction of sexual needs will be mutually shared. It is expected that this will create bonds of long-lasting affection that will serve to strengthen the marriage bond.

# THREE
# MARRIAGE AND
# FAMILY LIFE

*THE FAMILY IN AMERICAN LIFE*

Marriage and family life are not phenomena of recent development in human history but have existed in various forms since early times. The study of the social life of various primitive peoples still existing on our "civilized" planet brings out the fact that in every group some form of marriage exists with restrictions on the sexual freedom of the individuals involved. The marriage bond unites the male and female into a sexual union that provides protection, companionship, and the nurture of offspring following procreation.

Various theories exist as to how family life originated but it is commonly thought that the family is the result of natural selection, the family developing as an adaptive mechanism in relation to man's struggle for existence. The earliest type of marriage was monogamy or pair mating—the sexual union of a male with a female for an indefinite period of time. The family eventually developed out of the helplessness of the human infant and the need of both the mother and the young for protection over a relatively long period of time. Although polygyny (multiple wives) and polyandry (multiple husbands) have existed at various times, they are considered to be late social adaptations resulting from primitive man's adaptation to industrial and social changes.

The family has been a unique institution in America. During the nineteenth century, it was a patriarchal institution with its members held together by legal, economic, and religious bonds. It was ruled by a husband and father whose power and authority over the wife and children was unquestioned. He usually ruled with affection and love and the family provided comradeship and mutual help in times of stress. It also served as a source of educational and sometimes vocational training.

However, a number of changes have been occurring in recent years indicating that the role of the family in our social system is being modified. Instead of marriage being a lasting "until death do us part" relationship, it is becoming more a temporary union lasting "until I'm not satisfied with it," and then it is terminated. That marriage is not performing its expected function is indicated by the increasing number of divorces, the number in 1975 totaling over one million. Marital fidelity seems to be decreasing as indicated by the increased percentage of married men and women who admit to engaging in extramarital affairs and the increased incidence of mate swapping. There is also an increase in the number of common-law marriages or liaisons and living-together arrangements and an increased participation in group marriage involving communal living, sometimes including indiscriminate mating. Another factor is the increased acceptance of homosexuality as a way of life in which straight marriage is avoided.

The causes of the so-called breakdown of marriage as an institution in our society are many. The following are some of the important factors involved. (1) The change from a rural, nonindustrial society to an urban, industrial economy with accompanying reduction in the influence and authority of a patriarchal father and economic dependence on the home. (2) A decrease in the role of religion and an increase in secular influences on family life. The influence of the Christian church has declined as indicated by its changing views on male dominance, divorce, contraception, and even homosexuality. (3) The changing status of women from that of almost complete dependence on the husband to one of equality— intellectually, economically, legally, emotionally, and sexually. (4) The mobility of our society in which the establishment of a permanent home and stable family life is made difficult. (5) The lessening of the stigma connected with divorce and the liberalization of divorce laws. (6) The decreased role of the home from the educational and nurtural standpoint resulting from the large number of married women working outside the home and the reduced contact of parents with children.

However, in spite of the dire predictions that marriage is a failure, marriage and family life is still the predominant type of heterosexual association in our society today, approximately three fourths of all first marriages remaining intact. Just as marriage in the past has changed to adapt to the changes of the society of which it was a part, so marriage of today will adapt to present and future conditions. The roles of husband and wife will be modified, attitudes toward children will change, and new types of sexual relationships will be developed. Some look on these changes with despair and as indicators of decadence and the disintegration of our society. Others welcome the changes with hope and optimism, regarding them as long overdue. Greater individual freedom in love and sexual expression is now being granted and experienced and the excessive restraints of the patriarchal monogamy are being done away with, changes regarded as healthy and desirable in a society such as ours in a changing world.

In the meantime, average young people will grow up very much like their parents. They will experience puberty and adolescence. Their sexual instincts will develop, and they will engage in various types of sexual activities, possibly even intercourse. Petting will be engaged in, and finally they will meet the person with whom they fall in love. This may happen once or it may be repeated several times. Eventually the "one" will be found whom they want to marry. When marriage occurs, a legal and emotional bond will be established that can, and most often does, last throughout life. This relationship permits the freest of heterosexual expression and the development of a nuclear family in which children can grow up with affection and security. Trust, confidence, and fidelity are still worthwhile goals and their attainment usually results in a lifetime of satisfaction, companionship, friendship, and love. Above all, in a marriage, a feeling of personal value, a feeling of being needed, and knowledge that, in your family, you are a very important person, make life more meaningful. These are some of the rewards coming from a successful family, the product of marriage. A discussion of the various steps leading to marriage and some of the important factors in a successful marriage follow.

## DATING

A unique feature of American social life is dating. Contrary to common belief, dating is a fairly recent social custom. Before the 1920s, in the horse-and-buggy days, a boy did not begin to court a girl until he began wearing long pants (about the age of 16). Then when a boy started going with a girl, it was with the distinct approval of the girl's parents. Much of the lovemaking took place in the home in the family parlor. After going together a few times, the girl would then begin to be regarded as "his girl" and no other boy was supposed to intrude into the affair. Similarly, the girl was expected to limit her attentions to her boy otherwise she would be classified as a "flirt." After going together for some time, with the permission of the girl's parents, they would become engaged and after a time, the marriage would take place. They would then establish a new home and settle down usually near the homes of their parents.

But all of this changed in the early decades of the twentieth century, especially the "Roaring Twenties." The more important contributing factors were the development of the automobile, the motion picture, and the radio. These coupled with other changes following the conclusion of World War I brought about the emancipation of women and the development of a new concept in marriage. Marriage came to be regarded as a union involving companionship. As a consequence, a new type of relationship developed between boys and girls and young men and women. They started going together and participating in social engagements with the sole purpose of enjoying each other's company. No serious commitment was involved. They went together for fun. Often they would date once and never see each other again. This type of casual dating is a unique phenomenon of American culture.

Dating developed into a heterosexual association characteristic of the adolescent subculture. Dating represents a degree of maturity and heterosexual acceptability and is a mode of attaining status. After reaching a certain age, usually high school age, boys and girls expect to begin dating. Typically, girls start dating at an earlier age than boys. To fail to do so means being excluded from social activities, consequently pressure is brought on adolescents by their parents and peers to participate in dating.

Dating is usually engaged in first as a recreational activity, sex in itself not being a primary factor, however males may utilize it as a possible entry to a relationship permitting sexual expression. Couples start going together principally because they enjoy each other's company and their relationship is a source of pleasure. Dating enables couples to participate in social activities, and in the process more is learned about members of the opposite sex and the appropriate behavior pertaining to them. Dating is a means of achieving social status and of being included within and associating with others of one's peer group. It provides the opportunity of developing friendships that often play an important role in future mate selection. It is often the beginning of a courtship, which ends in marriage.

Dating usually begins as *casual dating* in which individuals of one sex occasionally associate with those of the other sex in an activity of some kind, usually for pleasure or entertainment. Each however remains free to withdraw from the association if he or she wishes. No personal commitment is involved. However if the association continues and the couple repeatedly appear together, then the association becomes that of regular dating with the development of an interpersonal commitment involving both an affectionate and erotic interest in each other. This usually leads to an exclusive relationship between the pair and *steady dating* or *going steady* becomes the pattern. This is characteristic of adolescent dating of today, together with a trend toward earlier dating in which boys and girls even below high school age are encouraged to form pairs in participation in social activities.

Although going steady is the usual pattern of behavior among adolescents, it is severely criticized by some of the older generation, their parents. It definitely reduces the number of persons of the opposite sex with whom a boy or girl may come into a close relationship with, and it often leads to a premature commitment to marry one who is not the most desirable partner. It tends to establish a monotonous association when exciting new options might otherwise be available. It usually leads to the development of intimate sexual relationships that have generally been restricted to marriage. Emotional involvement is almost inevitable when going steady, and sexual involvement is a common occurrence. Serial relationships in which a person is committed to one partner for a limited time and then shifts to another are increasing in frequency.

Regardless of one's views on steady dating, it is now a well-established pattern of behavior and most adolescents value it highly as it provides assurance of being accepted and being able to participate in the activities of one's peers. It tends to place a value on honesty and openness in sexual relationships, and if sexual activities are engaged in before marriage, they are with the one with whom the relationship is meaningful and lasting and often a prelude to marriage. The monogamous principle of marriage is still maintained in premarital behavior.

A recent innovation in dating techniques has been the utilization of the electronic computer in bringing about the matching of couples. Computer service organizations have been established which provide their clients with the names of prospects which have made their availability known and provided data concerning such personal characteristics as age, height, religion, education, and interests. This service is of special importance to individuals who have limited opportunities of meeting prospective dates. Evidence indicates that couples matched on the basis of common interests, education, and values are generally more compatable than those matched by chance encounters.

## ENGAGEMENT

An engagement is an agreement between two individuals to marry each other. In the early part of the century, an engagement was accompanied by a public announcement of an intent and promise to marry. It was regarded as a binding agreement and was not broken except for the most urgent reasons. Engagements today are less binding.

Courtship usually involves the following steps: regular dating, steady dating, private understanding to marry, and, finally, an engagement with a public announcement. At any stage in this series, the relationship can be terminated, and it frequently is, with no breach of faith involved. There is a hesitancy among American couples today to regard any human relationship, even marriage, as permanently binding, consequently the engagement is regarded more as the removal of the girl from the marriage market than a positive commitment to marriage.

Engagement for an average couple usually occurs in this way. After going steady for a period of time, they decide that they are in love and plan to marry. This may be publicized by the girl accepting informally his fraternity pin or more formally, an engagement ring, the traditional diamond. Among a girl's associates, this is a sign for congratulations. A public announcement usually follows, with the prospective bride's picture appearing in the local paper.

To the couple involved, engagement changes their relationship to each other from a fun and recreational association to the more serious relationship of a couple planning to get married. Many factors that were of little consequence during the early stages of courtship now begin to assume major importance. An engagement gives a couple the opportunity to become better acquainted with each other and to learn how each reacts to the other during an extended, close relationship.

During an engagement plans are made for the wedding, the honeymoon, and the establishment of a home. Relationships with future in-laws are established or amplified. Many questions arise that require answers. Often conditions that may affect a marriage are kept secret during early courtship; now they must come out into the open. Is the woman able to bear children? Are there health factors that need to be explored? What is the attitude of each toward having a family? Toward contraception? Toward the religious upbringing of their children? How will they get along economically? Is his income adequate or will she need to take a job? How will the finances be handled? These and countless other questions will arise. Most need to be resolved *before* marriage.

Disagreements often arise during an engagement but, in most cases, they are resolved without difficulty. However when disagreements involve severe personality conflicts or reveal marked differences in likes and dislikes or differences in patterns of living, serious thought should be given to termination of the engagement. Couples who experience trouble during their engagement are much more likely to experience trouble in marriage, consequently, if there is serious doubt as to whether the marriage will be a happy one, it is better to break the engagement than to enter into a marriage that is likely to turn out to be unsatisfactory.

How long should an engagement last? There is no definite length of time. Some engagements are short lasting only a few days or weeks; others are long lasting for months or years. An engagement should last long enough to enable a couple to get to know each other well enough to determine compatibilities or incompatibilities. Important factors in broken engagements center around these situations: the degree of emotional involvement, ability to stand separation from one another, difficulties with future in-laws, cultural differences, and personality conflicts. Testing these during an engagement helps to determine whether these situations are likely to cause difficulties during married life.

Sometimes difficulties encountered during an engagement can be resolved through proper counseling. Individuals trained in premarital counseling can, with a fair degree of accuracy predict the probability of success in a marriage. A lack of information or lack of understanding of the problems involved can often be resolved through the help of a sympathetic and understanding counselor.

*Sex Relations During an Engagement*
Although marriage has been generally regarded as a prerequisite for sexual intercourse, especially for the female, there has been a marked trend in recent years for couples to engage in the sexual act before the marriage vows have been declared or the marriage legalized. This has been the result of many factors. The liberalization of premarital sexual codes has led to the general acceptance of the "permissiveness with affection" standard in which intercourse is acceptable for both sexes when a stable, affectionate relationship exists. This, together with a greater knowledge of the nature of one's sexuality, the availability and use of contraceptives, and the belief that sexual intimacy before marriage will facilitate sexual adjustment and the development of a more enduring relationship in marriage, have increased the incidence of premarital sexual relations.

Whether sexual intercourse should be engaged in before marriage is a question each couple must decide for themselves. The long period of adolescence, the high emotional involvement, the sexual intimacies of an engaged couple, their lack of supervision and ready opportunity for sexual activity all make abstinence extremely difficult.

Does engaging in sexual activities lead to a happier and more stable marriage? Studies in general indicate that premarital sexual relations do not necessarily have an adverse effect on sexual adjustment following marriage, in fact it may be favorable to it. However, with respect to overall marital adjustment, abstaining from premarital intercourse seems to have a stabilizing influence, divorce occurring less frequently in such marriages.

## MARRIAGE

*Marriage* is a social institution in which a man and woman declare their intention to live together as husband and wife. It is usually formalized by a legal or religious ceremony. In the United States it is the culmination of heterosexual sex-pair associations and is one of the most highly valued relationships in American society. The majority of individuals today find their greatest satisfactions in life in the marital relationship as indicated by the fact that over 90 percent of all men and women marry. Furthermore, they marry earlier, fewer remain unmarried, and more remarry following death of spouse or divorce than in any other civilized country. A prime factor in American marriages seems to be the development of a nuclear family, however the desire to have children is usually not a motivating factor for marriage. Most couples marry for love and companionship. Some may marry for other reasons, for example, financial security or social status.

Marriage and family life, as it has been regarded in the past, is undergoing marked changes. Many factors point to the breakdown in family life as indicated by the increased incidence of divorce and desertion and the development of alternate patterns of sexual relationships. The question then arises, "What leads to a successful marriage?"

Success in marriage depends on many factors. Age and emotional maturity, social factors, race, religion, education, and other factors as compatability, understanding, and ability to adjust to each other are of importance. All of the above-mentioned factors play a role in success or failure in marriage.

*Age and Emotional Maturity*
Studies indicate that when girls marry before the age of 18 or men before the age of 21, the chances of success are much less than when marriage occurs at a later age. Men over 22 and women over 20 are much more mature in every respect. Personality traits are more firmly established and less likely to change and both are better prepared to meet the responsibilities of married life. They are usually much more adaptable to the demands of married life and more considerate of their partner as to material and emotional needs, consequently are less likely to resort to divorce when difficulties arise. Marriages of couples of greatly disparate ages, as older men with young women, or more rarely, vice versa, pose special problems although, not infrequently, such marriages turn out satisfactorily.

*Social Factors*
Similarity in cultural and economic backgrounds play an important role in marriage stability. When individuals are of the same cultural background with comparable interests and values, misunderstandings and conflicts are lessened and adjustments are more readily made. Marriages between individuals of different social classes usually encounter difficulties. Marriages among those of the lower social classes are less stable than those of the middle and upper classes. In cross-class marriages, the status of the husband is usually the determining factor, marriages of women to men of a lower social class usually resulting in maladjustment.

*Race, Religion, and Education*
Differences in race, religion, and education may be significant factors in marital success. In interracial marriages, differences in customs, standards of behavior, and attitudes of husband and wife toward each other may lead to internal conflict. This may be intensified by attitudes and prejudices of the families involved and the society in which they live. Further problems arise in connection with children and their upbringing and in their rela-

tionships with their friends and associates. A great amount of tolerance and understanding is necessary to overcome these difficulties.

With respect to interfaith marriages, the degree of disruption depends on the importance religion plays in the lives of those involved and the extent to which their religious differences were discussed and worked out before marriage. In general, a shared religion is a unifying influence and an important factor in holding couples together. Religious conflict can be disruptive. As in interracial marriages, children may be a disturbing factor as to the religion in which they are to be brought up. The intensity of religious beliefs held is usually the determining factor in the stability of such a marriage. When the religious faith of one or the other is not strong, religious differences can usually be worked out satisfactorily. It is noted however, that the least stable marriages of all are those in which no religious faith is held by either spouse.

Differences in education may or may not be of importance in marital success. Well-educated couples more generally have better adjusted marriages than poorly educated couples, but if the marriage is unsatisfactory, they are more likely to resort to divorce. Couples with similar levels of education are generally more compatible in marriage. Marriages in which the wife is more highly educated than the husband have a higher than normal divorce rate.

*Other Factors*

Of the many factors that determine whether a marriage will be successful or not, probably the most important is the desire and the ability to adjust to each other. This involves an understanding of each other's personalities and the willingness of each to make changes necessary to insure a harmonious life. The romantic love that exists during courtship and early marriage is likely to change to the less exciting, but more enduring and deeper, feeling of companionship and affection. A considerable amount of sexual adjustment is usually necessary based on the different sexual natures of husband and wife and different attitudes of each toward sex. To men, love comes or is expressed by sexual activity; to women, sexual activity comes as a result of love. Or as expressed by a poet in another way, "of men, love is but a part; 'tis woman's whole existence." Marital satisfaction, achieved when both have a mutual understanding of the needs and desires of the other, provides a cementing bond that will hold a wife and husband together through the difficulties usually encountered.

The goals of marriage have been changing in recent years. Formerly it was concerned principally with economic survival, emotional security, and the raising of a family. Now it is expected to provide both husbands and wives with the opportunities for intellectual growth, a warm and intimate companionship, satisfying sexual experiences, and allow for the development and expression of self. Whether such can be obtained without the loss of some basic ideals on which sound marriages are based, that is, commitment, responsibility, and sacrifice remains to be seen.

## MARRIAGE LAWS

Because the family is a social and legal institution, every state has laws pertaining to marriage, the rights and obligations of the parties involved, relations of parents to children, the dissolution of a marriage, and related matters. The laws vary from state to state and are constantly being changed in an effort to keep up with changing customs. The following are some of the representative laws.

### Minimal Age

Every state requires that a person reach a certain age before marriage. Without parental consent, the minimum age for males ranges from 18 to 21, for females, from 14–18. With parental consent, the minimum age for males ranges from 16 to 18, for females, from 14 to 18. Most states make exceptions in cases of girls who are pregnant.

### Who May Perform the Ceremony

After a license has been obtained, a marriage ceremony must be conducted. In most states, licensed or ordained clergymen and authorized judicial officials (certain judges, justices of the peace, certain civil magistrates) may perform the ceremony. A few states permit certain public officials as governors or mayors to officiate at weddings.

### Waiting Period; Physical Examination and Blood Tests

Many states now require that a period of time, usually three to five days, elapse between the application for a marriage license and its issuance. This waiting period is to discourage marriages from taking place on sudden impulse. Also most states require that a person seeking a marriage license submit a physician's report affirming that he or she has been examined and found to be free of certain communicable diseases, especially venereal diseases. This is usually determined by blood tests. A few states also require freedom from certain other diseases as tuberculosis and certain mental disorders.

## Prohibited Marriages

Every state has some prohibitions against the marriage of closely related persons. Incest or marriage between members of a nuclear family is not only prohibited by law but by strong social and religious taboos. Some states prohibit the marriage of first cousins and a few between distant cousins. Also a person judged mentally ill is usually prohibited from marrying and, in some states, the prohibition includes persons afflicted with other disorders of the nervous system, such as extreme mental retardation. The marriage of a person already married to a living spouse (bigamy) is prohibited.

All states have laws under which a marriage can be annulled or declared void. Marriages in which consent was obtained by force or duress, misrepresentation, or fraud may be annulled. Marriages that are prohibited by law can be declared void. The inability of either spouse to engage in normal sexual intercourse or sterility of either spouse discovered after a marriage are usually considered as grounds for annulment.

## INTERMARRIAGE

Intermarriage is the marriage of persons who differ with respect to their religious, ethnic, or racial backgrounds. *Interfaith marriages* are those in which persons of different religious faiths (Jewish, Catholic, Protestant) marry or in which anyone of these three Judeo-Christian groups marries a person of a different faith as a Mohammedan or Hindu. *Interethnic marriages* are those in which persons raised in different cultural and national environments marry. *Interracial marriages* are those in which persons of different races, especially races that differ in color, marry.

The incidence of intermarriage has been increasing in recent years and will probably continue to increase. The greatest increase has been in interethnic marriages followed by interfaith marriages. Interracial marriages show the lowest rate of increase.

A number of factors have been operating in the United States to increase the incidence of intermarriage. Probably the most significant single factor has been the increase in enrollments in colleges and universities in recent years. The majority of high school graduates enter institutions of higher learning today where they are subjected to influences that tend to break down barriers that separate groups. Of primary importance is the general attitude of liberalism and tolerance that is encouraged in nearly every aspect of our culture. To this is added the release of students from home influences, especially those of family and church, accompanied by greater permissiveness granted by parents; the influence of peer pressure and a general rebellion by students against customs, manners, and present-day life-styles; the greater opportunities afforded through various intellectual and social activities for students of diverse backgrounds to meet and socialize; the decline in the influence of religion together with the development of ecumenicalism and the decline of secularism; the elimination of barriers based on race, religion, color, or national origin with respect to housing, membership in national fraternities or sororities; the increased participation of all groups in aspects of college life; and finally the development of more tolerant and understanding attitudes toward individuals who do intermarry. To these may be added the changing attitudes of most young persons toward marriage, the trend toward delaying marriage, and the desire to try out new alternatives that may be offered.

Another important factor in the increase of intermarriages is the changing nature of the family. The older type of patriarchal, authoritative family has changed to one of greater freedom and less restraint. The family has become more mobile, increasingly more urban, and because many of the family functions have been taken over by the school or state, its role as a molding influence in our society has been weakened. The increasing mobility of families has lessened the influences of the local community on the family and the influence of the extended family (grandparents, in-laws) has tended to decline with the increased isolation and independence of the nuclear family.

Increasing liberalism among young people, especially college students, has led to a greater acceptance of intermarriage. The tendency of today's youth to rebel against acceptance of parental attitudes with respect to social, cultural, and religious values is widespread. As a result, bans against intermarriage as established by tradition and custom tend to be questioned and disregarded. The influences of the family, the church, and society in general have declined in recent years and, as a result, intermarriage between individuals of groups differing in religion, race, national origin, or color is occurring more frequently than in the past.

A number of additional factors have increased the incidence of inter-marriage. When a racial, ethnic, or religious group is in a minority, there is a tendency of individuals of that group to marry outside their own group. When sex ratios are unbalanced, there is also a tendency to intermarry. The trend toward rejection of standardized forms and patterns of behavior with accompanying development of individualized behavior favors the selection of a mate with little consideration for rules of custom and tradition. Often intermarriage is the result of the desire to escape parental domination, those involved regarding themselves as freed or emancipated individuals. The rebellious person avoids conformity and acts to dissociate from his or her group to demonstrate independence. The desire for new experiences in a new and different culture is also an incentive. Romantic love, which may develop in almost any situation in which two individuals are thrown into a close and intimate association, together with the idea that it is the inherent right of any individual to be happy, are probably the motivating factors in the majority of intermarriages.

In spite of the many influences that tend to encourage intermarriage, there are also factors that discourage it. These may be of legal, social, psychological, or cultural nature. The role of the parent and family is still strong in the selection of a mate, many children preferring to avoid bringing into the family one whom they feel might not be totally accepted. The clergy similarly exerts a strong influence on members of their respective faiths and their general disapproval of intermarriage acts as a deterrent. Membership in a group or subgroup characterized by a sense of belonging is an important part in the cultural development of an individual. Intermarriage consequently is, in a sense, a repudiation of or desertion of the group. As a result, those participating in intermarriage may be subject to penalties as isolation or ostracism from the group. Pride in race or nationality or disdain for another race or nationality are effective deterrents to marrying outside a specific group. Prejudice and bias against members of groups other than the one to which we belong are strong influences operating to discourage intermarriage. The tendency to conformity is difficult to overcome.

*Interfaith Marriages*
These principally involve marriages between persons of the three primary faiths, Catholic, Protestant, and Jewish. Generally speaking each of these groups oppose or have a negative attitude toward members of their faith marrying a person of a different faith. A part of this opposition results from the different attitudes toward sex and marriage held by the different faiths and the insistence by certain faiths that children of such a marriage be raised in their faith. Also it is generally assumed that lack of harmony on the part of the parents in the training of children can be a disrupting influence on the home increasing the chances that the marriage will turn out unsatisfactory. In the marriage of individuals of the same or similar religious faiths, children usually tend to develop a closer relationship with their parents and there seems to be a close correlation between this relationship and success or failure in marriage. Studies of divorce rates show that the rate is lowest among Catholics and Jews, higher among Protestants, higher still in mixed marriages, and highest among couples who have no religious affiliation.

While parents may reconcile their religious differences, children are often denied the opportunity of affiliating with or identifying with any religion. The ability of couples to reach agreement and settle disagreements in the field of religion is one of the more difficult adjustments to make for beliefs in creeds and doctrines are not readily subject to rational analysis and acceptance as most other matters are. Since a successful marriage usually depends on the possession of common interests and involvement in common activities, divisiveness brought about by belonging to different faiths makes it more difficult for an enduring relationship to be established. Depending on the strengths of the faiths of the couple involved, unhappiness may carry over to the families of each of the partners and tensions often develop leading to in-law problems. There seems to be common agreement that married couples who share a common faith tend to have a more harmonious married life than those who do not.

However, it must be recognized that interfaith marriages may turn out to be satisfactory and enduring. When faiths are not too strong, compromises can usually be worked out that are mutually satisfactory. The willingness of one partner to allow children to be raised in a faith other than his or her own is usually essential.

*Interethnic Marriage*
This is a marriage between persons belonging to distinct groups that differ in their culture and national origin. Ethnic groups usually have a basic pattern of family life with a common religion and language. They tend to have distinctive customs, patterns of dress and behavior, commonly recognized moral codes, and usually have a common center of loyalty as a country, religion, or language. They are often self-centered, priding themselves on those characteristics that differentiate them from others.

Ethnic differences were pronounced in the early development of this country when migration of peoples from all sections of Europe and other parts of the world brought to this country large populations of various nationalities that tended to settle in particular regions of America and there retain their cultural differences. Groups were often differentiated along national and religious lines. There were the Irish, Italian, and French Catholics, Protestant groups such as the Lutherans, Episcopalians, and Scotch-Irish Presbyterians, and Jews from Russia and Germany who formed distinct ethnic groups. However with America serving as a "melting pot," ethnic differences tended to recede and through intermixture of various groups and especially the use of a common language, the identity of the various groups tended to disappear and their members came to regard themselves as Americans.

Among the major ethnic groups are the Pennsylvania Dutch, the Amish of Northern Indiana, the Lutherans of Minnesota and Wisconsin, the French-Canadians of New England, the French peoples of Louisiana, the Spanish of the Southwest, the Indians of the Midwest and West, the Mexicans in Texas and California, the Chinese of New York and California, and the Japanese of the West Coast. In all of these groups, religion and national origin have been significant unifying factors. However separateness of ethnic groups has tended to decline as Americanizing influences have exerted their effects. As most members of these groups are now native-born Americans (about 95 percent) and speak English, the hold of the old country with its language and customs has diminished. Religious views have been liberalized, and while there is still a reluctance for members of the various groups to marry outside the group, interethnic marriages are occurring with greater frequency. As the Americanization continues and the "melting pot" concept continues to hold, as time passes, the differences that differentiate ethnic groups from each other and from the American population in general will tend to recede and their cultures and life-styles will blend in with the American scene.

*Interracial Marriage*
An interracial marriage is one in which the participants belong to different races. There is evidence that no pure "race" of mankind exists, all human beings belonging to the one species, *Homo sapiens*. As a consequence all humans are related to each other. However, throughout the history of mankind, as a result of migrations, wars, conquests and many other factors, humans have become distributed over the surface of the earth and as a result of hybridization, isolation, and adaptation to various types of environments, groups of humans that differed from other groups have developed. These we call *races*. Classification of races has been difficult. One classification is based on shape and size of the skull; another is based on color of skin and hair form. In general races are thought of in terms of skin color (white, brown, black, yellow). Interracial marriage includes crosses between any of the above groups.

Interracial marriages are accepted in various ways throughout the world. *Miscegenation* or the mixing of races is common in Hawaii and South American countries and intermarriage is common in most of the countries of Europe where it is generally accepted both legally and socially. However, in certain countries—especially England, South Africa, and the United States with predominantly white populations—a race consciousness exists that has prevented the free marriage of individuals of different races. White prejudice has been strong against intermarriage of whites with Orientals, Indians, Mexicans, and blacks. Of these groups, it is strongest against the blacks.

During the development of the United States, in the days before the Civil War, marriages between whites and black slaves or servants were prohibited by law. In spite of legal prohibitions, intermixture of races took place on an extensive scale usually between the white master and black women. Resulting offspring together with an admixture of Indian blood have given rise to many groups of racially mixed peoples which form pockets of hybridization scattered throughout various regions of the United States. Intermarriage between members of these groups is common.

Following the Civil War, special efforts were made to prevent the mixing of the black and white races. All Southern states and many other states passed antimiscegenation laws that forbade interracial marriage or cohabitation between whites and blacks or persons of any color. These laws were primarily the result of race prejudice in which the blacks were considered as inferior and sexual contact of black men with white women was especially feared and avoided. Black men were denied white females as sexual partners under penalty of death. Interracial marriage was prohibited as a means of maintaining the "purity" of the white race and preventing the mixture of the races. Whites in pre-Civil War days had feared revolt and reprisal by slaves; after freeing the slaves, fears of murder and sexual assault by blacks still persisted. Violence was often resorted to as a means of preventing racial intermixture. A social order in which separateness in nearly all aspects of life, such as housing, schooling, and social relations was established and maintained until the development of the civil rights movement of the 1960s.

As a result of the civil rights movement, barriers to the separation of blacks and whites have steadily fallen with resulting desegregation of the races and the increased integration of blacks into nearly every segment of our society—housing, education, business, politics, arts, recreation, and social life. One of the last legal barriers to fall resulted from a Supreme Court decision in 1967, which held all laws against miscegenation to be unconstitutional. As a result of the greater contact of the races with each other and the elimination of legal prohibitions, there has been a gradual increase in interracial marriages and with it a growing acceptance of black-white marriages.

Black-white marriages involve a white male and a black female or a black male and a white female. The most common type of black-white marriage is that between white brides and black grooms, accounting for 0.5 percent of all marriages in 1980. Marriages of black brides and white grooms are not nearly as frequent, accounting for 0.2 percent of all marriages in 1980. Marriages between a white person and a person of another race other than black accounted for 0.4 percent in 1980. However, the rarest marriage combination was that between a black person and a person of a race other than black or white; this pattern accounted for less than 0.1 percent of all marriages in 1980. In a recent survey in 35 states, the five states with the most interracial marriages were, in descending order: Hawaii, Alaska, Montana, South Dakota, and Kansas. Data on interracial marriages were not available for some of the most populous states: California, Maryland, Michigan, New York, Ohio, Oklahoma, Texas, and Washington. The total number of interracial marriages in the United States is estimated to be about two million. Of this number, less than half are black-white marriages.

Marriages between blacks and whites occur primarily for the same reasons that most couples marry—the participants fall in love with each other and want to share each other's lives. Persons of different color who are associated in social or business relationships or who have mutual interests in the arts, music, or the theater are most likely to marry. However, members of either race may exploit members of the other race and date and marry across racial lines for various reasons other than mutual respect and love. Both whites and blacks may date and marry to demonstrate lack of racial prejudice. For both it may be a sign of rebellion against parental authority. White females may date black males to show contempt for the establishment, to identify with the oppressed, to express liberalism and contempt for authority, for revenge against parents, relatives, or ex-husbands. Black males may exploit white females, especially liberal women, by using the woman's feeling of guilt over racial injustice to gain sexual favors, or for revenge for racial injustice. Blacks, both male and female, may marry whites to improve social status or to gain economic advantages. Some white females seek black companions because of the mistaken belief in the sexual prowess of black males. White males may be attracted to black females because of their belief in the heightened sensuality and responsiveness of the black female and also for the same reasons that white females seek black males as given above.

Whites who participate in interracial marriage are often divorced or widowed persons and are usually older than those who marry within their own race. They are also urban rather than rural, and more likely to be service workers or laborers than professional or white-collar workers. However, interracial marriage is occurring with greater frequency in the better educated and among the socially and professionally prominent in both white and nonwhite groups.

Many problems are encountered in interracial dating and marrying. Although much progress has been made in integration of blacks and whites, a considerable amount of hostility is still often encountered in both white and black communities against individuals who date and marry outside of their race. Among the problems encountered in interracial marriages are disapproval by the families of those involved, social disapproval by the community, difficulties in interpersonal relationships, and difficulties in the raising of children.

Disapproval by both nuclear and extended families of those involved is a common reaction. It may vary from mild unacceptability and grudging acceptance of the situation to outright hostility. This not only occurs in white families but is common and intense among blacks. Parents sometimes disown or disinherit children who marry against their objections; blacks, especially women, cannot understand why a black man would prefer a white woman to a black woman. Among both blacks and whites, rejection by in-laws is a common reaction and oftentimes family relations are strained or completely broken.

Social disapproval is still a common reaction to mixed dating and marriages. This is experienced in both black and white communities where a mixed couple receives looks of disapproval or looks that may convey hatred, contempt, or sometimes pity. Older blacks and whites especially express strong silent disapproval by intense staring. Mixed couples may encounter unfavorable reactions in restaurants, in getting taxicabs, in securing favorable housing, and in being accepted socially by either black or white groups. Mixed married couples frequently find that they are deserted by their former friends and associates.

Difficulties are often encountered by mixed couples in their interpersonal relationships, for each partner may read racial overtones in the behavior of the other. Racism has led to the development of both black and white stereotypes, and, in conflicts that may develop in any close relationship, latent racism may manifest itself in disparaging references to race. Blacks find it difficult to trust whites or to really understand what they are thinking: whites similarly fail to understand blacks, especially their intimate feelings and emotions.

Probably the most serious problem arising from interracial marriages is that which develops when children result from the marriage. A child of mixed parenthood encounters many problems not experienced by all-white or all-black children. He or she has the primary problem of identity. To which race does he or she belong? Inability to identify with both parents may lead to resentment against one or the other. The mixed child is invariably regarded as a black, and he or she is subjected to the same discrimination that blacks in general experience. This is especially trying for teenagers when parents of black peers or white peers reject the child's ambiguous racial background. At this time, the mixed child needs considerable parental support and understanding to minimize the psychological trauma that may result.

Interracial marriages experience greater difficulties than intraracial marriages, as is indicated by the higher divorce rate for mixed couples. Divorces of white husbands from nonwhite wives are more numerous than those of nonwhite husbands from white wives.

Attitudes toward interracial marriage are, however, slowly changing. Even though there is an increasing tolerance for and acceptance of interracial marriage, general approval by either whites or blacks is likely to be slow in coming and the rate of racial intermixture is likely to be slow in the foreseeable future.

*Homogamy and Heterogamy*

*Homogamy* is the marriage of persons with like characteristics; *heterogamy*, the marriage of unlike individuals. These terms are used interchangeably in the literature with endogamy and exogamy, *endogamy* being marriage within one's own group or social unit and *exogamy* marriage outside one's group. A vast amount of research on homogamous factors such as similarity in age, income, social class, ethnicity, moral values, religion, intelligence, education, and cultural background overwhelmingly supports the concept that the *greater* the similarity between two individuals, the greater the likelihood of marital success. Only in two areas are the results questionable; these are interracial marriage and personality needs. Interracial marriage has been previously discussed; with respect to personality "needs," some research indicates that complementary rather than similar personality "needs" might be of greater importance in a successful marriage.

## THE WEDDING AND EARLY MARRIED LIFE

Following the issuance of a marriage license, a wedding ceremony takes place, legalizing the marriage. In a typical monogamous marriage, the nature of the wedding and the details of the various events associated with it are usually arranged by the principles involved and their families. For some couples marriage is considered a sacred act with important religious meanings and the marriage vows are said according to ceremonial rites before a priest, rabbi, or clergyman. For other couples, marriage is a civil rite required by law binding a couple together legally. It can be performed by certain authorized officials as a judge or justice of the peace. Some marriages are simple ceremonies performed before members of the families involved or intimate friends. Others are elaborate public ceremonies preceded and followed by social events, such as showers, bachelor parties, and receptions. All weddings require official witnesses.

A modification of the traditional marriage is an *elopement* or a *secret marriage*. An elopement is a marriage in which the two participants run away and are married and make known the marriage after the ceremony. In a secret marriage, the wedding is not revealed until some time after it has taken place. Elopements and secret marriages occur for various reasons. Sometimes parents are opposed to a couple marrying or the participants may simply wish to avoid a public wedding. Pregnancy of the bride is a common reason. However the conditions that usually necessitate an elopement or a secret marriage are often such that the chances of a successful marriage are lessened. A much higher divorce rate is encountered in these marriages.

Following the wedding, the newly married couple usually takes a vacation trip before settling down and establishing a new home. This period, called the *honeymoon,* serves as a time in which the first steps in adjustment to family life take place. It is a period in which a couple, for the first time, can freely associate with each other in a way not possible before the wedding; they will begin to really learn about each other. It is the time which marks the transition from a single to a married state, a period characterized by separation of each from their families and the formation of a new family. The honeymoon may be spent in a quiet secluded place, usually a welcome relief from the hectic activities usually associated with a wedding, or it may involve extensive travel. When travel is a part of the honeymoon, overcrowded activities and a too-extended schedule are usually unwise during this important period of adjustment.

Sexual experiences on the wedding night and during the honeymoon are looked forward to usually with mixed feelings of anticipation and anxiety. As mutual satisfaction and sexual adjustment is stressed in most marriage manuals or books on newly married life, inability to experience exhilarating pleasure may be regarded as failure. Difficulties can be avoided by planning and preparation. Premarital consultations by both husband and wife with a gynecologist are highly recommended. Advice as to the most desirable contraceptive method and instructions as to its effective use can be obtained. A physical examination of the bride to determine the possible existence of a resistant hymen is desirable.

Oftentimes the first sexual encounters are unsatisfactory. Failure of the bride to experience an orgasm is common; premature ejaculation or inability to maintain an erection are common problems of grooms. The inexperience of the bride and the awkwardness and lack of tenderness and feeling of the husband are often causes of sexual disharmony. Mutual understanding and consideration is of primary importance. Often attitudes developed early in life that sexual activities are wrong or sinful are carried into married life. Changing to regard such as proper and essential is rarely accomplished overnight. Sexual hangups may last a long time and some require outside help to resolve.

*Posthoneymoon Life*
This is the period of true adjustment. Marriage is usually the culmination of a romantic love affair; the honeymoon a continuation of the courtship. Following the honeymoon, every couple settles down and begins to experience the realities of married life; the realities of everyday living. The idealized picture each has of the other is likely to be altered, sometimes shattered. True behavior, often concealed during courtship, begins to be revealed; problems that were once avoided or ignored must now be solved. However, if the couple is emotionally mature and if each was successful in choosing the right mate, with the aid of love and understanding,

most problems can be resolved and the basis of a permanent marriage established. In this process, the romantic love of courtship changes to the mature love of marriage.

An important part of this adjustment is that pertaining to sexual activities. Sexual satisfaction in marriage involves a considerable amount of sexual adjustment. Although most new brides and husbands are reasonably well informed on the basic facts of sexual anatomy and physiology, the emotional aspects are often ignored. These can only be learned through personal interaction and understanding. Sometimes a period of weeks or months may elapse before satisfactory sexual relations are established.

Sexual satisfaction and harmony increase as love and understanding of each other increase. As couples mature physically and emotionally, sexual feelings, actions, and expectations change. Sexual adjustment is necessitated throughout an entire married life as one's sexual desires and needs change with age, experience, physical conditions, and other factors. Patience and understanding, love and consideration for each other are of primary importance.

## PARENTHOOD

Most married couples expect to have children although the desire for children is not a motivating factor for most marriages. The reasons for such are complex. Some regard it as their religious duty to have children; most expect to receive satisfaction and pleasure and a sense of fulfillment in raising a family. With effective methods of contraception now available, having a child now is usually a deliberate, planned action.

Children have far-reaching effects on a marriage. For some couples, children may cement the bond between husband and wife, provide more common interests between the two, increase the general happiness in a home, and reduce the possibilities of discord. For other couples, children are a disrupting influence. A new child may lessen the feeling between husband and wife, the child's needs coming before those of the parents. A child may reduce a wife's sexual responsiveness and seriously interrupt sexual relationships. Instead of increasing happiness in a home, a child or children, by increasing responsibilities and difficulties, may have the opposite effect. Instead of being a preventative of divorce, they may be precipitating factors. Whether children make marriage more or less happy cannot be answered definitely.

As a consequence, the decision to have a child or children is an important decision. Most parents regard having children as a rewarding experience and derive pleasure from them. They enjoy playing with them, instructing them, and watching them grow up. They get much pleasure from their children's accomplishments and they have a feeling of accomplishment when their children mature, assume a useful role in society and begin to raise their own families. Some feel that, as parents, their purpose in life has been accomplished.

However many couples, for various reasons, voluntarily wish to remain childless. With the availability of new and improved contraceptive techniques, childlessness is now within the reach of all. Also there are many couples who desire to have children but, for various reasons, are unable to do so. For these couples, adoption is an alternative to conception. Oftentimes a presumably "sterile" couple will adopt a child, and then, not infrequently, conception will take place shortly after adoption. In such cases, psychological factors are probably the basic cause of the sterility.

*Adoption* is the process by which a legal relationship of parent and child is established between persons not so related. When finally legalized, the adopting parents and adopted children have the same rights, obligations, and relationships as those that exist between parents and their natural children. An adoption is generally permanent and can be voided only in exceptional cases. As adoption proceedings are generally of a complicated nature, it is recommended that the services of a lawyer be obtained to assist in the preparation of all necessary documents and to safeguard the interests of the adopting parents and the child.

Adoption is resorted to by couples who are infertile or by couples who are fertile but, for various reasons, do not wish to have additional natural children. It is also resorted to by couples, or sometimes individuals, who wish to provide care for children who have been orphaned, or whose natural parents have deserted them, or are unable for physical or mental reasons to care for them. Steps in adoption of a child involve the filing of a *petition* in an appropriate court giving information about the various parties involved. A judicial hearing is then held in which the qualifications of the adoptive parents are examined following which the petition is granted or denied. In most states, a *probationary* or *trial period* of six months to a year is required during which the ability of the adoptive parents to take care of the child and the child's ability to adjust satisfactorily to his new parents is checked. If conditions are satisfactory, the adoption is made permanent; if not the child is usually removed from the custody of the foster parents.

Most children who are adopted are illegitimate. The mother usually voluntarily gives up the child for a number of reasons: her inability to take care of it, the possibility that the child would interfere with future prospects of marriage, to avoid the stigma of illegitimacy for the child, or to enable the child to grow up in a more favorable home environment. In all cases, consent of the mother is required unless the child is abandoned or neglected.

Today there are more couples desiring to adopt a child than there are children available for adoption. This situation is the result of several factors that have reduced the number of available children. Fewer out-of-wedlock mothers are surrendering their children for adoption, abortion is more readily available, and the adoption of children holds little or no social stigma.

Legal adoptions are of two types; *private* or *independent* adoptions and *agency* adoptions. Independent adoptions are those which are arranged for privately through an agreement between the natural parent or parents and the adoptive parents, oftentimes a relative. Generally most adoptions are arranged through state-licensed private agencies or public welfare agencies to whom a child has been given by one or both of the natural parents. Since the agencies are regulated by law, these agencies offer the most complete protection to the natural parents, the adoptive parents, and the child. Because of the extreme demand for adoptive babies, there exists a "black market" in which the buying and selling of illegitimate or unwanted children exists. However, the payment of money other than for medical and hospital expenses and reasonable lawyers fees is strictly illegal and an adoption may be voided if it can be shown that the child was given up under duress or because of threats or misrepresentations.

Depending on the state and the agency, various eligibility requirements designed primarily to protect the welfare of the adopted child are required of persons who wish to adopt a child. Briefly, they are as follows. The foster parents must be married, living together, and both must consent to the adoption. They must be adults, usually 10 to 15 years older than the child, however they must not be too old, usually 35 or 40 being the maximum age. They must possess the character, physical and mental health, and social status necessary to provide a decent and adequate home for the child, and they must have the financial means to support and take care of the child. Most states require that adoptive parents be residents for a certain period of time in the territory of the agency.

Some states restrict adoption to married couples who can certify that they are unable to have children of their own. However, it is not unusual for a childless couple to conceive following the adoption of a baby, the previous infertility in such cases being the result of psychological problems. Psychogenic infertility, though, accounts for less than 10 percent of couples who are infertile, consequently conception following adoption is a relatively rare occurrence.

Adoptions are also generally limited to couples who are psychologically stable. Adoptions to save marriages that are falling apart or the adoption of a child to replace a child lost through illness or accident are not recommended. Single women, divorcees, widows, and single men are usually not permitted to adopt children except under exceptional circumstances, since it is generally considered that the optimal conditions for the normal development of a child are those encountered in a normal family situation. Singles wanting to adopt a child are often limited to taking an older child, a delinquent, or a physically or mentally handicapped child.

Some states require that foster parents and children be of the same race and, generally speaking, adoption agencies follow the policy of not crossing racial lines when placing a child. However, in areas where there are more minority children than adopting minority parents, this policy may be ignored. Also there is an increasing number of white families who desire to adopt black children as an aid to the process of desegregation and integration. Interracial adoptions, however, raise the problem of the adopted child losing his sense of racial identity and suffering an identity crisis in growing up in a family with a different cultural background. Similar problems may develop when the foster parents have a different religion from that of the child's natural parents. Consequently, adoption agencies generally try to place children in the homes of couples or families of similar cultural and religious backgrounds.

Because of the scarcity of adoptable children in the United States, many prospective parents are turning to the adoption of a child from a foreign country. This requires the approval of the U.S. Immigration and Naturalization Service. This agency investigates both the child and the circumstances of the prospective parents before deciding upon what action to take. Foreign children who are adopted must be registered as aliens for their first two years in the United States. Following that period, naturalization papers may be filed for citizenship.

A problem involving most adoptive parents is when to tell the child that he or she is adopted. Some say that a child should be told at as early an age as possible; others disagree maintaining that between the age of 6 and 10 is the best time. While there is no specific time that is applicable to all adopted children, the circumstances differing with each child, the child should at least be informed before he finds out from extrafamilial sources.

A problem frequently encountered in adopted children is, on reaching the age of 18 to 20, their desire to seek to learn the identity of their natural parents and to make an effort to find them. Because all adoption proceedings are confidential and records of the hearings unavailable except through a court order, an adopted child may have difficulty in learning the names of his or her real parents unless the foster parents choose to reveal the information. New birth certificates are usually issued to adopted children under the new family name. A new organization called the Adoptees' Liberty Movement Association (ALMA) has been developed to assist adopted children in locating their biological parents. One of the aims of this organization is to bring about a change in adoption laws so that records pertaining to specific adoptions that are now unavailable will become accessible to those who desire to know and seek to find their biological parents.

That adoption is now a widely accepted and socially approved procedure is indicated by the fact that in the United States from 1 to 2 percent of the population or about 2.5 million persons have been adopted. Such indicates that parenthood is primarily the responsibility and commitment to the care of a child and not restricted to a child's conception and birth.

## EXTRAMARITAL SEXUAL RELATIONS
Most marriages are entered into with the expectation that sexual activities are to be restricted to one's mate. When sexual relationships are satisfactory and as love and affection grow, such an arrangement may be mutually satisfactory. Many individuals tend to be faithful sexually and they usually expect the same of their spouse. However, for various reasons, extramarital affairs occur in many marriages. According to recent studies, 50 to 70 percent of males and nearly an equal number of females have experienced some sexual relations outside of marriage.

In many cases, it is a single affair involving only one other person. In others, it may be occasional affairs involving different persons, or it may be a chronic, often repeated situation. Affairs are usually carried on secretly, especially if the couple has children, although sometimes they may be carried on openly with the knowledge and consent of both parties.

The causes of sexual infidelity are numerous and varied. In most cases it is the inability to experience satisfactory sexual relations with one's mate, often due to lack of understanding or lack of concern for the partner involved. As a consequence, individuals seek to fulfill their sexual needs outside of marriage. Such conduct usually has a disruptive effect on a marriage, as is indicated by the number of broken marriages, and the fact that infidelity and sexual incompatibility are two of the major complaints in applications for divorce. However, the question arises as to whether adultery is the primary cause of marital disruption or a symptom of more deep-seated marital problems. Usually there is a general dissatisfaction with emotional and other aspects of the marriage.

A few years ago, a new form of adultery known as *mate swapping* or *swinging* came into vogue. In this activity, both husband and wife engaged in some form of sexual activity with one or more married couples with each other's consent. Both engaged in swinging at the same time and place and often in view of the other or others engaged in swinging. It was usually engaged in as a recreational activity with no emotional involvement between the participants. Proponents of swinging claimed that it counteracted boredom in marriage and increased sexual interest in their mate or partner, provided new sexual experiences, facilitated the shedding of sexual inhibitions, and broadened their sexual horizons. On the negative side, males often found that it was difficult to live up to their expectations of being extremely potent sexual performers, and personal jealousies often developed. Repulsive couples may have been encountered, exposure to venereal disease was always a possibility, and the hazards of being discovered and exposed to members of their families and associates were always present. Swinging was also criticized as being too mechanistic and lacking in any elements of personal commitment or involvement that are essential for the achievement of greatest satisfaction in sexual relationships.

Who engaged in swinging? Most swingers were "responsible" middle-age individuals of middle-class suburbia who did not regard their behavior as deviant but instead regarded themselves as the avant-garde of a new form of sexual behavior involving complete freedom in sexual relationships. By others, however, swinging was regarded as a deviant activity and it was condemned as being a manifestation of neurotic behavior.

## TYPES OF MARITAL RELATIONSHIPS

Although the typical monogamous marriage and its resulting nuclear family (father-mother-offspring) is still the norm throughout the greater portion of the world, various other relationships are being developed that are at variance with the family unit of the past. These have resulted from the changing role of the family in society and the development of a new sexual freedom among young people, especially in the United States and European countries with accompanying changes in sexual behavior patterns. Of importance also are changes resulting from the development of the Women's Liberation Movement with the elimination of sex-role differences and the granting of greater freedom to women. Some of the types of relationships are the common-law marriage, the consensual union, the single-parent family, the contract marriage, the renewable marriage, and the group marriage or commune.

### Common-Law Marriage

Recognized in many states, common-law marriage does not require a license or ceremony. Under common law, a boy or man over the age of 14 and a girl or woman over the age of 12 need only to take each as man and wife and live together for a period of time. In some states, even if the specified period of time is met, parental consent is required if the male is under 21 and the female under 18.

### Consensual Union or Arrangement

This is the practice of a couple living together, to all appearances, as a married couple but lacking legal sanction. This life-style of formal uncommitment has become common among college students, divorced individuals, and others who feel that legal ties are a contributing factor to the breakup of marriages. They further believe that freedom to terminate the arrangement leads to greater regard for and consideration for each other, qualities often lacking in a convential marriage.

Among college students, *cohabitation* or living together without formal marriage vows is becoming an increasingly viable alternative to the traditional sequence of steady dating through engagement to marriage. The arrangement may or may not be marriage oriented and it may or may not involve unrestricted sexual intimacies. It is estimated that between 20 and 30 percent of college students participate in this type of relationship.

Most of the relationships are of a temporary nature only about a third lasting longer than six months. This high failure rate might be taken as an indicator of lack of success in this mode of life, however it must be considered that most couples who enter into this relationship do not want a permanent union as provided by marriage. Usually the woman hopes and expects that it will lead to an eventual marriage whereas the man is more likely to regard it as an alternative to marriage.

Motives for this type of relationship are varied. The couple has a greater freedom than they would have in the traditional engagement or marriage situation; they are provided with a homelike atmosphere within the financial abilities of each; each has a readily available sexual partner under conditions of privacy not available when living separately; and each gets to know each other in a way not possible when living apart. There is consequently less likelihood of each developing an idealized image of each other as so often occurs when a couple enters marriage through the traditional route.

However, cohabitation carries certain disadvantages. A couple is sometimes socially isolated from disapproving parents and friends; the freedom of each to have varied dating experiences is limited; it permits the sexual exploitation of either or both of the participants; and its ease of termination may lead to severe emotional problems. The idealness of the setup is of course altered if the woman becomes pregnant.

Most cohabiting students report that they hope to get married at some future time but not necessarily to the person they are living with. The relationship is something of the nature of an *apprentice* or *trial marriage* in that it shows how each partner can adapt to the institution of marriage and it is also a test for compatibility. Many couples who live together for a time eventually end up getting married to each other.

Consensual union is becoming a more common practice among the retired segment of our society. Widows and widowers who are thrown together in retirement homes or communities often desire to get married but to do so would mean a loss of economic benefits under our present Social Security system. As a result, couples who would like to get married find it is more rewarding financially and sometimes actually necessary to live together unwed. Needless to say, children of these individuals often regard the consensual type of living arrangement quite unacceptable for their parents whereas for themselves, it is considered a desirable and acceptable practice.

### The Single-Parent Family

Now becoming commonplace, this arrangement usually results when a girl or woman becomes pregnant and chooses to bear and raise her baby rather than submit to an abortion or to marriage to a boy or man who might make an undesirable husband and father. It also occurs when an unmarried man or woman adopts a child to raise as is becoming a common practice. This type of "family" is increasing in numbers with the growing acceptance of out-of-wedlock pregnancies and the lessening of the stigma of illegitimacy. The increasing independence of women is also a factor in the increase in this type of family relationship.

### Contract Marriage

This is one in which the participants agree prior to marriage to definite commitments or stated responsibilities with respect to allocation of work, relations to friends, continuance or development of outside careers, sexual relationships, and other aspects of married life. This has led to the concept of an *open marriage* in contrast to the *closed marriage* as a possible new life style for couples.

All marriages essentially are contracts, legal and psychological, which bind a husband and wife together. Many couples are now examining more carefully the nature of the contract that will bind them together for life or until legally separated. The type of marriage that most couples enter into unwittingly is the *closed marriage* which is the conventional type of marriage. It basically involves the establishment of a couple unit in which each belongs to the other, possessiveness playing a pronounced role in the behavior of the couple. In such a marriage, the freedom of each is restricted, fidelity and mutual exclusiveness being expected. Adaptation to the desires of one's mate often requires the subjugation of one's identity and individuality. As a result, the growth and development of each is accomplished at the sacrifice or expense of the other.

To counteract the shortcomings of this type of marriage and to provide for greater individual freedom and intellectual growth, a new type of marriage called the *open marriage* is now being advocated. In this type of marriage, various unrealistic expectations of the conventional marriage, especially the idea that in one person, your mate, you will find the person who will fulfill all your physical, emotional, intellectual, and sexual needs is recognized, and more realistic expectations are established and steps taken to fulfill them. In an open marriage, the concept of mutual growth and development is encouraged in both, especially the right to independent development of intellectual interests and career activities. Through an open marriage, mutual giving and sharing are experienced and closer interpersonal relationships developed with a possible strengthening effect on the marriage bond.

### Renewable Marriage

This proposed type of marriage is one for a limited length of time with the option of renewal upon mutual consent of both parties at the expiration of a specified time, for example every four years. This type of marriage, as proposed by Dr. Margaret Mead, a well-known anthropologist, would actually be for young people, a "two-step marriage" in which the first step or phase would be a trial period during which they would have no children. If the relationship was satisfactory, a second license could be obtained and the marriage made permanent.

### Group Marriage or Commune

This type of marriage is a departure from the traditional, monogamous marriage in that it involves small groups of males and females with their children who live together either under one roof or in a closely knit community. The group goes under various names as "family," tribe, commune, or intentional community. In some groups legal marriages and monogamous sexual relationships prevail but most involve *polygamous* (multiple wives) or *polyandrous* (multiple husbands) relationships.

Members of communes are mostly individuals who are disillusioned with traditional monogamous family life and seek to avoid its monotony and loneliness. Indiscriminate sex helps to solve some of the problems of sexual boredom and collective parenthood helps to reduce the burden of the care of offspring. Being a part of the group provides the feeling of belonging and being needed.

Communes are successful only to a limited extent. Most last only a short while although a few have persisted for several years. The successful ones are those which have a strong, patriarchal type of leader, a selective membership, and a more or less regimented routine of daily life with definite economic and familial responsibilities. Difficulties encountered are those typical of most group-living experiments, such as the lack of harmony among members and consequent inability to develop stable, enduring, emotional relationships. The sexual freedom of newly established communes usually is replaced after a time with limited monogamous associations, much like those of society in general.

In the various forms of associations proposed to replace the monogamous family, one of the major problems is that of child care. Who shall be responsible for the nurture and welfare of children? The mother? The father? The mother and father? Or the state through its social agencies? What will be the effect of lack of close parent-child relationships on the development of children?

Studies in general have shown that children who are separated from their parents in early years and grow up deprived of parental love and affection suffer a higher incidence of emotional disturbances than those who grow up in stable families. The development of close bonds of affection with parents is important in normal emotional development. Children who are deprived of such an opportunity during the early years of development exhibit the most pronounced ill effects.

## FAMILY DISSOLUTION

Most couples marry with the expectation that their marriage will be successful and enduring. However, for various reasons, married couples often find themselves so incompatible and conditions of everyday living so difficult that to continue living together would result in harm to themselves or their children. When this point is reached, or before, outside advice should be sought. Various counseling agencies are available for consultation and advice. Family-life or marriage counselors, clergymen, psychologists, and others may be consulted. Frequently with the aid of wise counseling, a marriage can be saved. When differences are irreconcilable, the following courses of action are available: separation, annulment, or divorce.

### Separation

This is the situation in which a husband and wife part company and maintain separate living quarters. The marriage is still effective and marital rights or obligations are in no way changed. Sometimes the separation is voluntary on the part of both and a legal agreement entered into with respect to care of the children and division of property. When either the husband or the wife leaves home without the consent of the other, the situation becomes that of *desertion*.

### Annulment

This is a court order declaring that a marriage never legally occurred. It restores both parties to the status they enjoyed before the marriage occured. It is based on the principle that the marriage was the result of fraud or misrepresentation, hence the contract was illegal. Possible causes include concealing a previous marriage or divorce, misrepresentation of chastity, marrying under legal age, fraudulent intent not to perform marriage vows, and incapacity to consummate a marriage. Annulment is the only method recognized by the Roman Catholic Church for terminating a marriage, since it believes in the concept of an indissoluble marriage.

### Divorce

A divorce decree, when granted, legally terminates a marriage at the time when the decree becomes final and it restores both parties to the unmarried state. It is the most widely used method for terminating a marriage. The number of divorces in the United States totaled about 1.2 million in 1980. Divorce rates are highest in the poorly educated, low-income groups; they are higher for nonwhites than whites and higher for childless marriages than marriages with children. An exception to the latter is in marriages in which the participants are below the age of 21.

The number of divorces has been steadily rising in recent years, due primarily to the steady increase in population in which the number of marriages has been increasing. But many other factors are involved. Among those of significance are (1) changes in the nature of marriage in which the patriarchal monogamy dominated by the father has been replaced by a relationship granting greater freedom and independence to the female, and (2) the secularization of marriage with the consequent lessening of religious restraints on divorce. Other factors of importance are increasing urbanization with a lessening in the importance of the home as a family center, the increasing mobility of the population, the increasing boredom of marriage, the increasing demands of women for personal growth and development, a more tolerant attitude of society towards divorce, the lessening of legal restrictions, and a simplification in the processes of obtaining a divorce.

Most divorces occur between the second and fourth years following marriage. All states have a residency requirement that varies from six weeks in Nevada to five years in Massachusetts. For a divorce decree to be binding, both parties, or their representatives, must appear before the court issuing the decree. In most states proof of marital offense is required before a divorce is granted. The most common offenses acceptable as grounds for divorce are adultery, insanity, conviction for a felony, alcohol or drug addiction, desertion, nonsupport, impotence, and cruelty (physical or mental). In some states, irreconcilable differences that cause an irreparable breakdown of the marriage are considered to be justifiable grounds. Cruelty is the most common cause of divorce being the primary cause in 52 percent of all divorces. Desertion ranks second with 23 percent.

A divorce decree may be interlocutory or final depending on the laws of the state where the action is taken. An *interlocutory decree* is one that does not become final until after a period of time—30, 60, or 90 days or a longer period. During this time, the parties may not remarry. After the completion of the waiting period, the decree becomes *final* and the parties are free to remarry. A divorce not only terminates the marriage but it also settles problems that result from a divorce such as who will have custody of the children, division of property, support of the wife and children, and related matters.

Because of the realistic recognition of the actual causes of divorce, there has been a trend in recent years toward the adoption of *no-fault divorce laws* in which the court gives formal and legal recognition to the fact that a true and viable marital relationship does not exist. It replaces the traditional procedure that required that one of the married pair sue for divorce bringing charges of marital misconduct against the other. As a consequence, the defendant in the action, if judged guilty, was often subjected to various penalties and the plaintiff frequently unduly rewarded.

In practice, the no-fault principle has actually been in effect for many years as 90 percent of all divorce cases are not contested, the marriage partners agreeing to the divorce and the terms of settlement prior to the time the case comes to trial. And, in most cases that are contested, the problem is not the preservation of the marriage but settlement of related problems as support, custody of the children, and division of property.

At least 13 states have adopted no-fault divorce laws, most of which accept "irreconcilable differences leading to irrevocable breakdown of the marriage" as justifiable grounds. Existing grounds for divorce have been broadened and liberalized some now including simple incompatibility. In some states, no-fault "divorce kits" are available that enable marital partners who desire a divorce to obtain one for a nominal fee without requiring the services of a lawyer.

Divorce has brought into existence an alternative to lifetime, monogamous marriage. It is called *serial monogamy* and involves divorce and remarriage a number of times. While such is not unfamiliar in the United States, the more usual pattern is for the divorced parties, both men and women, to remarry, and in a surprising number of cases, the second marriage turns out to be more successful than the first. The reasons for this are that both parties are older and more mature and they can often see what caused the first marriage to fail and seek to avoid failure in the second. Furthermore, the expectations of marriage are more realistic and sex is usually relegated to a lesser role than in the first marriage.

The welfare of children is an important factor to be considered in a divorce. It is often assumed that children are severely harmed emotionally by the breakup of a marriage and parents are often encouraged to stay together for the sake of the children. This however is not recommended.

Studies indicate that, in general, although divorce does present a crisis for a child and children from broken homes do experience emotional and adjustment problems to a greater extent than children from unbroken homes, they still grow up better adjusted than the children of parents whose marriages remain intact but whose homes are shattered by discord and unhappiness.

# FOUR
# THE MALE AND FEMALE
# REPRODUCTIVE SYSTEMS

The development of a new individual results when a male reproductive cell, a *spermatozoan* or *sperm,* comes into contact with and penetrates a female reproductive cell, an *ovum* or *egg.* The male and female reproductive systems comprise the organs and structures that are involved in the production of these two cells, the bringing of these two cells together, and the development of the egg if it should become fertilized.

In addition to its reproductive functions, certain of the reproductive organs produce hormones that have pronounced effects on the entire life of an individual. These hormones (androgens, estrogens, and progesterone) influence bodily development, behavior, and emotions involving the whole psychosomatic complex of an individual.

The *male reproductive system* (Fig. 2) includes (1) the *testes,* which produce the spermatozoa. Each testis is suspended from the body wall by a *spermatic cord* and both are enclosed within a sac, the *scrotum;* (2) a system of paired ducts, the *epididymis, ductus or vas deferens,* and *ejaculatory duct,* which convey spermatozoa to a single *urethra;* (3) accessory glands, which include the *bulbourethral (Cowper's) glands, seminal vesicles,* and *prostate gland;* and (4) the *penis,* a copulatory organ through which the urethra passes.

The *female reproductive system* (Fig. 3) includes (1) the *ovaries,* which produce the *ova;* (2) two *uterine tubes* or *oviducts,* which convey ova toward the uterus and in which fertilization occurs, (3) the *uterus,* in which development takes place if the ovum is fertilized; (4) the *vagina,* which serves as a copulatory organ and birth canal; and (5) the *external genitalia* or *vulva* (labia majora, labia minora, clitoris, vestibule, and mons veneris), structures at the entrance of the vagina. The vagina and uterus also serve as a pathway for spermatozoa in their course to the uterine tube where fertilization occurs.

## EARLY DEVELOPMENT OF THE REPRODUCTIVE SYSTEMS
During the fifth and sixth weeks of intrauterine development, the genital organs make their appearance. At this time, the sex of an embryo cannot

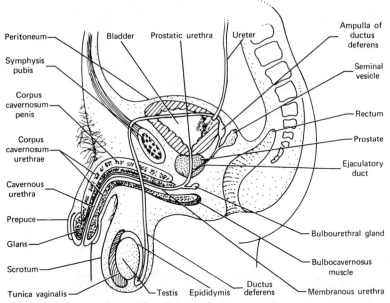

**Figure 2** Diagrammatic sagittal section of the male pelvis showing reproductive organs and related structures. (From E. Steen, *Laboratory Manual and Study Guide for Anatomy and Physiology.* **Wm. C. Brown Co., Dubuque, Iowa.)**

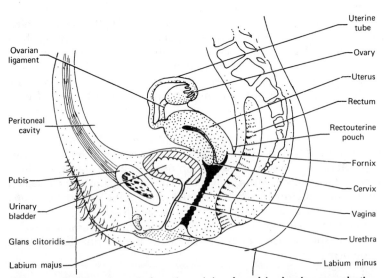

**Figure 3** Diagrammatic sagittal section of female pelvis showing reproductive organs and related structures. (From E. Steen, *Laboratory Manual and Study Guide for Anatomy and Physiology.* **Wm. C. Brown Co., Dubuque, Iowa.)**

be determined, hence the period constitutes an *indifferent stage* (Fig. 4). The reproductive system consists of a pair of generalized sex glands (*gonads*) and a double set of sex ducts. These are capable of developing into either a male or female set of organs. Basically, however, the system is essentially female.

If the genetic sex is *male,* as determined by the egg being fertilized by a Y-bearing spermatozoon, the primordial germ cells migrate from their point of origin in the embryonic yolk sac to the gonad and establish themselves in the inner portion or medulla. This occurs about the sixth or seventh week. There they produce a substance that induces the development of this structure into an organ capable of becoming a functional testis. It begins even in the embryo to elaborate androgens that act to suppress the growth of female structures and stimulate the development of male structures. The male embryo does not have any ducts intended primarily for the conduction of sperm from the testes so it appropriates a pair of ducts, the *mesonephric* or *Wolffian ducts,* which are primarily excretory ducts, and transforms them into sperm ducts. The urethra, the duct from the bladder, is also utilized for sperm transport. The testes during development shift their position migrating downward. During the seventh to ninth months, they pass from the abdominal cavity through the inguinal canal and assume their final position in the scrotum.

If the genetic sex is *female* as determined by the egg being fertilized by an X-bearing spermatozoon, the primordial germ cells, on arriving in the primitive gonad, establish themselves in the cortex. There they induce

**Urogenital homologies**

| MALE | EARLY INDIFFERENT STAGE | FEMALE |
|------|-------------------------|--------|
| Testes | Gonad | Ovaries |
| *Vestigial structures* | Genital ligaments | Suspensory ligament of ovary<br>Broad and round ligaments of uterus |
| Efferent ducts of testes | Mesonephric tubules | *Vestigial structures* |
| Epididymis<br>Ductus (vas) deferens<br>Seminal vesicles<br>Ejaculatory duct | Mesonephric (Wolffian duct) | *Vestigial structures* |
| *Vestigial structures* | Mullerian duct | Uterine tubes<br>Uterus<br>Vagina (upper portion) |

**Figure 4.   Homologies of the male and female internal urogenital organs.**

**Urogenital homologies**

| MALE | EARLY INDIFFERENT STAGE | FEMALE |
|------|-------------------------|--------|
| Urethra | Urogenital sinus | Vestibule |
| Prostate gland | | Vestibular glands |
| Bulbourethral glands | | |
| Urethral glands | | |
| Penis | Phallus | Clitoris |
| Scrotum | Genital swellings | Labia majora |

Figure 5.    Homologies of male and female external genitalia.

neighboring cells to surround them and the primary follicles of the ovary, each containing a single ovum, are developed. These increase greatly in number during late fetal life and at birth the number in both ovaries totals about 400,000. These comprise the entire number of follicles produced by a female. Most of these undergo regression and, of the total number, only about 400 (one a month) develop to maturity and liberate a functional ovum during a woman's life.

Present in the embryo in the indifferent stage, is a pair of ducts, the *Mullerian ducts*. In a female, the upper ends of these ducts become the *uterine tubes*; the lower portions fuse and develop into the *uterus* and upper end of the *vagina*. In the male, these ducts degenerate and only minute vestiges of them remain in the adult (Fig. 4).

Following development of the gonads and internal sex structures, the external genitalia develop appropriate for each sex. As with the internal organs, an indifferent stage persists for about eight weeks. A structure called a *phallus* becomes a *penis* in the male and a *clitoris* in the female. Two folds, the *genital swellings,* become the *scrotum* in the male and *major labia* in the female. Accessory glands, the *prostate* and *seminal vesicles,* develop in the male; a *vestibule* with an opening into the vagina develops in the female (Fig. 5).

It should be noted that the development of male structures involves marked modification of basically female embryonic structures and requires a masculinizing hormone acting during fetal development. The development of female structures is autonomous, no female hormones being required for differentiation.


## MALE ORGANS OF REPRODUCTION

### Testes

The *testes* or *testicles* are the primary male sex organs as they are the source of the male germ cells, the *spermatozoa*. Each is an ovoid organ about an inch and a half in length and an inch in diameter and contained within a sac, the *scrotum*. Each testis is divided into compartments that contain coiled *seminiferous tubules* in which spermatozoa are formed, a process called *spermatogenesis*. The structure of a tubule is shown in Figure 6a and 6b. Sperm are formed when cells from the outermost layer multiply and move to the center of the tubule (Fig. 6c). During this process, highly specialized reproductive cells are developed from generalized cells, a process called *maturation*. This involves two important changes. A reduction in the number of chromosomes occurs, the number being reduced from 46 to 23. One of the pairs of chromosomes is the XY pair, therefore, when reduction occurs, half of the sperm formed will possess an X chromosome and half a Y chromosome. These are called *sex chromosomes* as they are primary factors in the determination of sex as shown in Figure 1. The second change of significance is in the shape of the cell. The sperm acquires a tadpolelike form with a long, vibratile *flagellum* or *tail,* which endows the cell with motility.

In addition to sperm production, the testes are the source of *androgens* or male hormones. These hormones are produced by the *interstitial cells* (of Leydig) (Fig. 6b), which lie outside of and between the tubules. Androgens, of which *testosterone* is the principal one, control the development of secondary sex characteristics (distribution of hair, quality

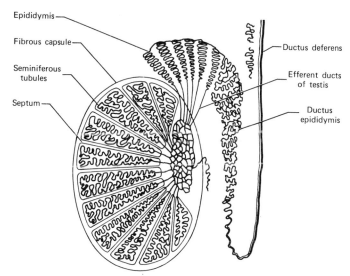

**Figure 6a** **Arrangement of the seminiferous tubules and the excretory ducts in the testis and epididymis. (From King and Showers** Human Anatomy and Physiology. **W. B. Saunders Co., Philadelphia.)**

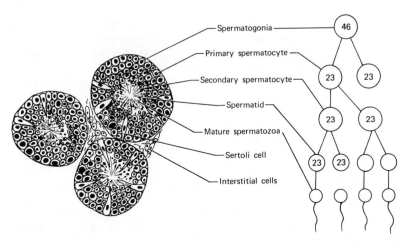

Spermatogonia — 46

Primary spermatocyte —

Secondary spermatocyte — 23   23

Spermatid — 23   23

Mature spermatozoa — 23  23

Sertoli cell —

Interstitial cells —

**Figure 6b** **Cross section of three seminiferous tubules showing structure and stages in spermatogenesis. (From E. Steen,** *Laboratory Manual and Study Guide for Anatomy and Physiology.* **Wm. C. Brown Co., Dubuque, Iowa.)**

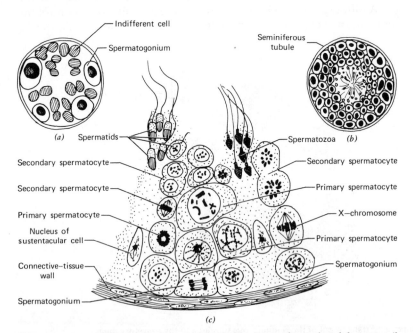

Indifferent cell

Spermatogonium

Seminiferous tubule

(a)   Spermatids —

Spermatozoa   (b)

Secondary spermatocyte —

Secondary spermatocyte —

Secondary spermatocyte —

Primary spermatocyte —

Primary spermatocyte —

X—chromosome

Nucleus of sustentacular cell —

Primary spermatocyte —

Connective-tissue wall —

Spermatozoon

Spermatogonium —

Spermatogonium

(c)

**Figure 6c** **Human testis tubules in transverse section. (a) Newborn. (b) Adult. (c) Detail of a small section of (b). (From Arey,** *Developmental Anatomy,* **W. B. Saunders Co., Philadelphia.)**

of voice, skeletal form, sebaceous gland activity). They also promote protein anabolism and the development and functioning of the accessory
sexual glands (bulbourethral, seminal vesicles, prostate) and they are primarily responsible for the male sex drive. Considerable quantities of
androgens are also produced by the adrenal cortex.

The production of spermatozoa is primarily under the control of hormones called *gonadotrophins* produced by the anterior lobe of the
pituitary. The follicle-stimulating hormone (FSH) and luteinizing hormone
(LH) are both essential for normal spermatogenesis. LH also is essential for
androgen production by the interstitial cells hence it is called *interstitial-
cell stimulating hormone* (ICSH).

Sperm production begins at puberty and, barring accident or disease,
continues until old age. The capacity of the testes to produce spermatozoa
is unlimited. Tremendous numbers are produced as the average number of
spermatozoa in each ejaculate is about 300 million. When sexual activity is
limited, excess sperm may accumulate, sometimes causing discomfort in
the testes. This is relieved by ejaculation as occurs in sexual intercourse,
masturbation, or nocturnal emissions (wet dreams).

Each testis is suspended from the body wall by a *spermatic cord* that
contains arteries, veins, lymph vessels, and nerves going to and from the
testes. The veins form an extremely convoluted network or *plexus,* the
*pampiniform plexus,* which plays an important role in regulation of the
temperature of the testis. Surrounding the cord are fibers of the *cremaster
muscle* whose fibers extend down and around the testis. Contraction of this
muscle draws the testis upward and closer to the body as occurs when the
testis is subjected to cold. Upon warming, the muscle relaxes and the testis
descends away from the body.


*Scrotum*
This pouch, which contains the testes, consists of an outer layer of skin and
an inner layer of smooth muscle, the *dartos.* Its surface varies under different conditions. Under conditions of warmth and in elderly or debilitated
persons, it is flaccid and stretched; in cold temperatures and in young and
robust persons, it tends to be tightly applied to the testes and has a corru-
gated surface. The presence of testes in a pouch outside of the body cavity
is a mechanism for reducing the temperature of the testes since scrotal
temperature is 2-3°C lower than body temperature. This is essential for
normal development of spermatozoa. If, in a fetus or infant, the testes fail
to descend and remain in the body cavity, a condition called *cryptorchism*
results; spermatogenesis does not occur and the testes fail to produce viable spermatozoa. Androgen production is not affected. Other conditions
that increase intrascrotal temperature as prolonged fever or the wearing of

tight-fitting, poorly ventilated shorts or jockstraps may result in impaired sperm production with resulting subnormal or complete infertility.

Cryptorchism in a young child can be corrected sometimes through the administration of gondotrophins. When this is ineffective, surgical intervention (orchiopexy) may be necessary. If the testes remain in the body cavity until puberty, irreversible changes take place that render the testes incapable of producing sperm. The optimum time for correction is by the fifth birthday.

### Epididymis

The epididymis (Fig. 6a) is a comma-shaped structure lying alongside each testis. It consists of a narrow tube, 18 to 20 feet long, but condensed into a structure about 2 inches in length. Its upper end or *head* is broad and receives the efferent ducts of the testis; its lower end or *tail* is narrow and continuous with the vas deferens; its central portion is the *body*.

Sperm, continuously produced by the seminiferous tubules of the testis, move passively through the efferent ducts into the epididymis. Here the sperm undergo maturation in structure and acquire an increased capacity for motility and fertility. Sperm from the testes are nonmotile and infertile.

The sperm slowly move through the epididymis being moved by peristaltic contractions of the tube. Normal passage requires about two weeks but may be faster or slower depending on frequency of ejaculation. The maturation of sperm in the epididymis and their survival are dependent on androgens produced by the interstitial cells of the testes. Spermatozoa not discharged as in prolonged periods of abstinence may die or show degenerative changes as reduced motility and an increase in abnormal forms. Sperm that die disintegrate and are resorbed or they are phagocytized.

Sometimes sperm may penetrate the walls of the tube and invade surrounding tissue. This may follow injury to the epididymis from inflammation or trauma. When this occurs, autoimmunization against spermatozoa may result, which may bring about the agglutination of sperm in the ejaculate with resulting infertility.

### Ductus Deferens and Ejaculatory Duct

The ductus deferens or *vas deferens* (Fig. 2) is a thick-walled muscular tube, 16 to 18 inches long, extending from the tail of the epididymis with which it is continuous. It passes upward from the testes in the spermatic cord and enters the body cavity through the inguinal canal. It continues backward and passes downward on the posterior surface of the bladder to where it joins the duct from the seminal vesicle to form the *ejaculatory duct,* which continues through tissue of the prostate gland to open into the urethra. The vas deferens possesses a thick, muscular coat capable of strong, peristalic contractions. It bears a dilated region, the *ampulla,* just before joining the ejaculatory duct.

### Seminal Vesicles

Each seminal vesicle (Fig. 7) is an outpocketing of the vas deferens at its junction with the ejaculatory duct. Each is a long, much-coiled, convoluted tube that forms a compact structure about $2\frac{1}{2}$ inches long. It is a gland whose secretory product contains *fructose,* a sugar, which is the principal source of energy for the sperm and a proteinaceous material essential for coagulation of semen after ejaculation. The secretion is a thick, yellow, alkaline fluid of sticky consistency.

### Prostate Gland

The *prostate* is a bilobed structure that surrounds the urethra near its origin from the bladder. It is about the size of a large chestnut and consists of some 30 to 40 individual glands whose ducts, 16 to 32 in number, open into the prostatic portion of the urethra. These glands produce a thin, slightly acid fluid that is responsible for the peculiar odor of semen. The secretion contains citric acid, acid phosphatase, and a number of enzymes that are present in semen. The prostate gland is unique in its tendency to enlarge especially in men over 60. The cause is thought to be the secretion of excessive estrogens by the adrenal cortex. Prostatic enlargement usually results in urinary obstruction and resultant infections.

### Bulbourethral Glands

These glands, also called Cowper's glands, are two small glands (Fig. 2) each about the size of a pea located behind the membranous urethra. Their ducts open into the bulb of the cavernous portion of the urethra. They, with other urethral glands, produce a clear, viscid, mucoid secretion that serves for lubrication.

### Penis

The penis (Figs. 2 and 7) is the male copulatory organ. Through it passes the *urethra,* which conveys both urine and spermatozoa. However, during sexual intercourse or at other times when the penis is erect, reflex mechanisms occur, resulting in the contraction of the internal sphincter muscle at the urethral orifice of the bladder, which makes it impossible for urine to be discharged.

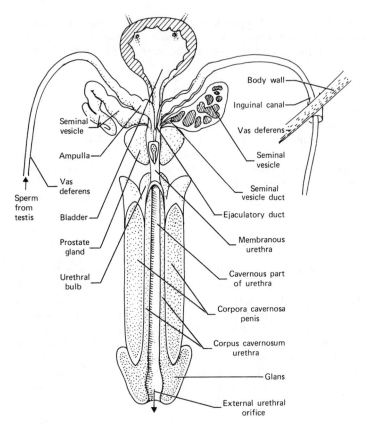

**Figure 7    Male reproductive and urinary organs. (Modified from King and Showers,** *Human Anatomy and Physiology,* **W. B. Saunders Co., Philadelphia.)**

The penis is a cylindrical organ composed principally of erectile tissue. In a flaccid state, it averages 3 to 4 inches in length and 1 inch in diameter. In an erect state, it averages 6 to 7 inches in length and 1½ inches in diameter. Its size, however, is extremely variable and has little relationship to masculine vigor or potency. In erection, a small penis increases in size to a greater extent than a larger penis.

Structurally, the penis consists of three parts, a root, body, and glans penis. The *root* is that portion by which it is attached to the pelvic floor. The *body* consists of three masses of erectile tissue, two cavernous tissue bodies, the *right* and *left corpora cavernosa penis,* which form the dorsal portion of the penis, and the *corpus cavernosum urethra (spongiosum),* which is ventrally located and encloses the urethra. (Fig. 7). The erectile tissue contains large blood spaces which, under sexual stimulation become filled with blood causing the penis to become hard and erect. The *glans penis* is the cone-shaped, distal end of the penis and is a continuation of the corpus cavernosum urethra. The skin, which is loosely attached to the body of the penis, projects over the glans forming a loose fold, the *prepuce* or *foreskin* (Fig. 2). The inner surface of the prepuce and the neck of the glans contain modified sebaceous glands, the *glands of Tyson,* which secrete a material with a peculiar odor. This substance, together with shed epithelial cells forms a white, cheeselike material called *smegma* that tends to accumulate beneath the prepuce. It decomposes readily and, unless removed regularly, may be a source of irritation.

## Circumcision

This is an operation in which the prepuce is surgically removed. It is an ancient operation of unknown origin but often done for religious purposes, when it is usually performed on infants. However it is currently resorted to primarily for hygenic reasons.

Sometimes the opening of the prepuce is excessively constricted preventing retraction of the prepuce over the glans penis, a condition called *phimosis*. This may be a congenital condition or the result of inflammation or trauma. Or once retracted, it may form a tight band about the glans penis preventing venous return resulting in a painful swelling of the glans, a condition called *paraphimosis*. Circumcision prevents the accumulation of smegma beneath the prepuce, which is often a source of inflammation with resulting development of adhesions. Circumcision is a preventative measure against cancer, cancer of the penis occurring exclusively in uncircumcised men and cervical cancer in women occurring more frequently in women whose husbands are uncircumcised.

## Erection of the Penis

Erection of the penis may occur in infants and young children before puberty. In adults it commonly results from sexual stimulation and is involuntary in nature, although it can be induced voluntarily by stimulation of various structures, expecially the sex organs. It commonly results from the action of higher nervous centers involving thought, memory, fantasies, or erotic activities. Erection is a reflex action the reflex centers being located in the lower spinal cord. The action however can be facilitated or inhibited by the functioning of the higher brain centers hence psychic factors play an important role in its occurrence and maintenance.

In an erection, the arteries to the penis are reflexly dilated resulting in an increased flow of blood to the erectile tissue. As a consequence, the spaces in the corpora cavernosa become filled with blood while at the same time the veins are constricted preventing the outflow of blood. The resulting turgidity of the cavernous bodies causes the penis to become rigid and assume an erect position.

Although most erections are the result of erotic stimuli, other stimuli may induce the response. A full bladder, phimosis, inflammation of the bladder or prostate gland, tight-fitting garments, or pressure of bed covers may initiate the reflex. Erections may occur at the beginning of sleep or during sleep, and are a common occurrence on awakening. Erections may occur quickly in a matter of seconds as is characteristic of young males. ' With age, erections tend to occur more slowly but they can be maintained for a longer period of time. Following ejaculation, detumescence quickly occurs and the penis assumes a flaccid state.

## FEMALE ORGANS OF REPRODUCTION

### Ovaries

The ovaries (Fig. 8) are the primary female sex organs as they are the source of the female germ cells, the eggs or ova. The two ovaries lie within the body cavity, one on each side of the uterus. Each is an ovoid body about an inch and a half long and three quarters of an inch in diameter and consists of a central medulla and an outer layer, the cortex (Fig. 9). The *medulla* consists of connective tissue containing blood and lymphatic vessels and some smooth muscle fibers; the *cortex* consists principally of *follicles* (Fig. 9) in various stages of development. Each follicle consists of an immature *ovum* or oocyte surrounded by one of more layers of follicle cells. At birth, each ovary contains many thousands of primary follicles each containing a cell capable of becoming a mature ovum. However, during a woman's lifetime, only a few, approximately 200 in each ovary, come to maturity and release an egg capable of being fertilized. The remainder degenerate a process known as *atresia*. The ovaries are thought to alternate in the release of eggs.

Before puberty, the follicles lie dormant and little change occurs in them. But at puberty (age 10–14), the ovary is stimulated by a *follicle-stimulating hormone* (FSH) from the pituitary (Figs. 13 and 14) and follicles are induced to develop. Normally only one follicle develops to maturity each month. The developmental processes are shown in Figure 9. Briefly they are as follows. The cells surrounding an ovum multiply and a *primary follicle* with a single layer of cells becomes a *growing follicle* with several layers. As the follicle increases in size, a cavity is formed within it and the ovum is pushed to one side where it is attached to the inner wall by a slender strand of cells. The follicle when mature is known as a *vesicular* or *Graafian follicle* and may form a bulge on the surface of the ovary. About the middle of the menstrual cycle, the follicle ruptures and the ovum, surrounded by a layer

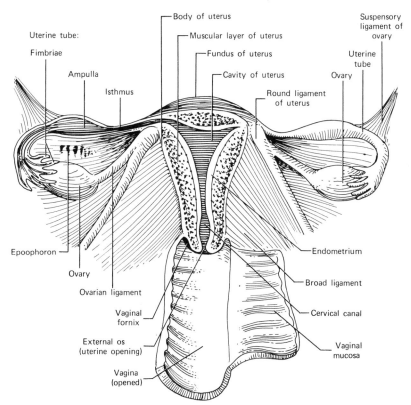

**Figure 8** The internal female reproductive organs. (**Modified from E. Steen,** *Laboratory Manual and Study Guide for Anatomy and Physiology.* **Wm. C. Brown Co., Dubuque, Iowa.)**

of follicle cells and its surrounding fluid, is expelled from the surface of the ovary into the body cavity, a process called *ovulation.*

Although there is no direct connection between the uterine tube and the ovary, the egg normally enters the expanded, fimbriated end of the tube that lies close to the ovary and partially encloses it. The egg is slowly propelled toward the uterus by the action of the cilia lining the tube and the contraction of muscle fibers in the wall of the tube. If sperm are present in the tube, fertilization usually occurs; if none are present, the egg degenerates and disappears. An unfertilized egg does not enter the uterus.

During the development of the follicle and its enclosed ovum, two important processes take place. First, the ovum, like the sperm, undergoes essential changes (maturation) preparing it for fertilization. These include a

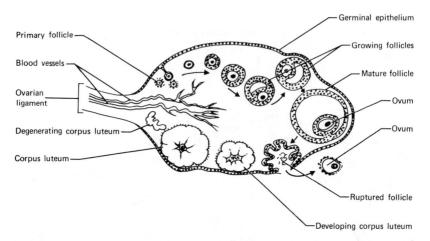

**Figure 9   The ovary showing development of follicles and the corpus luteum. (From E. Steen,** Laboratory Manual and Study Guide for Anatomy and Physiology, **Wm. C. Brown Co., Dubuque, Iowa.)**

reduction in the number of chromosomes from 46 to 23, a process not completed until the egg is fertilized. However, as the pair of sex chromosomes in the female is an XX pair, all ova, when mature, possess an X chromosome. Furthermore, the size of the cell increases slightly with the accumulation of a minute amount of yolk so that the mature egg is about 100,000 times the size of a spermatozoon. While markedly different in volume, the egg and the sperm contain the same amount of chromatin and contribute equally in inheritance. Although the egg is the largest cell produced by the body, it is still minute, being about 200 micra in diameter or about the size of a period on this page.

The ovary, in addition to producing ova, is involved in the production of hormones (estrogens, progesterone, and androgens). The follicle is the principal source of *estrogens,* which are the female sex hormones. These are responsible for the development of female secondary sexual characteristics (hair growth, quality of voice, body form, weight, and mammary gland development) and they have marked effects on the structure and functioning of the accessory sex organs, especially the uterus and vagina. They also affect behavior, especially sexual behavior; however, sexual drive and interest is only partially dependent upon ovarian estrogens.

A second hormone produced by the ovary is *progesterone,* secreted principally by the *corpus luteum,* a small, yellow mass of cells that forms within a ruptured follicle. It plays a special role in changes that occur in the uterus during the menstrual cycle and during pregnancy (Fig. 10).

A third hormone produced by the ovary is *testosterone,* an androgen.

Although produced in small amounts, it is important as it serves as an intermediate substance in the synthesis of estrogens and it is primarily responsible for female libido or sexual desire. Its source is not definitely known but it is thought to be produced by cells located in the hilus of the ovary, the region where blood vessels enter and leave.

### Uterine Tubes

Also called *fallopian tubes* or *oviducts* (Fig. 8), these are two muscular tubes, each about four inches in length, which extend laterally from the upper portion of the uterus. The outer end of each terminates in an expanded portion or *infundibulum* that bears *fimbria,* fingerlike processes that partially enclose the ovary, especially at the time of ovulation. Within the fimbria is an opening, the *ostium,* through which the egg enters. From the infundibulum, the tube expands to form the *ampulla,* then becomes short, straight, and narrow as the *isthmus* that joins the uterus. It passes through the thick wall of the uterus as the *intramural* or *uterine portion* and opens into the uterine cavity. Its inner surface is thrown into many folds and its lining cells bear *cilia,* minute, vibratile, hairlike processes.

### Uterus

Also called the *womb,* (Figs. 3 and 8), this is a hollow, pear-shaped structure which lies in the pelvic cavity above and behind the bladder. In a woman who has never been pregnant, it is about 3 inches long and 2 inches wide. It possesses a thick, muscular wall (*myometrium*) and its inner lining, the uterine mucosa, or *endometrium,* is a special type of tissue that undergoes cyclic changes during the menstrual cycle. Its outermost covering is a thin serous membrane, the *perimetrium.* The central portion of the uterus is the *body*; its lower, narrow portion is the *cervix* or neck. These two regions are separated by a slight constriction, the *isthmus.* The rounded portion of the uterus between the openings of the two oviducts is the *fundus.* The cervix, which projects slightly into the upper end of the vagina, contains the *cervical canal* through which sperm may enter or the menstrual fluid be discharged. The opening of the cervical canal to the vagina is the *external os*; the internal opening to the uterus is the *internal os.* The cervix contains mucus-secreting glands.

### Vagina

This is a muscular tube or *barrel* (Figs. 3, 8), 3 to 4 inches in length, which extends from the uterus above to the vulva below. It opens into a space, the *vestibule,* located between the minor labia. It is normally collapsed with its walls in apposition. The upper end of the vagina forms a circular groove which encircles the tip of the cervix of the uterus. This groove is divided into the *anterior, posterior,* and *lateral* (right and left) *fornices.*

The vagina lies between the urethra (tube from the bladder) and the rectum (digestive tube), consequently its external opening, the *vaginal orifice* or *intoitus* lies between the opening of the urethra (*urinary meatus*) and the *anus.* The vagina serves as the female organ of copulation and as a birth canal through which the fetus passes from the uterus to the outside. It also serves in the discharge of menstrual fluid.

*External Genitalia or Vulva*
This includes those structures that can be seen from the exterior (Fig. 10). On each side of the urogenital openings are two prominent, longitudinal folds, the *labia majora* (major lips). If these are separated, two lesser folds, the *labia minora* (minor lips) can be seen. At their anterior ends and partially concealed by them is a small, extremely sensitive, erectile structure

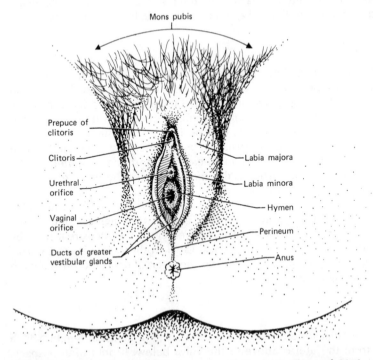

**Figure 10  Female external genitalia or vulva. (From B. G. King and M. J. Showers,** *Human Anatomy and Physiology,* **W. B. Saunders Co., Philadelphia.)**

the *clitoris*. Between the minor labia is a region, the *vestibule*, into which the urethra and vagina open. A perforated mucous membrane, the *hymen*, may partially close the vaginal opening in young females. It is broken during the first intercourse or by the insertion of a vaginal tampon or other object. Its presence however, as an indicator of virginity has been greatly overemphasized. Overlying the pubic symphysis is a rounded, elevated cushion of fat, the *mons veneris*. After puberty its surface is covered by pubic hair.

## MAMMARY GLANDS

Strictly speaking, these glands, the *breasts* or *mammae*, are not reproductive organs but, because they are closely related to the reproductive system and function only on the production of offspring, they will be described here.

The *breasts* or *mammae* (Figs. 11a and b) are two rounded bodies present on the thorax of females. They are composed principally of adipose (fatty) tissue in which the secreting structures that produce milk are embedded. These are saclike cavities called *alveoli* from which ducts lead and connect with *lactiferous ducts* that open on the *nipple*. Surrounding the nipple is a pigmented area, the *areola*.

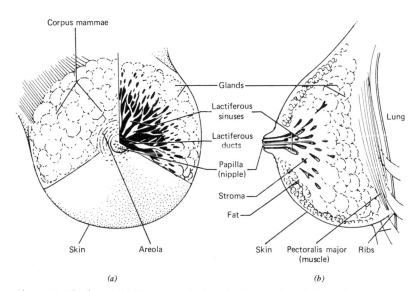

Figure 11    The breast. (a) Dissection of a lactating breast. (b) Relation of breast to chest wall. (From *Anatomical Studies for Physicians and Surgeons* by permission of Camp International, Inc. Jackson, Mich.)

Immature mammary glands consisting of a few relatively simple ducts are present in prepuberal individuals of both sexes. At puberty, in females, the glands, under the influence of estrogens, enlarge and the duct system develops. During pregnancy, under the influence of estrogen and progesterone, the duct system becomes extensively branched and secretory alveoli develop. Upon parturition, *lactation* is induced by pituitary hormones, prolactin and oxytocin (Fig. 12). Continued production of milk depends on the combined action of several hormones from the pituitary, thyroid, and adrenal glands plus the stimulus of sucking.

In the male, the mammary glands usually remain in a rudimentary state of development. However, occasionally, a male develops breasts which more or less resemble those of a female, a condition called *gynecomastia*. This results from altered androgen-estrogen balance as occurs in eunuchs or men given large doses of female hormones.

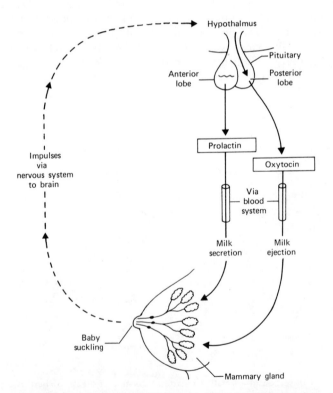

**Figure 12  Factors involved in lactation initiated by suckling at the breast. (Modified from C. J. Avers, *Biology of Sex*, Wiley, New York.)**

Following delivery of a baby, milk yield increases and then slowly decreases until the child is weaned. Milk production ceases if the glands become engorged with milk from failure of suckling or milking. Strong emotional states such as worry, fright, grief, or embarrassment may inhibit the production and ejection of milk. This probably results through the action of epinephrine (adrenalin), which has an inhibitory effect on milk ejection.

Under certain conditions, the mammary glands may secrete in the absence of pregnancy as mothers of adopted children are sometimes able to nurse these babies from the breast. In such cases, the sucking stimulus, when continued over a period of time (10 weeks or more) is sufficient to eventually initiate the hormonal production necessary to bring the glands to functional activity with resulting milk secretion. However, this can only occur in women who have had a previous pregnancy.

# FIVE
# MENSTRUATION AND
# THE MENSTRUAL CYCLE,
# MENOPAUSE

The human species is unique among the various species of mammals in that the adult female experiences a cyclic process known as *menstruation* in which there is a bloody discharge from the vagina. It is commonly thought that this process, which occurs once each month, is to bring about the discharge of an unfertilized egg. Such, however, is not the case because an unfertilized egg normally never reaches the uterus. What, then, is the significance of menstruation? Why do menstrual cycles occur?

## MENSTRUATION

This is the discharge of a fluid that contains blood, the secretions of glands, and cellular debris from the disintegrating lining (*endometrium*) of the uterus. It recurs at intervals of about 28 days, the period from the beginning of one menstrual period to the beginning of the next constituting a *menstrual cycle*.

The length of the menstrual cycle is variable. For most women, it falls within the range of 21 to 34 days but longer or shorter cycles may occur. Variations in the length of the cycle may be caused by age, endocrine disorders, illness, emotional disturbances, climatic factors, or other conditions. For only a relatively few women (10-15%) is the cycle exactly 28 days in length.

Menstrual cycles begin with the *menarche*, the first menstruation, which occurs shortly after the onset of puberty (about the age of 11 or 12). The cycles continue to about the age of 45 to 50, when they become less frequent and finally cease, a time called the *menopause*. During this interval, they occur regularly except during pregnancy and lactation when they cease temporarily.

75

Both the menarche and the menopause are related to the functioning of the ovary. The menarche indicates that the production of eggs has begun with the resultant possibility of pregnancy; the menopause indicates the cessation of egg production by the ovary and consequent inability to bear children. What then are the relationships? The connection between the two can be seen by noting the changes that occur within the ovary and the related changes that occur in the uterus as shown in Figure 10.

Activities in the ovary are regulated by hormones produced in the anterior lobe of the *pituitary gland* (Fig. 13). In turn, the pituitary gland is regulated by hormones or factors produced by the hypothalamus. The *hypothalamus* is a region of the brain lying directly above and connected to the pituitary by a slender stalk. It produces specific *releasing factors* (RF) or *hormones* (RH) (Fig. 14), or *inhibiting factors* (IF or IH). These are carried by the blood stream to the anterior pituitary where they act in stimulating or inhibiting the secretion of a number of specific hormones. The hypothalamus is also involved in the production of posterior lobe hormones. These hormones (*vasopressin* and *oxytocin*) are produced in nerve cells in the hypothalamus and pass by way of their axons to the posterior lobe where they are stored before their release. The production of releasing factors by the hypothalamus is dependent upon hormones reaching it by way of the bloodstream and nervous impulses arriving by way of the brain stem, the olfactory brain, or the cortex, or both.

The regulation of ovarian functioning is accomplished in the following manner. The hypothalamus produces a *releasing factor* (FSH-RF) (Fig. 13) which causes a *follicle-stimulating hormone* (FSH) to be produced and released by the anterior pituitary. This passes to the ovary where it initiates development of ovarian follicles. As a follicle develops, it produces estrogens which stimulate the uterine endometrium. Cells multiply, glands develop, and the endometrium becomes thicker and more vascular. It is being prepared to receive the fertilized egg. Then ovulation occurs, brought about by the action of another pituitary hormone, the *luteinizing hormone* (LH) acting with FSH. This hormone continues to act on the cells of the collapsed follicle and within its cavity a small, ovoid, yellow body, the *corpus luteum,* is formed. This structure is a minute endocrine gland that secretes *progesterone.* This hormone with estrogen, which is still being produced, continues to act on the endometrium causing it to become thicker. Its glands become more pronounced and begin to secrete a fluid, at first watery but later mucuslike. The endometrium is brought to its maximum thickness of about one-fifth of an inch (Fig. 14).

The changes that occur within the uterus are essentially to prepare the lining for the reception of a fertilized and developing egg. In most instances, however, the egg following ovulation is not fertilized and degenerates before reaching the uterus. When this happens, the corpus luteum begins to regress (*luteolysis*) and the secretion of progesterone

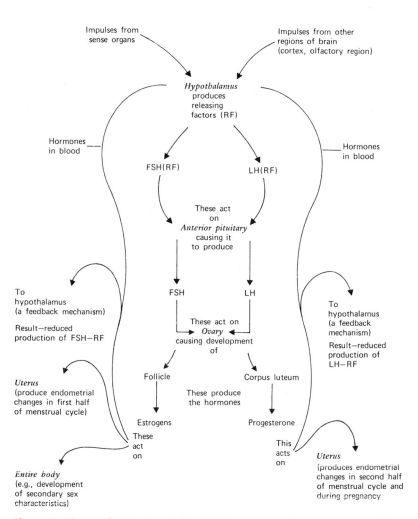

**Figure 13  Diagram showing interrelationships between central nervous system, pituitary gland, ovary, and uterus. (FSH = follicle-stimulating hormone, LH = luteinizing hormone, FSH(RF) = FSH releasing factor, LH(RF) = LH releasing factor.)**

declines. This is thought by some to be due to a substance from the uterus or to reduced LH production by the anterior pituitary or both. However, evidence from recent studies in reproductive physiology points to the local production of estrogens as being the principal factor in the regression of the corpus luteum. Since progesterone is the principal stimulus for the maintenance of the uterine endometrium, with its reduction, blood supply

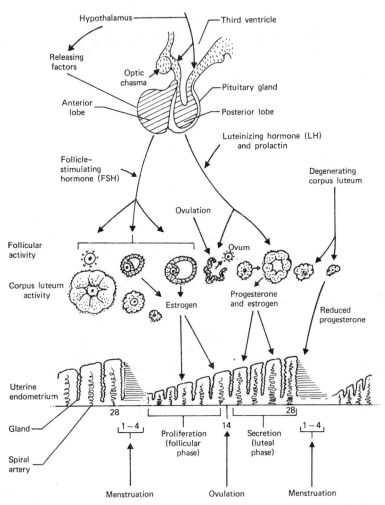

**Figure 14  Diagram showing relationships between the pituitary, ovarian, and uterine (endometrial or menstrual) cycles.**

to the endometrium is altered and cells begin to die. The disintegrating endometrium, now having no function, is discharged bit by bit, together with some loss of blood and glandular secretions. This periodic discharge from the vagina, which usually lasts four or five days, constitutes the *menses* or *menstruation*. It is a cyclic event occurring monthly. The cycles may vary in length from 21 to 34 days, most averaging 27 to 30 days. Following menstruation, the endometrium repairs itself and the cycle is repeated unless pregnancy occurs.

The events of a menstrual cycle of 28 days are summarized in the following table:

| DAYS | ACTIVITY OF ENDOMETRIUM | RELATED ACTIVITY IN OVARY |
|------|------------------------|---------------------------|
| 1–5 | Menstruation | |
| 6–7 | Repair | Development of follicle |
| 8–14 | Buildup of endometrium; proliferative phase | Increase in secretion of estrogens |
| 15–28 | Continued buildup of endometrium; secretory phase | Development of corpus luteum; increase in secretion of progesterone |
| 26–28 | Premenstrual phase | Decline in secretion of progesterone and estrogens |
| 1–5 | Menstruation | |

It is sometimes said that the uterus must be the most disappointed organ of the body. Every month it gets ready to receive a fertilized egg and prepares to take care of a developing embryo. But in 99–100 percent of the cases, the egg is not fertilized and the prepared materials have to be discarded. Every month from menarche to menopause this process is repeated unless interrupted by pregnancy or another condition, such as a pathological disorder.

Let us note what occurs if an egg is fertilized as happens if sperm are present in the uterine tubes. This usually occurs about the fourteenth or fifteenth day of the cycle. The developing egg (*blastocyst*) makes its way to and enters the uterus just prior to the anticipated time of menstruation. The lining (endometrium) has been prepared to receive the fertilized egg by the action of estrogens from the ovarian follicle and progesterone from the corpus luteum. The blastocyst on entering the uterus embeds itself within the endometrium, a process called *implantation*. The outer layer of the blastocyst, the *chorion* produces a hormone called *human chorionic gonadotrophin* (HCG), which maintains the corpus luteum. As a result, the corpus luteum does not degenerate but enlarges and continues the production of progesterone. Progesterone acts on the pituitary by way of the hypothalamus (a feedback mechanism), inhibiting the production of FSH and LH so that no more follicles are formed. It also acts on the uterine endometrium maintaining its integrity, which is essential for the completion of implantation and the formation of the placenta. Consequently, menstruation does not occur and the cycles are interrupted for the duration of pregnancy.

You might assume, from the above description, that actions that occur within the body are the result of "need." Such is not the case. Events happen because, first, there is a mechanism by which such can be accomplished, and second, there is a regulatory mechanism (nervous or endocrine) that controls these activities. Many of the activities involved in reproduction and development are under endocrine control. For example, when menstrual disorders occur, such as the absence of menstruation (amenorrhea), too frequent, delayed or painful menstruation (dysmenorrhea), the source of the trouble may be (1) the ovary, which produces the hormones that regulate the uterus, (2) the pituitary gland, which produces the hormones that regulate the ovary, or (3) the hypothalamus, which regulates the pituitary gland. However the nervous system through nervous impulses conducted over nerves also plays an important role in various reproductive activities. Most reproductive activities are reflexly controlled and, if uninhibited, take place naturally. Emotional states can markedly affect sexual activities as is indicated by the importance of psychogenic factors in conditions such as impotence and frigidity.

## MENSTRUAL DISORDERS*
Menstrual disorders include absence of menstruation (amenorrhea), premenstrual tension, and painful menstruation (dysmenorrhea).

### Amenorrhea
The absence of menstruation, this is a normal condition before the menarche, after the menopause, and during pregnancy and lactation. If it occurs at any other time, it is considered *pathologic*. Possible causes include (1) *congenital abnormalities,* an incomplete development of various organs of the reproductive system or the endocrine glands controlling them, (2) *disorders of the central nervous system* both organic and functional, (3) *various systemic disorders,* especially chronic diseases and endocrine malfunctioning, (4) *ovarian disorders,* either structural or functional, and (5) *disorders of the uterus,* as malposition, uterine tumors, or cervical stenosis.

### Premenstrual Syndrome (PMS)
It appears that most women experience some fluctuations in mood and/or physical symptoms over the various phases of their menstrual cycle. The variations in physical symptoms are most likely due to fluctuations in hormone levels, especially estrogens and progesterone. However, some wom-

*See also the section "Toxic Shock Syndrome" in the Preface to the Dover Edition, p. vii.

en who believe that "normal" women experience symptoms prior to and during their period may report such symptoms or magnify their intensity in part to make themselves appear normal. Women in whom the fluctuations in mood and physical symptoms seem to be more severe than minor changes are said to have premenstrual syndrome (PMS). The symptoms of PMS seem to vary and there is no single pattern. The research literature indicates that from two to four days before a woman's period she may develop some the following feelings in PMS: tension, irritability, anxiety, depression, nervousness, and low self-esteem. Other variable physical symptoms may include fatigue, headache, constipation, bloating, breast tenderness, and weight gain.

The incidence of PMS is uncertain. It has been reported by various investigators that from 25 to 75 percent of women have some of the aforementioned premenstrual symptoms. However, only 5 to 10 percent experience premenstrual distress of such a severe nature as to interfere with their normal functioning.

The cause or causes of premenstrual syndrome (also called premenstrual tension) are not definitely known, but most of the symptoms are thought to be related to increased water and sodium retention within the body primarily related to hormone (estrogen and progesterone) imbalance. Contributing factors such as lack of exercise and psychological disturbances may play a role. There has been a movement to set up PMS clinics for the treatment of victims but their effectiveness is questionable.

*Dysmenorrhea*

Difficult or painful menstruation (menstrual cramps), is common in adolescent girls but is not present at the menarche or for several months after the onset of menstrual cycles, these cycles comprising anovulatory cycles in which ovulation does not occur. Dysmenorrhea is associated with ovulatory cycles and usually ceases with the first pregnancy.

The pain of dysmenorrhea is usually most severe during the first 24 hours of flow. In most cases it subsides during the second day although it may persist throughout the entire menstrual flow. The pain consists of cramps, mild or severe, localized in the lower abdomen. Pain may also be felt in the lower back and thighs.

The causes of dysmenorrhea are still unknown. It may be associated with endocrine imbalance although endocrine therapy is of questionable value. It may also result from uterine abnormalities or pelvic disorders as endometriosis or pelvic inflammatory disease. Contributing factors are inadequate physical activity, poor posture, and possibly psychogenic factors. A negative attitude toward menstruation is thought to contribute to the severity of the pain. A recent theory postulates that the presence of prostaglandins in the uterine fluids may intensify uterine contractions causing vascular changes that would reduce the supply of oxygen. Reduced blood supply (ischemia) and muscle spasms are two common causes of pain.

Some women experience low abdominal discomfort associated with ovulation. This pain, called *mittelschmerz*, is a dull pain lasting only a few hours and usually affecting the lower right quadrant of the abdomen. Slight uterine bleeding may accompany, precede, or follow the abdominal discomfort. The exact cause of the pain associated with ovulation is unknown.

The usual treatment for premenstrual tension and dysmenorrhea involves the use of pain-relieving drugs or sedatives, application of hot packs, and rest. In severe cases, limitation of salt in the diet and the taking of a diuretic to decrease fluid content of the blood are resorted to, treatment beginning several days before the expected onset of menstruation. Tranquilizers may reduce nervousness and irritability. The severity of menstrual cramps is often reduced when women go on oral contraceptives or after they experience their first pregnancy.

Many women use over-the-counter (OTC) medical preparations for relieving menstrual discomfort. These commercial preparations for example, Cope, Midol, Pamprin, and Trendar are widely used. All contain a pain-relieving agent, usually aspirin combined with other components such as diuretics, antihistamines, and antacids. Whether any of these combination products is more effective than plain aspirin is questionable.

Aspirin is generally recommended as the cheapest and most effective agent. Its pain-relieving effect in dysmenorrhea is thought to be associated with its action in inhibiting the synthesis of prostaglandins by the uterus. Prostaglandins stimulate uterine contractility.

## MENOPAUSE

This is that period in a woman's life characterized by permanent cessation of menstrual activity. The menopause is considered complete after a full year of *amenorrhea* (absence of flow). Menstruations may cease abruptly but usually the menstrual flow gradually becomes less and less, the interval between the periods increases, cycles become irregular, and finally, when ovarian function ceases entirely, the menstrual cycles cease.

### Natural Menopause

This occurs usually between the ages of 45 and 50 but may occur anytime between the ages of 35 to 55. It is the result of declining activity of the ovary, which occurs as a result of the aging process. With cessation of ovarian function, the cyclic changes in the uterine endometrium also cease and eventually menstruation fails to occur. In many individuals the change is gradual lasting one or two years; in others it may occur in a much shorter time. There is a correlation between the time at which the menarche (beginning of menstrual cycles) occurs and the onset of the menopause. In general, the earlier the menarche, the later the onset of menopause; conversely, the later the menarche, the earlier the menopause.

When ovarian regression is gradual, the menopause may be without significant symptoms, however in some women, annoying and disturbing symptoms may occur. These include nervousness and anxiety, hot flashes, irritability, fatigability, depression, crying spells, and insomnia. Most of these symptoms result from endocrine imbalance that results in an unbalanced functioning of the autonomic nervous system. Through hormone therapy, especially injection of estrogens, judicious use of tranquilizers, and, in some cases, short-term psychotherapy, most of the symptoms can be alleviated.

Physical changes usually include gain of weight, a coarsening of the skin and growth of hair, and involution of the mammary glands. The uterus and ovaries atrophy, length and width of the vagina are reduced, the vaginal epithelium becomes thin and dry, and the size of the introitus is reduced. The latter changes sometimes result in *dyspareunia* (painful or difficult intercourse).

As the age of childbearing and fertility come to a close, some women fear that their sexual desire and sexual activities will also come to an end. This is usually not the case—sex drive and the capacity to enjoy sex may persist for many years. In fact, some women become more interested in sex after the menopause than before, because the fear of pregnancy is removed and they are free of the monthly recurrence of menstruation with its attendant tensions.

*Premature Menopause*
This occurs before the age of 35. It may be due to premature aging of the ovaries, development of ovarian cysts or tumors, or it may result from infectious diseases or other pathological conditions especially debilitating disease.

*Artificial Menopause*
This follows surgical removal of the ovaries, or it may result from irradiation of the ovaries or occur after radium transplants in the abdominal cavity. Menopausal symptoms are usually more severe than those following natural menopause.

# SIX
# SEXUAL INTERCOURSE
# AND SEXUAL BEHAVIOR

In sexual intercourse (coitus, coition, copulation, mating), there is a physical union between the male and female and the erect penis is inserted into the vagina. In lower mammals, females are receptive only during certain periods called *heat* or *estrus*. It is during these periods that the ovary releases the egg or eggs, consequently conception and the production of offspring are usually the result of such matings. In humans, however, there is no specific period for sexual activity, hence intercourse can be engaged in at any time during the menstrual cycle. Usually it is abstained from during the menses for esthetic reasons but there is no physiological reason for doing so. No harm can result.

Usually preceding sexual intercourse, *sexual foreplay* is engaged in, such as petting, fondling, kissing. Various parts of the body are especially sensitive to sexual stimulation. These are the so-called *erogenous zones* that include the genital organs, especially the penis and clitoris and surrounding areas, the breasts and nipples, the neck and ear lobes, the lips and tongue. All the senses are involved in sexual activity, especially the sense of touch (feeling and being felt), the sense of sight (seeing the naked body or sex organs), the olfactory sense (the smell of the body, especially sexual regions), the gustatory sense or sense of taste (touching of tongue and lips to parts of the body). All of these activities increase sexual excitement and are a prelude to bringing a male and female together in the final sexual act.

## Coital Positions
Various positions (Fig. 15) may be assumed to accomplish the coital act. The most common position is face to face with the man above and this position is generally regarded as the "normal" position and, in certain religions, the only permissible one. However the desirability of this position may be questioned and another position be considered more favorable depending on various factors. These include the size and weight of individuals

Face-to-face, man-above position

Face-to-face, woman-above position

Face-to-face side position

Rear-entry position

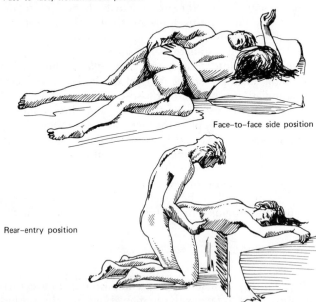

**Figure 15  Four basic coital positions. (From J. L. McCary,** *Human Sexuality,* **D. Van Nostrand Co., New York.)**

involved; conditions under which coition occurs, whether spontaneous or planned; likes and dislikes of coital partners; responsiveness of each to coital activity; adaptation to special conditions, such as pregnancy, fatigue, and minor pathological and other conditions. Usually most married couples experiment with various positions, eventually using the one or two positions that suit them most favorably. The following are some of the positions that may be assumed with some comments concerning each.

### Face to Face, with the Man Above

This is the most common position and, for most couples, the preferred position. A man's legs may be within or without a woman's thighs and his weight is usually supported on his elbows and knees. A woman may elevate her pelvic region by use of a pillow thus facilitating deep penetration. Clitoral contact may be increased by the woman closing her thighs together. This is erotically stimulating to both partners. The face-to-face position also facilitates mouth-to-mouth contact and lip-and-tongue kisses add greatly to the pleasure of intercourse.

The disadvantages of this position are that a woman's movements are restricted, male penetration may be too deep and uncomfortable, and manual manipulation of the clitoris difficult. Also free movement of the male tends to lead to premature ejaculation. It is also a difficult position for an obese male or for a woman in the late stages of pregnancy.

### Face to Face, with the Woman Above

This is the reverse of the previously described position and the woman may, and usually does, assume the active role. Her entire body may be in contact with his or she may assume a sitting position with her knees outside his thighs. In this position she can regulate coital movements and by changing her position, increase or decrease pressure on the penis or clitoris. By the male playing a passive role, ejaculation may be delayed. Because of the freedom of his arms, the male is able to caress almost any part of the female body.

The disadvantages of this position are that the male's freedom of movement is restricted and the physical demands put on the woman may not be acceptable. Sometimes the male feels that his masculinity is reduced by the assumption of a passive position and the dominating role of the female may not be acceptable.

### Face to Face, Lying on Sides

In this position the man and woman lie on their sides facing each other. Neither has to assume a passive position and each is free to caress each other as much as they might desire. The strain of physical activity is reduced for both and the position is especially desirable in late stages of pregnancy. The disadvantages of this position are that body movement is limited, clitoral contact reduced, and deep penetration almost impossible.

A modification of this position is the *lateral position* considered by Masters and Johnson as the most effective position for mutual satisfaction. In the side-to-side position, the woman lies with one leg between the man's thighs and the other leg thrown across his body. Mutual freedom of pelvic movement is possible and the male can readily maintain control of ejaculatory activity.

### Rear-Entry Position

In this position, the woman usually assumes a knee-chest attitude either in a face down position or lying on her side. The male then lies against her with her buttocks between or against his thighs and penetration of the vagina is from behind. The man's arms are free to encircle her body and his hands may caress any body area. Disadvantages of this position are that face-to-face closeness is lacking and deep penetration is difficult.

### Other Positions

In addition to the positions mentioned, there are others that may be assumed, as face-to-face, both standing; face-to-face, both sitting; and numerous others, most of which are variations of the basic positions described. Some involve gymnastic abilities beyond the range of most couples.

Generally speaking, any coital position that is mutually satisfactory to both partners is acceptable. A couple should experiment and feel free to assume the position that brings the greatest satisfaction to both. There is no right or wrong method of sexual intercourse.

### Ejaculation

During sexual intercourse, mechanisms are brought into play that bring about an ejaculation by the male. This is the forceful discharge of *seminal fluid* or *semen*. Semen consists of *spermatozoa* produced by the testes and *seminal plasma* that consists of the secretions of the accessory sex glands. These secretions serve as a vehicle for the sperm. They also contain substances that activate the sperm, nutrient materials for energy, and buffering agents that protect the sperm against the deleterious effects of acid vaginal secretions. The seminal plasma comprises the major portion (about 90 percent) of the ejaculate.

Spermatozoa, which are produced continuously, accumulate in the epididymis and the ductus deferens, especially the ampulla. Nerve impulses arising from various sexual stimuli reflexly stimulate the sex glands (bulbourethral, prostate, and seminal vesicles) increasing their rate of secretion. When stimulation reaches a certain threshold level, responses are initiated which bring about an ejaculation. This occurs in two stages. In the *first stage,* contractions beginning in the efferent ducts of the testes and continuing as peristaltic contractions in the epididymis and vas deferens force the sperm toward the urethra. The prostate gland expels its fluid into the prostatic portion of the urethra. To this is added the sperm from the vas deferens and the secretions of the seminal vesicles. At this time the internal sphincter of the bladder contracts preventing semen from entering the bladder (*retrograde ejaculation*) or urine from contaminating the semen. In the *second stage,* the semen is expelled forcibly through the penile urethra by contractions of the external sphincter of the bladder and two pairs of muscles, the *bulbocavernosus* and *ischiocavernosus* located at the base of the penis.

When an ejaculation occurs in the male, there is a brief period of intense nervous and emotional excitement accompanied by marked physical activity. This occurs at the climax of the sexual act and is known as the *orgasm.* An orgasm (one or several) also may occur in the female with comparable feelings but without any associated discharge. Sexual tension is usually slower in building up in the female and the female may not experience an orgasm as readily as a male. However, the female can experience longer orgasms and orgasms can occur repeatedly in rapid succession, whereas in the male, once an orgasm occurs, it requires an interval of time, sometimes minutes, sometimes hours, before another orgasm can be experienced.

## SEXUAL RESPONSES

The sexual responses of the male and female have been extensively studied by Masters and Johnson and described in their book entitled *Human Sexual Response.* In both sexes they fall into four categories or phases, designated the *excitement phase, plateau phase, orgasmic phase* or *orgasm,* and *resolution phase* or *recovery.* The principle responses or activities occurring in each of these phases are summarized in Table 2.

### Generalized Body Reactions

During sexual intercourse, a number of generalized body reactions occur. Muscles in various parts of the body contract, especially those of the arms, legs, abdomen, and buttocks. The heart rate increases with resulting

Table 2   Phases in sexual response

| MALE | FEMALE |
|---|---|
| *Excitement Phase* | |
| Erection of the penis resulting from increased blood flow (vasocongestion). Penis increases in length and diameter. Partial elevation of testes and increase in size. | Moistening of vagina by appearance of beads of moisture on inner surface. Increase in size of clitoris. Elevation of uterus. |
| *Plateau Phase* | |
| Increase in circumference of glans penis. Full elevation of testes. Appearance at tip of penis of mucoid material from bulbo-urethral glands. | Engorgement and swelling of tissues in outer third of vagina (orgasmic platform) and major labia. Ballooning of inner two-thirds of vagina. Further elevation of uterus and cervix. Enlargement of uterus; elevation of clitoris. Appearance of mucoid material from glands of Bartholin. |
| *Orgasmic Phase (Orgasm)* | |
| Ejaculation of semen resulting from contractions of the vas deferens and accessory organs. Contraction of anal and urethral sphincters. | Contraction of uterus, orgasmic platform, anal and urethral sphincters. (Nothing in the nature of an ejaculation occurs.) |
| *Resolution Phase* | |
| Reduction in vasocongestion. Loss of penile erection. Refractory period sets in (i.e. inability to have another orgasm). May last for a few minutes or hours. | Reduction in vasocongestion. Reduction in orgasmic platform. Decrease in size of clitoris. Lack of a refractory period with ready return to orgasm. |

increase in pulse rate and blood pressure. Engorgement of blood vessels occurs especially in the pelvic region. The rate of breathing increases and perspiration appears over much of the body. Other changes include swelling of the areola and erection of the nipples and a reddening of the skin over much of the body (sex flush).

*Postorgasmic Reactions*
Following the orgasm, the resolution period sets in and the various physiological changes that took place to enable a sexual union to materialize are now reversed. The male experiences a feeling of relief with the discharge of semen and detumesence of the penis occurs. Withdrawal of the penis usually follows. A feeling of lassitude is experienced often followed by sleep. Comparable feelings occur in a woman if she has experienced an orgasm. As a woman is usually more slowly aroused than a man, she often

fails to achieve orgasmic satisfaction. Often a repeat performance a short time later will provide a more satisfying experience for both.


### Emotional Reactions

The preceding description of sexual intercourse and the various responses deals almost entirely with physiological activities that are associated with or result from the act. Of equal or greater importance are the *emotional* responses. Man differs from other animals in his ability to experience emotions, the degree to which emotions are felt, and his reactions to them. Emotions involve feelings and accompanying these feelings are responses. These responses constitute behavior.

During and following every sex act, emotions are experienced. Emotional responses are variable. For some, intercourse may simply provide physical relief and satisfaction; for others, because of the intimacy, closeness, and oneness of the relationship, it may constitute the ultimate expression of love and affection. The pleasures of the act are increased by the knowledge that each is giving pleasure to the other as well as satisfying individual emotional and physical needs.

In sexual intercourse, some experience joy and happiness; for others it may give rise to feelings of anxiety, fear, or guilt. Taboos, inhibitions, and repressed feelings concerning sex may seriously interfere with natural and free participation in the act. Suppression of natural responses and inability to experience an orgasm may lead to frustration. Lack of knowledge and understanding of the emotional aspects of sexuality in both men and women are often important contributing factors in sexual maladjustment.

Sex and sexual activity are an integral part of growing up and maturing. Successful sexual relations between couples do not always come easy. They involve knowledge, a sympathetic understanding, patience, persistence and, often, a reeducation. The education a child gets in sexual behavior is not the education for the fulfillment of the sexual needs and desires of an adult so continuous growth and development are essential for the realization of the ultimate in sexual maturity.


### SEX AND AGING

As the human life span has become longer, there has been an increasing concern about the quality of life and not just the quantity of life. Gerontologists have long been concerned with the problems of reduced income, increased physical disabilities, inadequate housing, and poor nutrition. Now researchers are paying more attention to the psychosocial needs of the elderly.

Society has been indoctrinated with the idea that sexual attractiveness and love are more or less an exclusive property of youth. Society has stereotyped late adulthood as the "sexless older years" or the "over-the-hill" age. Coitus is regarded by many as unnatural, unbecoming, or childish for older people. Since sexual behavior is largely a reflection of individual and cultural expectations, this generally held belief about the sexless older years has become for many a self-fulfilling prophecy.

The widespread denial of sexuality in the older years often complicates and distorts interpersonal relations late in life. Serious conflicts may develop between adults and their older parents who may be thinking of remarriage or the development of a new personal relationship. Older men who express any interest in sex are often characterized as "dirty old men" and their activities regarded as childish foolishness. A common myth is the notion that child molestation and sexual deviance are more common among older men.

Sexuality is not necessarily limited to intercourse for the aged anymore than it is for younger individuals. It can be expressed by continued closeness, intimacy, and affection and it may provide some romance in otherwise dull and uninteresting lives.

As people age, all physiological activities tend to slow down and there is an actual decrease in the size and weight of most organs. It is not surprising then that sexual desire is less pronounced in older people and that sexual relations occur less frequently on an average than among younger persons. The decrease in sexual activity can be attributed to a number of factors. Among them are a general decrease in sexual desire or libido that occurs in most individuals. This may be associated with or the result of mental or physical fatigue, preoccupation with business or recreational activities, excessive consumption of food leading to obesity, excessive indulgence in alcohol, or boredom with one's mate as a sexual partner. Sexual boredom commonly develops when older people fail to maintain their attractiveness or develop their sex appeal to the fullest extent. There is a general reluctance among older people to alter their pattern of sex activities and to adopt new and innovative patterns. As a result the sex act often becomes a mechanical and a repetitious act with little variety.

Physical disabilities, some minor and some of a major nature, may have a serious effect on sexual interest and ability. During most illnesses, interest in sexual activities is generally reduced or absent. Following heart attacks, men are often reluctant to resume sexual relations for fear of precipitating another attack. However heart specialists today generally consider that, in the absence of chest pain, it is safe to resume sexual activities after an interval of six to eight weeks following release from a hospital after an acute coronary attack. They regard the tension resulting from abstinence, especially when accompanied by fear of failure or fear of possible physical harm, as more harmful than any condition that might result from coitus. It is important that in nearly all cases of illness sexual activity be resumed as soon as possible, such always being consistent with the patient's health.

Masters and Johnson have found that the key to a continuous, productive sex life in old age is the establishment of a good and active sex life in younger years. One of the most generally believed myths is the erroneous idea that one "uses oneself up" by early frequent sexual intercourse and masturbation; that the supply of germ cells is limited and once exhausted, cannot be replaced. This is contrary to fact. The testes and accessory sex glands are capable of producing an unlimited amount of seminal fluid and, in general, continuous functioning improves their productivity. The research of Kinsey and his associates supports this view; they found that those who began their sexual activity early in life did not end it earlier but later.

In women, a major transitional phase in reproductive physiology occurs in the form of the *menopause* or *climacteric,* which usually occurs between the ages of 45 and 50. During this period, ovarian function, that is, the production of ova and ovarian hormones (estrogens and progesterone), ceases. This is due to the failure of production of the follicle-stimulating and luteinizing hormones by the pituitary gland or the failure of the ovary to respond to them. As a result, follicles are not produced and corpora lutea do not develop. Consequently, eggs are not released from the ovary and uterine endometrial changes resulting from the release of ovarian hormones fail to occur. This brings about the cessation of menstrual cycles, the primary indicator of the menopause. The changes may occur suddenly but in most cases they occur over a period of months or years. With the cessation of the production of ova, a woman becomes infertile.

A number of symptoms, both physical and mental, are sometimes associated with the menopause. Physiological symptoms include hot flashes resulting from dilatation of skin arterioles giving a feeling of warmth. Other symptoms may include fatigue, breast and joint pains, headaches, dizzy spells, and heart palpitations. Emotional disturbances are not uncommon resulting from the disturbing nature of the entire period, resulting principally from the realization that the period of childbearing is over. Only about 30 to 40 percent of the women going through menopause have any symptoms of significance and less that 15 percent are disturbed enough to seek medical help.

Physical changes associated with the menopause include a decrease in length and width of the vagina, a thinning of the vaginal wall, and a decrease in size of uterus, breasts, major labia, and uterus, and a general decline in the elasticity of the connective tissues of these organs. All of these changes, brought about by the diminished production of estrogens (Fig. 16) may make intercourse painful (dyspareunia). However, despite these changes, the organs retain their ability to respond to sexual stimulation. Fortunately, most of these conditions can be corrected by estrogen replacement therapy, which can take the form of pills, creams, or injections. Although the extent to which estrogen therapy should be utilized

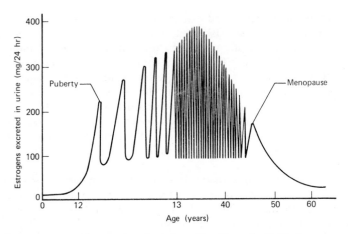

**Figure 16  Estrogen secretion throughout life. (From A. C. Guyton,**
*Textbook of Medical Physiology,* **W. B. Saunders Co., Philadelphia.)**

following menopause is debated, there is no question that sex-steroid replacement therapy is essential for many women in order for them to function as an effective sexual partner. The specific changes that occur in sexual response are listed in Table 3.

In males, total cessation of testicular function does not occur comparable to the cessation of ovarian function in the female; consequently marked changes, like those that constitute the female climacteric, are not observed in most men. However with aging, the amount of sperm and the secretions of the accessory glands are reduced in quantity and sexual activity tends to diminish. The level of testosterone declines gradually with age (Fig. 17) however fertility is maintained in spite of the low level of testosterone. It is consequently possible for elderly men to father children. It has been found that in the male, as well as the female, early and regular sexual activity combined with adequate mental and physical well-being during the later years provide a positive sexual climate that can continue into the seventies or eighties.

Among the effects of aging in the male are reduced frequency in involuntary morning erections and a reduced sex drive. Some women in the later years of marriage may desire intercourse more frequently than their husbands. This increased demand by the wife for sexual performance by the husband may have the effect of inducing impotence in the male as some males withdraw from coital activity rather than face the ego-shattering prospects of repeated episodes of impotence. For some males, threatened impotence induces a panic that may lead to extramarital affairs in an attempt to retain their virility and masculinity. However any beneficial

**Table 3  Differences in sexual response cycle of the aged**

| MALE | FEMALE |
|---|---|
| *Excitement Phase* | |
| Erection of penis takes longer. | Lubrication delayed and reduced. |
| Erection without ejaculation can be maintained longer. | Uterus only slightly elevated. |
| | Sex flush not common. |
| *Plateau Phase* | |
| Slight elevation of testes. | Delayed onset of orgasmic platform. |
| Sex flush is uncommon. | Reduced elevation of uterus and cervix. |
| *Orgasmic Phase* | |
| Inevitability sensation of ejaculation lost. | Orgasmic phase shortened. |
| Less semen ejaculated with less expulsive force. | Contractions of uterus, orgasmic platform, anal and urethral sphincters reduced in number and intensity. |
| Contractions of accessory organs, anal and urethral sphincters reduced in number and intensity. | |
| *Resolution Phase* | |
| Rapid reduction in vasocongestion. | Rapid reduction in vasocongestion. |
| Loss of erection extremely rapid. | Rapid return of vagina to normal size. |
| Refractory period much longer. | |

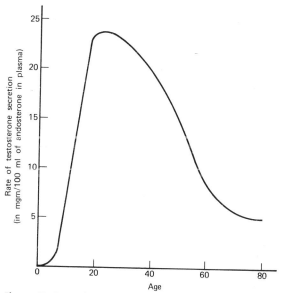

**Figure 17   Rate of testosterone secretion in the male, by age. Note peak at age 17–18. (From A. C. Guyton,** *Textbook of Medical Physiology,* **W. B. Saunders Co., Philadelphia.)**

change is usually short lived and old habits of performance are usually resumed in a short time.

The effectiveness of hormone replacement therapy by injection of androgens is highly questionable. While androgen therapy seems to produce a generalized feeling of well-being and enhances a feeling of increased sexual capability, its effects seem to be largely of a psychological rather than a physiological nature.

Other psychosexual problems involve the elderly who are "singles" (divorced, widowed, widowers). The woman who is widowed is most often involved with these problems since 75 percent of American women become widows due to their longer average length of life. The problem is the "double standard for the aged," comparable to that which exists in youth. Aged single men and widowers have only slightly less frequent sexual activity than married men of a comparable age, however single and widowed elderly women have sexual relations much less frequently than men of a similar age. This may be a factor in the increased frequency in masturbation reported in women of this age group. The double standard can also be seen in the social approval men receive in dating and marrying younger women while the association of older women with younger men is generally disapproved.

Finally it must be recognized that some men and women may look forward to having a justifiable reason for ending their sexual relationships. For some, engaging in sex is regarded almost exclusively as an obligation or duty rather than a pleasurable activity hence freedom from participation in it may be welcomed.

In summary, an active sex life for post-menopausal women and for men of a similar age is not only possible but desirable. As the gerontologist, Victor Kassel, recently mentioned to a group of hospital and nursing home administrators:

"It is a well-established fact that for many people sexual orgasm brings immediate relief from anxiety. And it is a fact that most patients in nursing homes suffer from chronic anxiety. What most nursing home operators do about anxiety is zonk their patients with tranquilizers so that they walk around all day like zombies and don't make trouble for the staff."

According to Kassel, nursing home patients, (the elderly in general) should be allowed and even encouraged to engage in sexual activity. Continuation of sexual relations in old age is of prime importance as a source of psychological reinforcement. Kassel suggests that the policies that have led to almost total segregation of the sexes and the separation of even husbands and wives in old age homes and mental institutions be reexamined since such practices are recognized as contributing to the problems of the aged.

## SEXUAL FANTASIES AND DREAMS

A sexual fantasy is a mental picture or image of a sexual phenomenon or activity created when one is awake. In everyday language, it is a daydream. Fantasies are usually unrealistic but they are a means of providing satisfactions that are oftentimes impossible to attain in reality. Sexual fantasies occur in both sexes and in all ages. They enable individuals to reenact past experiences with pleasure and to experience joys that were actually missed. They often provide a substitute activity giving temporary pleasure in anticipation of a future indefinite and uncertain happening and they often provide an interest that makes the boredom and tediousness of everyday life much more bearable.

Sexual fantasies or reveries usually revolve around the generally accepted norms of sexual behavior but sometimes fantasizing may involve activities such as homosexuality, sadism, or incest. In these cases, instead of being pleasurable, the fantasies may cause mental distress and feelings of guilt. In extreme cases they may lead to antisocial behavior.

Males tend to have fantasies that involve a high order of genital activity. Fantasies usually accompany and are an integral part of male masturbation. They may include intercourse in unusual positions or places or intercourse with unavailable females as movie stars or pinup girls. They may include sex relations with animals, oral sex, anal sex, homosexual, or incestous relationships. They often involve efforts to overpower a female toward whom aggressive action would not normally be taken but who, on being taken, willingly accepts his advances. Or the opposite fantasy may occur in which the male is aggressively pursued and sexually subdued by an attractive young female.

In women, however, fantasies tend to be less orientated genitally and to have as their major component romantic relationships. Two common fantasies are those of being with another man such as a famous actor, musician, or even a casual acquaintance and being forced to surrender. Women often fantasize when having intercourse with their marital mate. Such is not a sign of marital maladjustment but a means of increasing the pleasure of the sexual act with her partner with whom she has become too familiar and with whom sexual relations have tended to become routine and boring. Sexually responsive orgasmic women active in seeking sexual relations are more likely to fantasize than women who have a low level of sexual response.

Not all fantasies are pleasant. Women may fantasize that their vaginas are too large, or that they are too small and are likely to be torn. They may picture a man's penis as penetrating and injuring their internal organs. Men may picture a vagina as a trap capable of holding the penis indefinitely or they may fear that it is full of sharp objects capable of injuring any penis that enters. Fantasies of this type, if temporary and fleeting, are of little significance but if they are persistent, they may be indicative of a disordered personality.

Sexual dreams are closely related to sexual fantasies in that both are frequently indicative of repressed desires. A dream is an attempt at wish fulfillment. In daydreams the activity is generally consciously limited to that which might logically be readily accomplished, and it is recognized as a daydream or fantasy. In nocturnal dreams, however, conscious restraints are lacking, consequently the dream becomes the means of expressing repressed desires or wishes. Nocturnal dreams are often characterized by a sense of reality, so real sometimes that the dreamer is relieved on awaking to find that the dream incident had not actually occurred. Sexual dreams in males are often accompanied by full or partial erections however erections during sleep are not always accompanied by sexual dreams. Orgasms frequently accompany sexual dreams.

Many dreams are fleeting and disconnected and can only be sketchily recalled. Psychoanalysts, through the analysis of dreams, by interpreting their symbolic meaning rather than their manifest content, can often gain an understanding of a person's mental and emotional problems. Dreams may reveal repressed wishes, hidden motivations, hatreds, fears, frustrations, and feelings of guilt that could not be discovered in any other way. Through dream analysis, the opportunity is given for the release from tension and anxiety that often accompany repressed desires.

# SEVEN
# PROBLEMS OF
# SEXUAL RESPONSE,
# SEXUAL DYSFUNCTION

*SEXUAL INADEQUACY OR DYSFUNCTION*

Theoretically, the sexual act and the orgasmic responses resulting from it are simple, cut-and-dried procedures that result in ecstatic pleasure to the participants. However, such is often not the case. The complete fulfillment of the sexual response is often lacking and a condition of *sexual inadequacy* or *dysfunction* results. This may involve either the male or female or both.

The first experiences in sexual intercourse, whether premarital or following marriage, are often unsatisfactory. Sexual intercourse involves three possible types of experience. Basically, it is an act of reproduction resulting from instinctive drives that bring male and female together for the purpose of producing offspring. This, called *reproductive sex*, is basic sex, a characteristic of all organisms that utilize sexual reproduction as a means of perpetuating the species. However, among humans, sexual intercourse is also a means of expressing love, a process by which two individuals fuse their bodies and their feelings into one, each giving to the other and each receiving and sharing the experience of becoming one. This is emotional or *love sex*. It may be experienced within or outside of marriage. It varies with the age and experience of the individuals involved and their commitment to each other; it may be of short duration or it may be permanent and last a lifetime. The third type of sexual relationship is that in which intercourse is engaged in for physical pleasure and satisfaction and without emotional involvement. This is recreational or *fun sex*.

These three types of sexual expression are often experienced simultaneously. The persons involved are frequently married and sex activities satisfy the needs of both parties. While sex for reproduction was once essential for the survival or economic well-being of families, tribes, or communities, the problem confronting the human race today is overpopulation, hence the emphasis has changed from production of offspring to limitation of population. However, as there is always the possibility that conception may occur, engaging in sex for emotional or recreational satisfaction always necessitates a knowledge of and utilization of contraceptive measures.

Although, theoretically, sexual intercourse should prove satisfactory to both sex partners, in many marriages or in sexual relations outside of marriage, satisfactory relationships may fail to materialize. Sexual inadequacy or sexual dysfunction may exist.

*Sexual dysfunction* is the condition in which the ability to perform the sexual act is impaired or unsatisfactory responses during or following the act result in one or both parties. The most common forms of sexual dysfunction are, in the male, premature ejaculation, impotence, and ejaculatory incompetence; in the female, vaginismus, dyspareunia, and frigidity or orgasmic impairment. To these may be added sexual apathy that may apply to either sex.

The causes of sexual dysfunction are many and varied. Some are of a superficial nature and easily corrected; others are deep seated and may require special counseling, medical treatment, or sometimes psychiatric treatment. As the nature of the sexual response is conditioned by early training and experiences, religious attitudes, moral instruction, parental attitudes, and peer viewpoints, any or all of these factors may be operative in sexual dysfunction. Satisfactory sexual relations seldom result in the first attempts at intercourse. Sexual harmony and mutual satisfaction often fail to materialize during early encounters. Lack of experience, inadequate knowledge of the basic functioning of the reproductive systems, and especially a lack of understanding of the emotional needs and expectations of the parties involved are important factors in limiting satisfaction obtained from the sexual act. To these may be added fears of pregnancy or venereal disease, religious or moral inhibitions, and unrealistic expectations for sexual satisfaction. Sexual harmony and adequate sexual performance are usually not acquired overnight nor should they be expected to result at any and all times. The cooperative efforts of both partners involved are essential and, as in all complex activities, repeated experiences, more knowledge of the problems involved, and an understanding of all the factors related to more successful performance almost invariably result in improved sexual relations. An understanding of each partner's physical, mental, and emotional needs coupled with a sincere desire to mutually satisfy each other is most helpful in solving many of the difficulties encountered.

## SEXUAL DYSFUNCTION IN THE MALE

### Premature Ejaculation

This is the condition in which ejaculation occurs before penetration or within two minutes after penetration, a time usually inadequate for a woman to be sufficiently aroused to experience an orgasm. Masters and Johnson define the condition as that in which ejaculation cannot be delayed for a sufficient length of time to satisfy the coital partner in half of the coital acts. Premature ejaculation or lack of ejaculatory control is common in many males, especially among the young and inexperienced. Among married couples, during the first year or for a longer period, premature ejaculation is common but with practice and experience, delaying techniques can be developed and most men learn to delay the orgasm until the appropriate time, that is, when the coital partner is experiencing her orgasm.

However for a large number of males, ejaculatory control is difficult or impossible to attain. As a result, marital difficulties, often of a serious nature may develop. The woman, not experiencing an orgasm, although highly stimulated, obviously is unfulfilled and sexually frustrated. Physically the man is satisfied. He has become sexually excited, had an erection, penetrated, ejaculated, and experienced an orgasm. However, he is deprived of the satisfaction derived from repeated thrusts and the pleasure of vaginal containment. He may feel that he is a failure as a lover. To the woman involved, she has not been satisfied and she tends to place the blame on him. She may feel that she is being used merely as a sex object to satisfy her husband's sexual desire.

Efforts are usually made to bring ejaculation under control but successive attempts usually bring the same results as the husband is fearful that premature ejaculation will occur and it usually does. As both become concerned over possible failure, the pattern of their sexual life tends to change. At first the wife tends to be tolerant and both feel that the situation will correct itself. However, if the condition persists, sexual intercourse becomes less frequent and that makes it more difficult for the male to hold back when intercourse does occur. The male becomes convinced he is a poor lover and that he is a second-rate individual as far as sexual activity is concerned.

The specific causes of premature ejaculation are not known. It has been attributed to physical factors as a hypersensitive glans or to infections of the urethra or prostate gland, but these are factors only in a limited number of cases. Masturbation and circumcision have been considered as possible causes but studies have shown that premature ejaculation is not caused by masturbation and there is little or no difference in the sensitivity of a circumcised penis as compared with an uncircumcised one.

Psychological and emotional factors seem to play a primary role in most cases of premature ejaculation. Early sexual behavior especially during adolescent years may set a pattern that persists throughout life. Common experiences often involve excessive petting that leads to sexual contact or intercourse in which rapid ejaculation by the male and lack of sexual satisfaction by the female are the usual results. When first sexual experiences are with prostitutes, the action of quick entry and ejaculation, which is encouraged by the prostitute, may establish a pattern of sexual reaction that will persist in later life. Sexual relations during adolescence, which are often carried on where exposure to adults—often parents—is threatened, tend to develop a pattern of rapid entry and ejaculation. The practice of withdrawal as a method of contraception may also be a factor in the development of premature ejaculation. This technique rarely provides sexual satisfaction and, since withdrawal usually must take place hurriedly and without regard to the partner's sexual response, it tends to facilitate premature ejaculation. It is also a poor method of contraception.

Correction of premature ejaculation can, in many cases, be accomplished if there is a sincere desire on the part of the male with the cooperation of a sympathetic and understanding mate. Ejaculation is a reflex activity, and like all reflexes, is subject to a limited degree of control through the action of the higher nervous centers. By conditioning, most males can learn to delay ejaculation until the appropriate time, that is, when his mate is ready to experience an orgasm. For inexperienced males, the use of a heavy latex condom tends to dull the sensitivity of the penis. This together with slower thrusting movements may delay ejaculation. Some resort to thinking of nonsexual subjects or to diversionary actions as pinching themselves or biting their lip.

However, for many premature ejaculators, none of these procedures are effective. For some, repeated failure bcomes a persistent way of life unless special methods are employed. Successful treatment usually requires both partners desiring and seeking professional help. Clinics have been established to treat various types of sexual dysfunction, one of the best known ones being the Reproductive Biology Research Foundation under the direction of Dr. William Masters and Mrs. Virginia Johnson. Through utilization of the "squeeze technique" in which ejaculation is retarded by the wife applying a firm pressure on the erect penis just below the glans, the need for ejaculation disappears although the erection persists. This, combined with a female-above coital position (later changed to a lateral position, considered to be the most desirable position for coitus), has resulted in successful results in a high percentage of cases treated.

Self-treatment utilizing the squeeze technique is sometimes effective. Success will depend on the cooperation of both husband and wife and a thorough understanding of the method employed in delaying an ejaculation, details of which can be obtained in various publications dealing with sexual inadequacies.

*Impotence (Erectile Dysfunction)*
This is the inability to achieve an erection or to maintain an erection of sufficient strength long enough for coitus to be cosummated. All males experience impotence at various times during their life. In childhood and old age, impotence is a normal condition. Immediately following coitus or masturbation, impotence persists for a variable length of time (from minutes to hours).

**Organic Impotence.**  Impotence is of two types, organic or physiological. *Organic impotence* results from, or is associated with, a structural disorder. Structural defects such as phimosis (constriction of orifice of prepuce), hypospadias or epispadias (urethral opening on lower or upper surface of penis) resulting from maldevelopment, or disorders resulting from pathological conditions as prostatitis, lesions of the penis, orchitis (inflammation of testes), or other conditions may result in impotence.

**Physiologic or Functional Impotence.**  This results from pathological or psychogenic factors. *Pathological conditions* that are associated with or may cause impotence are systemic disorders such as diabetes mellitus, various debilitating diseases, endocrine disorders especially those involving the pituitary gland, chronic poisoning as in alcoholism or drug dependency, and excessive X-ray irradiation.

*Psychogenic impotence.*  This is, by far, the most commonly encountered form of impotence. This type may be divided into two categories: primary and secondary. *Primary impotence* is that in which the subject has never been able to achieve an erection or to maintain one long enough to engage in coitus. *Secondary impotence* is that in which a previously potent man loses the ability to attain an erection.

The causes of primary impotence are often deep-seated. Excessively strict religious beliefs, an extremely domineering mother or mother-son sexual relationship, unfortunate or inept first coital experiences, homosexual encounters, especially seduction, or encounters with an unsympathetic prostitute are common experiences of men experiencing primary impotence. Fear of failure or fear of not being able to perform as most men think other men are capable often results in complete failure to obtain an erection.

In secondary impotence, successful intercourse has been accomplished a variable number of times. However, at some time, sooner or later, a man may fail to have an erection. It may be due to fatigue, illness, overindulgence in alcohol, or any of a number of causes. Ordinarily such would be of little consequence. However if a man is ridiculed or compared unfavorably with other men, he may begin to suspect that he is losing his coital ability. He becomes anxious and begins to worry about how he will be able to perform at the next encounter. When that occurs the fear of failure may lead to failure and a pattern of sexual incompetence is established.

Society has helped the male ego to develop the notion that masculinity and personal worth are directly related to potency. As a consequence, any failure in ejaculatory performance becomes an affront to a man's ego. Therefore the goal of therapy must be to correct the causes of impotence and not its symptoms. Once a male has been impotent, the problem tends to develop into a state of severe performance anxiety.

**Treatment of Impotency.** When impotence is due to organic or pathological causes, medical treatment or correction of the cause may in some cases be effective. But as most impotence is due to psychogenic factors, treatment by trained individuals who have a thorough understanding of the condition may be essential. Impotence may be absolute or relative. A man may be impotent with his wife but capable with another woman. A man may be impotent with women of certain types or personalities, as women who are virgins, women they respect, or women whom he may identify with his mother or sister but still be able to function sexually with other women.

Treatment is primarily directed toward eliminating the fear of failure in the male and redirecting his interests in becoming an active participant in the sexual act. The wife's attitude and actions are of primary importance in his overcoming his inadequacy. For her to indicate impatience or disappointment over her husband's failure usually aggravates the situation by increasing his fear of failure.

A procedure that is, to some degree, effective is for the couple to engage in sexual play not for the purpose of intercourse but simply for the pleasure involved. Usually the male will be able to relax and eventually manipulation of the penis will usually induce an erection. However coitus is not attempted immediately but usually delayed for several days. Then, with the wife initiating the coital attempt in a female-superior position, when an erection has been attained, she inserts the penis and, if the erection is maintained, she then allows the male to engage in thrusting movements. If the penis should become flaccid upon insertion, it should be withdrawn and the procedure repeated until success has been achieved.

*Ejaculatory Incompetence (Retarded Ejaculation or Inhibited Orgasm)*
A third type of dysfunction involving the male is the rare condition in which a man is unable to ejaculate during coitus. The erection may persist for a considerable time but ejaculation and experiencing an orgasm fails to occur. This condition, like impotence, may be absolute or relative. When relative, ejaculation may occur in nocturnal emissions, masturbation, homosexual encounters, or in intercourse with particular women.

Though the causes are usually psychologic, pathological conditions such as diabetes, or multiple sclerosis or the use of certain drugs, especially hypertensives, may be causative factors.

When it involves a wife, the causes are nearly always of a psychogenic nature. It may be from the desire to avoid contamination, the fear of inducing pregnancy, or it may be a means of punishing a wife for undesirable or unfaithful behavior. Because nonejaculators have a prolonged "staying power" and are capable of inducing multiple orgasms in their coital partners, they are sometimes regarded as being superpotent. However the situation usually becomes unsatisfactory or unbearable for the wife and it is she who usually seeks help.

This form of dysfunction is generally more easily corrected than other forms. Usually it involves the use of manual manipulative techniques to induce an ejaculation extravaginally. First attempts are usually unsuccessful but with patience, an ejaculation can be brought about. Once this is accomplished and capable of being repeated, the male, when brought close to the point of ejaculation, then inserts the penis into the vagina and if ejaculation occurs, the primary psychological block has been overcome. Repetition of this precedure usually results in the restoration of normal ejaculatory competence.

## SEXUAL DYSFUNCTION IN THE FEMALE

### Vaginismus

This is the condition in which there is a painful spasm of the muscles at the entrance of and surrounding the vagina. The spasm is involuntary and results in a partial or total blocking of the introitus thus preventing penetration in coitus or, if penetration is accomplished, severe pain may result.

The causes of vaginismus are almost entirely psychological. Impotence of the husband has been found to be one of the primary factors involved. Vaginismus may be either the cause or the result. Repeated failures at penetration may cause the wife, out of frustration, to involuntarily close the entrance to the vagina. On the other hand, vaginismus, by preventing penetration, may induce impotence in a male as a protective mechanism against his attempting intercourse.

Many other situations have been established as possible causes. An initial painful coital experience, the trauma of rape, the fear of pain or possible pregnancy, strict religious training and the association of sexual relations with sin, and homosexual experiences have all been factors in various cases of vaginismus.

Most cases of vaginismus can be corrected, in general, without much difficulty. The cause or causes should be ascertained and then both partners should be provided with a detailed explanation of the anatomical, physiological, and psychological factors involved. Treatment involves the dilation of the vaginal opening by the use of dilators of increasing size. The containment of the largest dilator for several hours each night usually results in relief of the spasms within four or five days. In difficult cases, counseling to relieve sexual tensions, unwarranted fears, and feelings of guilt or shame may be desirable or essential.

*Dyspareunia*
This is painful intercourse. The pain may occur on entry, during coitus, or following coitus. Its causes may be physical or psychological and it may involve members of either sex. In women, physical causes of dyspareunia are a resistant hymen, vaginal-vulval or urinary infections, irritation resulting from intravaginal contraceptives, both chemicals and the rubber of condoms or diaphragms, or irritation from substances used in douching. Pain may result from organic conditions as a displaced uterus, a shortened vagina following hysterectomy, excessive tightness following episiotomy repair, abnormal growths, or other pathological conditions. The most common cause is inadequate lubrication. This is especially common in women following the menopause due to a thinning of the vaginal mucosa resulting from hormonal imbalance. It can usually be corrected by estrogen treatment. In younger women, adequate foreplay usually results in adequate lubrication. If it is insufficient, a lubricating jelly is recommended.

In spite of the many physical conditions that may cause dyspareunia, in the majority of cases, coital or postcoital pain has its origin in psychogenic factors. This is interpreted as being a defense mechanism by which sexual intercourse can be avoided or at least reduced in frequency. It may be caused by vaginismus, which may result from the fear of genital injury during coitus; fear of pregnancy; or fear of acquiring a venereal disease. A painful initial sexual encounter, rape, or homosexual experience may result in an aversion to sexual intercourse. Guilt due to a repressive religious background, resentment toward males in general or an aversion to a specific male are possible causative factors. Inconsiderateness on the part of unknowledgeable and inept males and lack of sexual foreplay as a preliminary to the sexual act often leaves the woman sexually unsatisfied with pelvic organs chronically congested. Such may result in a more or less chronic low abdominal pain.

Treatment for dyspareunia is dependent upon the causes. For excessive dryness, a lubricating ointment used freely is usually adequate, or in older patients with senile vaginitis, estrogenic suppositories may be employed. However for most physical conditions, the correct diagnosis followed by the proper medicinal or surgical treatment is essential. For dyspareunia of psychological origin, special counseling or psychotherapy may be necessary. In all cases the husband should be involved in the treatment as the condition is often the result of faulty coital techniques. Instruction in improvement of sexual techniques, alleviation of unreasonable fears, and a better understanding of physical and emotional factors involved in sexual relationships often bring marked improvement.

Dyspareunia may also be experienced by the male but it is less common. In uncircumcised males, lack of cleanliness may result in irritation of the glans. *Phimosis,* a condition in which the foreskin cannot be retracted over the glans, may make coitus very painful or impossible. Pain may also result from hypersensitivity of the penis to chemicals used for contraception or feminine hygiene. *Priapism* characterized by a persistent and painful erection of the penis and *Peyronie's disease,* in which the shape of the penis is abnormal, are both characterized by painful coitus. Both necessitate medical and sometimes surgical treatment.

*Orgasmic Dysfunction; Inhibition of Sexual Excitement; Anorgasmia*
This is the inability of a female to experience an orgasm during sexual intercourse. Sometimes the term "frigidity" is applied, which means "coldness" and implies a lack of sexual responsiveness that is often assumed by the male to be deliberate and willful. The term "orgasmic impairment" is preferred because this term indicates that the condition is basically the inability of a female to respond to sexual stimulation. Masters and Johnson consider that a woman is orgasmically dysfunctional when she cannot go beyond the plateau phase in sexual response. They recognized two major types of orgasmic dysfunction: primary and situational. *Primary orgasmic dysfunction* includes women who have never experienced an orgasm; *situational orgasmic dysfunction* includes women who have experienced at least one orgasm by some means, either by coitus, masturbation, or other method of stimulation but are no longer capable of responding.

The causes of orgasmic dysfunction are of organic, physiological, or psychological origin. *Organic causes* include abnormalities of the sex organs, pathological conditions especially infections and lesions of the genital tract or diseases involving the central nervous system. *Physiological causes* include vaginal anesthesia, endocrine disorders, disorders of the nervous system, systemic disease, excessive use of alcohol or drugs, and changes occurring as a result of aging.

The predominant causes of female orgasmic dysfunction are of a *psychologic nature.* Faulty sex instruction in childhood in which sexual thoughts and feelings were considered to be shameful or sinful may result in such feelings being carried over into marriage. The fear of pregnancy, hostility to men in general, overt or unconscious homosexual inclinations, or disinterest in intercourse other than for procreation may be factors in lack of sexual response. The lack of preliminary lovemaking and sexual foreplay and the actions of an inexperienced and inconsiderate male are frequent causes of a lack in female response.

The effects of female orgasmic dysfunction are often serious and far reaching. Sexual intercourse should be mutually satisfactory to the participants and marriages in which sexual relations result in physical and emotional satisfaction to both parties are generally much more stable than those in which sexual disharmony exists. Women who fail to experience the pleasure of sexual intercourse often tend to become neurotic and it is thought that such may be a factor in premenstrual tension or in menopausal symptoms in later life. The inability of a woman to respond to her husband is a common cause of either or both seeking sexual satisfaction in extramarital relationships usually with unsatisfactory results. Lack of orgasmic response may be a factor in the development of dyspareunia or vaginismus because a woman may consciously submit to coitus as a "duty" but subconsciously seek to avoid it by responses that make entry difficult.

### Treatment of Female Orgasmic Dysfunction
As in other disorders, this depends primarily on its causes. If the causes are of an organic or physiological nature, proper medical diagnosis and treatment should be secured. If psychological factors are the principal ones involved, as is usually the case, the condition can sometimes be corrected through self-therapy. If this is not effective, counseling by a marriage counselor, a clinical psychologist, or a physician may result in improved response. When causes are deep seated, psychotherapy may be necessary.

The first essential of therapy is a thorough understanding of the anatomy of the sex organs and the physiology of the sexual act, together with an understanding of the emotional factors involved. Sexual activity is an essential part of male-female relationships and when the female experiences sexual pleasure and the male makes a special effort to see that his partner responds, mutual pleasure results and the possibility of orgasmic response is enhanced. Some specific actions that may be taken by an unresponsive woman are as follows: removal of negative attitudes and inhibitions resulting from early training and indoctrination; altering coital techniques and encouraging stimulation of the most sensitive bodily areas through adequate foreplay; selection of the proper time and conditions for the sexual act, and finally, the acquiring of a attitude of confidence and a feeling of relaxation.

Occasionally a woman does not know whether she has ever experienced an orgasm. For such women, masturbation is recommended, especially through the use of a vibrator. This enables her to know what the feeling is like and what initiates it. Sometimes the use of a vibrator preceding intercourse to bring a woman almost to the point of orgasm will result in an orgasm during coitus. Sometimes, a second coitus a short time after the first will induce an orgasm that failed to occur in the first encounter. Exposure to erotic literature or indulgence in erotic thoughts or fantasies often facilitate the experiencing of an orgasm. In any case, no attempt should be made to "will" an orgasm.

When marital counseling is resorted to, both partners involved should participate since lack of sexual response by the female may be due to actions, or lack of actions, on the part of the male. Lack of knowledge of how to arouse erotic feelings and responses is common. Rough, inconsiderate treatment, and failure to express love or tender feelings, so important to women, are often contributing factors.

In severe cases of sexual disharmony and dysfunction, a team approach may be necessary. This type of treatment promoted especially by Dr. Masters, a gynecologist, and Mrs. Johnson (now Mrs. Masters), a psychologist, enables both members of a marital unit to have someone to identify with, eliminates sexual involvement of patient and therapist, improves accuracy of therapist's diagnosis, and enables better communication between therapists and patient through elimination of sex-linked biases. Their work at the Reproductive Biology Research Foundation in St. Louis, in which a two-week intensive therapy program is utilized has resulted in successful treatment in about 80 percent of cases.

*SEXUAL APATHY*

This condition, although rare, may exist in individuals of either sex. In its extreme form, little or no interest in members of the opposite sex exists and the sex drive is missing or extremely low. In the male, there may be a total absence of sex desire and inability to attain an erection, as in primary impotence. In females, the desire for coitus may be lacking and there may be little or no desire to associate with males or to participate in sexual activities.

Sexual apathy may be the result of various factors. Basic sex drive in humans, as in all other animals, has a genetic basis and its fundamental nature, whether strong or weak, is probably determined by a person's genetic makeup. Genetic factors acting through the organization of the nervous and endocrine systems that control the body's activities determine the fundamental nature of the sex drive.

Early training and experiences have a marked effect on the sexual performances of animals, including humans. Repeated frustration will cause domestic animals to lose interest in sexual activities. Environmental conditions play an important role, wild animals often failing to mate in artificial environments. Young primates reared in isolation fail to develop normal mating responses. Among humans, experience, training, and education all have an influence on how an individual responds to his innate sexual drive. Ridicule, laughter, threats of punishment, fear, or guilt feelings may hamper one's psychosexual development to the extent that sexual activity is shunned or avoided altogether.

Physiological or pathological factors may induce sexual apathy. In most chronic or systemic diseases, sexual desire and activities are dulled or lacking. The same is true in most acute illnesses. During the acute phase and recovery, interest in sex is at a minimum. In old age and senility and in bodily conditions in which there are marked changes in bodily metabolism as in starvation or extreme fatigue, sexual desire and activities are reduced to a minimum or are lacking entirely.

Attitudes toward sexual activity may contribute to sexual apathy. Among celibates in the clergy, sublimation of sexual feelings and activities to so-called higher ethical, cultural, or moral values may result in reduced interest in sexual activity. Finally, more or less continued sexual activity, once it has been engaged in, is an important factor in maintenance of sexual drive. Among married couples, a long period of abstinence due to illness or absence of a mate, may result in marked impairment of interest and response in sexual activities. Among widows and widowers, sexual desire tends to decline following cessation of sexual stimulation of any kind. Steady or continuous functional activity acts to maintain the regulatory functions of the nervous and endocrine systems in their highest state of effectiveness.

*SEX THERAPY*

A recent development in the field of sex has been the development of sex therapy centers in which treatment is offered for the solution of various sexual difficulties such as impotence, premature ejaculation, frigidity or orgasmic impairment, and painful intercourse. Pioneers in this field were Dr. William Masters and Virginia Johnson who, through their book *Human Sexual Inadequacy* published in 1970, publicized their work done at the Reproductive Biology Research Foundation which was marked by considerable success.

As a result of their work sex therapy clinics have sprung up throughout the country at an alarming rate. It has been estimated that from 4000 to 5000 "Clinics" or "Treatment Centers" now exist for the treatment of sexual problems. Because sex therapy is just beginning to be recognized as a specialized field, a person or couple seeking help might find it difficult to know to whom they should go. There are no certification requirements nor have any standards for the training of sex therapists been established. Medical and counseling organizations have not been able to control in any way those who enter the field, hence the majority of sex clinics established are of a questionable nature. Of the total number in operation in the United States today, only about 100 are considered as being legitimate. Dr. Masters estimates that the number of sex clinics operated by persons adequately trained in the field to be no more than 50, the remainder being of questionable or fradulent nature operated by untrained persons and, in many cases, by outright charlatans and quacks.

Many of the sex-counseling clinics that have been established claim that their staff members use the "Masters and Johnson Method." All too often this claim simply means that one or more of their "therapists" have read *Human Sexual Inadequacy* and, in most cases, this reading has been of a most superficial nature. Sex therapists who are trained by Masters and Johnson must complete a course of training at their clinic in St. Louis, this course requiring from four months to a year to complete. At the present time, only eight teams have been trained by Masters and Johnson at their Reproductive Biology Research Foundation.

Among the principles followed by Masters and Johnson in dealing with sexual inadequacy is the use of a male-female therapy team in the treatment of a wife-husband marital unit. Patients are much more receptive to therapy by members of the same sex, but treatment of couples is essential since sexual disorders are frequently the result of faulty interaction between members of a pair. Furthermore, in treatment, sexual activities of couples are not observed by the therapists and no sexual contacts are made by the therapist. The couple's sexual activities remain entirely private.

Unqualified "therapists" often resort to questionable procedures such as nude encounter groups or actual intercourse with the patient or they may provide surrogate partners for their patients. Clinics employing these procedures should be avoided as these activities may intensify the problem instead of alleviating it. Moreover, sexual problems are often defense mechanisms for more deep-seated problems and their correction without provision of a substitute could cause problems of a more serious nature.

At present, no state has any law requiring that a sex therapist be licensed, consequently, individuals with no medical or counseling experience are free to set themselves up as "sex therapists." The American Association of Sex Educators and Counselors is preparing a pamphlet titled "The Professional Training and Preparation of Sex Counselors" that, hopefully, will help to standardize the type and amount of training necessary for a person to become a sex therapist. Various psychiatric associations are attempting to establish standards for sex therapy. In the meantime, a person seeking help in the solution of a sexual problem should be extremely careful in the selection of a therapist. Usually a physician, marriage counselor, or minister can refer one to a qualified therapist. Sex therapists who advertise are to be regarded with suspicion and their credentials should be carefully checked.

Sex therapy may be offered by physicians, psychiatrists, marriage counselors, psychologists, or others who have a thorough understanding and knowledge of the field. Sexual problems and marriage problems are closely related and the solution of both often comes through correction of misinformation about sex and improvement in communication between the participants.

# EIGHT
# CONCEPTION AND
# DEVELOPMENT

## INSEMINATION; FERTILIZATION

During intercourse, semen is deposited in the upper end of the vagina near the cervical opening a process called *insemination*. The amount of semen discharged averages about 3 or 4 milliliters (about a teaspoonful) and contains on an average about 300,000,000 spermatozoa. Some of the sperm enter the uterus through the cervical canal and pass to the region of the openings of the uterine tubes. Sperm are propelled by the lashing movements of their whiplike tails but their movement through the uterus is principally the result of muscular contractions of the vagina and uterine wall. Movement through the uterus is rapid usually only a matter of minutes. On entering the uterine tube, sperm move to the region of the ampulla where fertilization usually occurs. Some may traverse the entire length of the tube and even enter the body cavity.

Relatively few sperm of the millions deposited in the vagina reach the fertilization site. Studies indicate that less than 1 percent of the total number survive the hazards of the journey. Many sperm are immobile; some are defective in structure; others fail to survive the acid environment of the vagina, and some fail to penetrate the viscid cervical mucus or to enter the constricted opening of the uterine tube.

Once in the uterine tube or oviduct, the narrow, constricted isthmus may impede their progress. Once past the isthmus, sperm may reach the ostium or opening of the oviduct and be lost in the peritoneal cavity. While in the uterine cavity, spermatozoa are subjected to the phagocytic action of macrophages and leukocytes and many are destroyed. The presence of spermatozoa antibodies may reduce their motility and ability to penetrate the cervical mucus. For many the energy supply is inadequate.

In their passage through the uterus and uterine tube, the sperm of most animals, and presumably humans, must undergo a physiological process called *capacitation* in which they acquire the ability to penetrate the corona radiata, a layer of follicle cells and the zona pellucida, a thin membrane, which surround the released ovum. Capacitation, while enhancing the fertilizing ability of the sperm, reduces their vitality. Their length of life in the female genital tract is not definitely known (it may be several days) but their fertilizing power is generally considered to be limited to 24 to 48 hours.

If an egg is liberated from the ovary and if sperm are present in the uterine tube, as the egg enters the tube and moves toward the uterus, *fertilization* (penetration of the egg by the sperm) usually occurs (Fig. 18). Fertilization accomplishes these results; it initiates development and it brings into the egg the male complement of chromosomes. These chromosomes contain the determiners of hereditary traits and, of the chromosomes, the sperm will contribute one, an X or a Y chromosome, which will determine the sex of the individual. Although the egg and the sperm differ markedly in size, each contributes equally (or nearly so) in the hereditary makeup of an individual.

What happens to the nonfertilizing spermatozoa? Some are lost through the semen that drains from the introitus. Most, however, are destroyed through processes that take place within the female reproductive tract. Some are destroyed by or through the action of cytolytic enzymes. Most are eliminated through the action of phagocytic cells, cells that ingest and destroy foreign matter. Many die and disintegrate.

What happens to an egg if no sperm are present or if it is not fertilized? It slowly moves along the tube but degenerates before it reaches the uterus. Unfertilized eggs are rarely, if ever, found within the uterus.

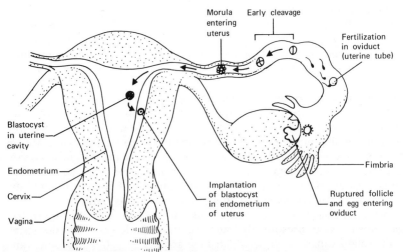

**Figure 18 Diagram showing ovulation, fertilization, early developmental stages, and implantation of the blastocyst in the endometrium of the uterus.**

## EARLY DEVELOPMENT AND IMPLANTATION;
## THE BEGINNING OF PREGNANCY

Following fertilization, the fertilized egg or zygote, now containing 46 chromosomes, undergoes cleavage (Fig. 19), that is, it divides into 2 cells, then 4, 8, 16, 32, and so on resulting in the formation of a morula, a solid ball of cells (third day after ovulation). At this stage it enters the uterus. The morula then hollows out and becomes a blastocyst (fourth day after ovulation). The blastocyst (Fig. 19) consists of a small group of cells, the inner cell mass, which is attached to the inner surface of its wall, the trophoblast. In this stage it is free within the uterus for two or three days, then if the endometrium is prepared to receive it, it embeds in the uterine lining, a process called implantation (Figs. 18, 19). This occurs on the sixth or seventh day after fertilization and is accomplished through the action of enzymes secreted by the trophoblast. With implantation, pregnancy is established.

### Signs of Pregnancy

Determination as to whether or not a state of pregnancy exists depends on symptoms observed by the subject and signs noted by the physician or examiner. Positive determination depends on laboratory tests.

Early presemptive signs include the cessation of menstruation, the occurrence of morning sickness or nausea, an increase in size and fullness of the breasts, and an increase in the pigmentation in the areola about each of the nipples.

Probable signs of pregnancy include a change is consistency and size of the uterus, the occurrence of uterine contractions, softening of the cervix, increased leukorrhea, enlargement of the abdomen, increased tenderness of the breasts, and a positive pregnancy test.

Positive signs of pregnancy include hearing and counting the fetal heartbeat, ballottement or detection of fetal movements, and the appearance of the fetal skeleton in an X-ray film. X rays should, however, be avoided except when absolutely necessary. To these have been added the recording of the fetal heartbeat by a fetal electrocardiograph and the detection of the fetal heart action by ultrasonic waves transmitted to the mother's abdomen through a transducer applied to the skin. These new techniques enable a more positive diagnosis of pregnancy, the presence of twins, the viability of the fetus, and the condition of the fetus during delivery.

### PREGNANCY TESTS

The first indication of pregnancy is the missing of an expected period, that is, the failure of menstruation to occur. However, a missed period should not be the cause of unnecessary concern for delayed or missed periods can

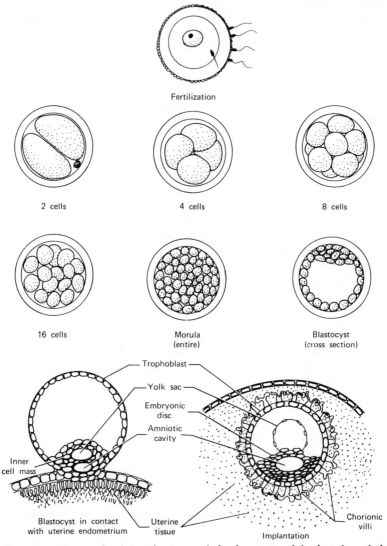

Fertilization

2 cells

4 cells

8 cells

16 cells

Morula
(entire)

Blastocyst
(cross section)

Trophoblast

Yolk sac

Embryonic
disc

Amniotic
cavity

Inner
cell mass

Blastocyst in contact
with uterine endometrium

Uterine
tissue

Implantation

Chorionic
villi

**Figure 19  Diagram showing early stages of development and implantation of the blastocyst.**

result from a number of causes. They may be due to altered ovarian function, which can be the result of disease or pathological conditions, the effects of certain drugs, disturbed pituitary function, or nervous stress or tension, especially marked emotional disturbances.

However, pregnancy is the most common cause for a missed period. Consequently, if a woman has had the opportunity to become pregnant and hasn't had a period for six weeks, that is, the period is two weeks overdue, an arrangement should be made for a pregnancy test. Advice concerning pregnancy tests can be secured from a physician, a university or community health service, or from a planned parenthood clinic.

Tests for pregnancy depend principally on the presence of chorionic gonadotrophin in the urine or blood. When pregnancy occurs, the embryo implants in the uterine lining. Its outermost membrane, the *chorion,* together with some uterine tissue produces a hormone called *human chorionic gonadotrophin* (HCG), which is excreted in the urine. Tests are made to see if this hormone is present.

Pregnancy tests are of four types: biologic, immunologic, hormonal, and radioimmunoassay. In *biologic tests,* the urine of the person being tested is injected into a laboratory animal, such as a frog, toad, mouse, rat, or rabbit. The presence of HCG in the urine will cause the release of eggs or sperm from the gonads or vascular changes in the gonads of the injected animals. These tests are not positive until about six weeks after the last menstrual period. The embryo has by this time been implanted about three or four weeks. Between this time and the third month the tests are 95 percent or more accurate.

Biologic tests have been largely replaced by *immunologic tests* that are less difficult to perform, less expensive, and provide results in a much shorter period of time. Among the tests now widely used are latex agglutination or latex agglutination inhibition response, which can be obtained in a two-minute slide test or hemagglutination inhibition results, which are obtained in two hours. These tests have an overall accuracy of 95–98 percent.

Coming into use now is the *radioimmunoassay,* which is characterized by extreme sensitivity. It involves taking a few drops of blood from a woman's finger and within an hour, results are available. It is effective as early as six days after conception and even before a menstrual period is missed or before implantation has occurred. Tests indicate close to 100 percent accuracy. This type of test is considered to be superior to any of the presently used tests.

*Hormonal tests* involve the administration, either orally or by injection, of hormones, especially progesterone. When given for three days, on withdrawal, bleeding will follow within a few days if the woman is not pregnant. There is evidence, however, that the use of hormones especially progestins during the first 12 weeks of pregnancy is associated with the occurrence of congenital malformations and cervical cancer, consequently their use is not recommended.

A variety of home pregnancy test kits are sold over the counter in the U.S. These kits measure the presence of HCG in the urine. However, they are not as accurate as physician-administered tests. If used to detect pregnancy up to ten days after a missed period, the average accuracy is about 66 percent. False positives (indicating you are pregnant when you are not) run about 45 percent. False negatives (indicating you are not pregnant when you are) run about 15 percent. We recommend instead that a young woman go to her physician, a university health service, or the local office of Planned Parenthood for pregnancy testing.

## DEVELOPMENT OF THE EMBRYO

With the implantation of the blastocyst within the endometrium of the uterus, the conditions are now favorable for the continuation of embryonic development. Within the *inner cell mass*, two cavities develop, one the *amniotic cavity*, the other the *yolk sac*. (Fig. 19). Between these two cavities lies a disc of cells, the *embryonic disc*. It is from the embryonic disc that the embryo develops. The walls of the amniotic cavity grow down and around the embryo to form a protective sac, the *amnion*. This becomes filled with a fluid, the *amniotic fluid*, which forms a protective cushion for the embryo during its intrauterine life. The yolk sac enlarges and persists as an out-pocketing of the digestive tract but it serves no nutritive function as it contains no yolk. In its walls some early blood cells are formed and primordial germ cells are thought to arise there. However it is a temporary, transient structure and is lost with the placenta at birth.

The outermost layer of the blastocyst, the *trophoblast*, grows rapidly. It is transformed into the *chorion*, which develops fingerlike processes called *chorionic villi* (Fig 19). These grow into the uterine endometrium and it is from these structures, together with some structures from the uterus that the *placenta* is formed. The placenta is the organ through which the embryo will get its nourishment. It is attached to the embryo by the *umbilical cord* (Fig. 22).

In the development of the embryo, the cells of the embryonic disc (Fig. 19) arrange themselves into three layers, the ectoderm, mesoderm, and entoderm. From these layers the various systems of organs of the body develop. The *ectoderm* gives rise to the epidermis of the skin, the nervous system and sense organs; the *entoderm* gives rise to the digestive and respiratory systems. The remaining systems (skeletal, muscular, circulatory, urinary, and reproductive) develop from the *mesoderm*.

By the end of the eight weeks of development, all of the primary organ

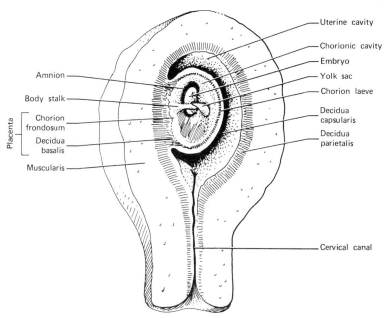

Uterine cavity
Chorionic cavity
Embryo
Yolk sac
Chorion laeve
Decidua capsularis
Decidua parietalis
Cervical canal

Amnion
Body stalk
Placenta
Chorion frondosum
Decidua basalis
Muscularis

**Figure 20   Longitudinal section of a gravid uterus to show relationships of the embryo (age five weeks) and placenta to uterus. [From B. G. King and M. J. Showers,** Human Anatomy and Physiology, **(after Arey). W. B. Saunders Co., Philadelphia.]**

systems have come into existence. Nails begin to form and ossification (bone-forming) centers appear. Externally, the embryo, now about an inch long, begins to acquire a human form; the human face is recognizable. (Fig. 21) The period of the embryo ends and that of the fetus begins.

Keep in mind that during the first two months of development, when the various complex organ systems of the body are coming into existence, any significant change in the intrauterine environment might possibly alter the developmental processes and be the possible cause of malformations sometimes present at birth. It is recommended that, as far as possible, drugs of any kind that might alter the body's blood chemistry be avoided. A classic example illustrating the serious consequences is that of the drug, thalidomide, a "harmless" sleep inducer which, in Europe, was the cause of hundreds of babies being born with short and deformed arms and legs, a condition called phocomelia.

## DEVELOPMENT OF THE FETUS
From the beginning of the third month to parturition, the developing individual (Fig. 21) is called a fetus. Some of the changes that occur during this period are as follows:

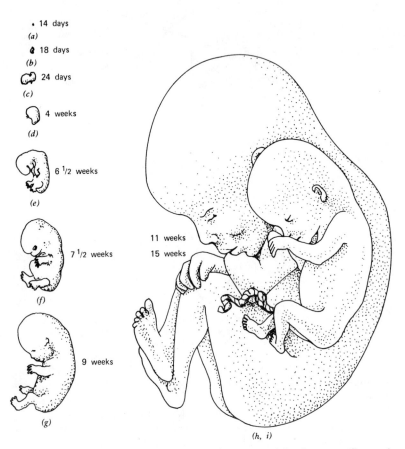

- 14 days
(a)
18 days
(b)
24 days
(c)
4 weeks
(d)
6 ¹/₂ weeks
(e)
7 ¹/₂ weeks
(f)
9 weeks
(g)
11 weeks
15 weeks
(h, i)

**Figure 21  Human embryos and fetuses at various stages of development. All natural size. (From L. B. Arey, Developmental Anatomy, W. B. Saunders Co., Philadelphia.)**

### End of Third Month
The fetus is about 3 inches long (from head to toe) and weighs 1 ounce; it has a distinct human appearance although the head is still exceptionally large; umbilical herniation becomes reduced as intestine is withdrawn into the body cavity; fingernails and toenails take form and ossification centers for most bones appear; the face is distinctly human.

### End of Fourth Month
Fetus is 6½ inches long and weighs 4 ounces; a fine, downlike hair (lanugo) covers the body; a few head hairs appear; faint fetal movements called *quickening* may be felt.

## End of Fifth Month

Fetus is 10 inches long and weighs about 8 ounces (Fig. 21); quickening movements are definitely felt; fetal heart beat may be detected; eyebrows and eyelashes appear.

## End of Sixth Month

Fetus is 12 inches long and weighs about 1½ pounds. It resembles a miniature baby but skin is wrinkled and red with little subcutaneous fat; skin is covered with a greasy *vernix caseosa,* which forms a protective covering; eyelids open. A fetus born prematurely at this age has a fair chance of survival.

## End of Seventh Month

Fetus is about 15 inches long and weighs about 2½ pounds. Subcutaneous fat begins to be deposited; testes begin to descend into the scrotum.

## During Eight and Ninth Months

Fetus acquires a weight of 6½ to 7½ pounds and is approximately 20 inches long (Fig. 23); skin loses red color and becomes white or pink; body becomes plump; the downy hair covering the body disappears; nails project beyond tips of fingers and toes, testes are usually in the scrotum and labia majora are in contact.

## DEVELOPMENT OF THE PLACENTA

The *placenta* in late pregnancy is a disklike structure about 8 inches in diameter and 1 inch in thickness (Fig. 22). It is formed in early development by the union of chorionic villi of the embryo with the uterine endometrium of the mother (Fig. 20) and contains blood spaces that receive blood from the mother, bathing the chorionic villi that project into these spaces. The villi contain capillary blood vessels of the fetus and on the surface of these villi, the exchange of substances between the mother and fetus takes place.

The placenta serves a number of important functions. Among them are:

**1 Nutrition**   Carbohydrates, proteins and fats (or their derivatives), vitamins, mineral salts, and water pass from the mother's blood to the fetus.

**2 Respiration**   Oxygen diffuses from the mother's blood into fetal blood and carbon dioxide passes in the reverse direction.

**3 Excretion**   Waste products of fetal metabolism, as urea, pass from fetal blood into the mother's blood, the placenta acting as a kidney.

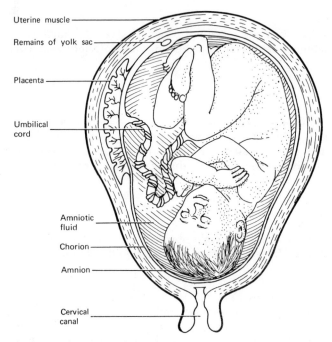

Uterine muscle

Remains of yolk sac

Placenta

Umbilical cord

Amniotic fluid

Chorion

Amnion

Cervical canal

**Figure 22   Diagrammatic longitudinal section of the uterus illustrating the relation of an advanced fetus to the placenta and other membranes. (From L. B. Arey,** Developmental Anatomy, **W. B. Saunders Co., Philadelphia.)**

**4   Barrier**   The placenta acts as a barrier to many substances that might be injurious to the fetus as pathogenic bacteria, some viruses, and certain large macromolecules. However some chemical substances as anesthetics, antibiotics, and certain drugs, as heroin, may cross the barrier.

**5   Endocrine Organ**   The placenta during early pregnancy takes over the function of the ovary and produces estrogens and progesterone. It also secretes a chorionic gonadotrophin of importance in the maintenance of pregnancy.

**6   Miscellaneous Functions**   The placenta also transmits various substances as hormones (for regulation), antibodies (for protection) and enzymes (for catalysis).

The placenta is connected to the fetus by the *umbilical cord,* which contains arteries through which fetal blood is pumped by the fetal heart to the placenta and a single vein by which blood is returned to the fetus. Note that there is no direct connection between the circulatory systems of the mother and fetus.

Also there are no direct nerve connections between the mother and

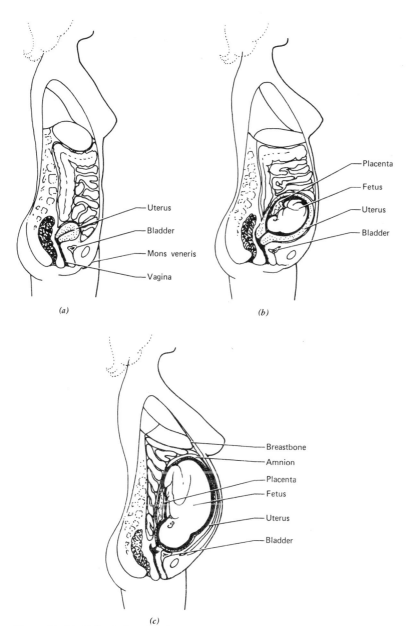

Figure 23 (a) Abdominal cavity of a woman at the end of the first lunar month. (b) Abdominal cavity of a woman at the end of the fifth lunar month. (c) Abdominal cavity of a woman at the end of the ninth lunar month. (From K. L. Jones, et al., *Sex,* **Harper & Row.)**

fetus. There is no basis for the commonly held view that prenatal influences or "maternal impressions" as a mother's thoughts, experiences, or actions can influence the physical development of an unborn child and cause various types of deformities. However it must be recognized that the emotional state of the mother may affect a developing fetus through hormones produced by the mother's glands that pass through the placenta into the fetal blood. These may have a significant effect on fetal development.

### MATERNAL CHANGES

As the fetus increases in size, the uterine muscles hypertrophy, the muscle cells increasing in number and length. Until the fourth month, the uterus with its contained fetus occupies the space within the pelvic cavity. As the fetus continues to grow, it pushes the uterus upward into the abdominal cavity where it causes a considerable displacement of the internal organs (Fig. 23). The abdominal wall begins to be distended and pregnancy can usually be detected externally. The skin of the abdomen becomes stretched and marked with reddish or bluish streaks. Mammary glands enlarge with the development of secretory tissue. Postural changes occur and the structure of the pubic symphysis and sacroiliac joints is modified, the tissues becoming softer. Pigmental changes may take place in the skin with the development of a yellowish-brown pigment, giving rise to the "mask of pregnancy" noticeable especially in the neck and face. Changes may occur in the bones and teeth especially if there is a dietary deficiency in calcium. There is an increase in blood volume due principally to the increase in water content. The increased blood volume necessitates increased work on the part of the heart. Urine production is usually increased and disorders in urination may occur due to increased pressure on the bladder. Respiratory function is increased and often shortness of breath occurs.

Other changes are occurring. The nervous system may be affected with resulting changes in mental and emotional states. The endocrine organs also respond with the production of hormones that are primarily responsible for the changes occurring during pregnancy. These and other phenomena associated with pregnancy and childbirth will be discussed in the next chapter.

# NINE
# THE MONTHS OF
# PREGNANCY; CHILDBIRTH

## PRENATAL CARE

As soon as it is determined that a condition of pregnancy exists, a woman should place herself under the care of a specially trained physician, an obstetrician, gynecologist, or a specially trained nurse-midwife who works under the direction of an obstetrician. This should be done within the first two or three weeks after the first menstrual period is missed. During the first session, the physician will usually obtain a complete medical history. This will include information about her menstrual cycles, data about previous pregnancies, miscarriages, or abortions; a record of all diseases she has had and their after effects, if any. He will ask about her family and the possible existence of any familial or hereditary conditions and he will seek similar information from her husband. He will calculate the probable time of delivery figuring 280 days from the onset of the last menstrual period before pregnancy (actually 266 days from conception).

Following the initial period of getting acquainted and establishing a condition of good rapport with the patient, the next step is a physical examination. This is a thorough examination covering all parts of the body. Included will be a thorough pelvic examination to determine the condition of the internal sex organs. The vagina and the size, shape, position, and consistency of the uterus are checked. The condition of the ovaries and uterine tubes is noted. The cervix of the uterus is carefully examined and usually a Pap smear is made to be sure that no cancerous or precancerous condition exists. The cervix is tested for gonorrheal infection. The bony structure of the pelvis is checked to be certain that no abnormalities exist that would interfere with normal delivery.

In addition to the physical examination in which data on height, weight, blood pressure, lung and heart conditions are obtained, several laboratory tests are made. Blood tests and urine examination are routine. Serological tests for syphilis are usually mandatory even though the premarital test was negative. Blood type (A, B, AB, or O) is determined in case of the possible necessity of a transfusion at delivery. Also the determination of whether the blood is Rh positive or Rh negative is made. This is of special importance in the treatment or prevention of erythroblastosis, a serious disease that may develop if a Rh- woman becomes sensitized following the transfusion of blood from an Rh+ donor or from an Rh+ fetus from a previous pregnancy.

Following the initial visit, for the next five months, the prospective mother should visit the physician once a month; in the ensuing months, more frequently and in the final months, every week or two. During these visits, the physician will check weight and blood pressure and examine the urine. He will check on the development of the fetus, its position, and the possible existence of a multiple pregnancy. A careful watch will be maintained for symptoms of any possible complications of pregnancy. He will give advice on the various problems and difficulties of pregnancy and instruction on how to meet various situations that may arise. A few of the concerns of most pregnant women are as follows.

*Weight Control and Diet*

A pregnant woman invariably gains weight resulting from the development of the fetus and associated structures, enlarged uterus and breasts, increased amounts of body fluids, and the tendency to accumulate body fat. The total amount of weight gain is normally between 20 and 25 pounds. Over the period of pregnancy, this averages 2 to 3 pounds a week. This increased weight, while often undesired, is inevitable and essential for the well-being of the mother and fetus. While the physique of the average woman is considerably altered by pregnancy, following birth of the baby, the excess weight is usually lost within a period of two to three months after delivery and the general structure of the body is little changed.

Excessive weight gain, however, should be avoided as excess weight tends to increase the likelihood of and the seriousness of any complications of pregnancy. Strenuous dieting to lose weight during pregnancy is to be avoided. Special attention should be given to the diet, which should be adequate and contain all the essential nutrients, especially proteins, minerals, and vitamins. Fluid intake should be adequate at all times.

*Breast Feeding*

The question as to whether to breast feed or bottle feed a baby must usually be resolved. This is an individual decision that depends on a number of factors, often emotional or psychological factors playing a dominant role. In general, nursing is most advantageous to the baby. Mother's milk is a natural food and contains the essential nutrients plus immune bodies that protect the newborn infant against disease. It is always available when needed, easily digested, and always at the right temperature. It is inexpensive and always fresh, clean, and sterile. The psychological value of nursing to both mother and infant is well-established. The close contact between mother and infant provides for the infant a feeling of security and being wanted and cared for that has a lifelong impact on character development. For the mother, nursing is usually a pleasurable experience and through nervous and endocrine responses initiated by suckling, involutional processes involved in the return of the uterus and other organs to normal are accelerated.

Further advantages of breast feeding are a lower incidence of cancer of the breast in women who have nursed their babies. For the baby, mortality and morbidity rates are lower for breast-fed babies than for those artificially fed. Respiratory infections and gastrointestinal disturbances occur less frequently and allergic disorders are less common in breast-fed babies. The protein of a mother's milk is of a more favorable composition and considered to be of higher nutritional value than that in cow's milk. The fluid (colostrum) secreted during the first three days following birth is thin and watery. About the fourth day, the milk assumes its normal composition and color. Weight gain is slower in breast-fed babies hence there is less likelihood that the baby will become obese. Obese infants are less healthy than normal-weight babies and tend to develop obesity in adulthood. Malocclusion of the teeth is less common in breast-fed babies and there is some evidence that the sudden infant death syndrome (SIDS) occurs less frequently in breast-fed infants.

However, some mothers, for various reasons, are not able to, or prefer not to nurse their babies. Such babies must be bottle fed. For mothers who are working or who psychologically prefer not to be tied down to the necessity of nursing or for whom nursing is an unpleasant experience, bottle feeding may be substituted. If the mother, when giving the bottle can hold the baby under conditions that simulate nursing, some of the emotional benefits of nursing to both mother and child can be provided.

Bottle feeding may be advantageous in some respects. Cow's milk is richer than human milk as it contains about twice the amount of protein and four times the amount of mineral matter, especially calcium and phosphorus. Fat content is about the same but carbohydrate content is lower. Advocates of bottle feeding also cite the high levels of DDT sometimes present in human milk, as much as two to six times that allowed in milk sold commercially. Drugs taken by women during the period of lactation may be passed to the baby through the mother's milk, such as barbiturates, heroin, amphetamines, alcohol, various laxatives, and nicotine. Maternal smoking of a pack of cigarettes or more a day not only may cause a significant reduction in milk production but may give rise to digestive disturbances as nausea, vomiting, abdominal cramps, and diarrhea in the baby.

*Various Concerns*

A pregnant woman is often concerned about the extent to which conditions of everyday life or various activities should be altered to adjust to her condition. Some concerns involve the following.

**Clothing.**   In general, the wearing of special supporting apparel in the form of maternity corsets and girdles is not required except in special cases of weak abdominal muscles or low back pain. A brassiere to support the enlarged breasts is usually worn. Loose clothing and properly fitted shoes with medium or low heels are recommended.

**Physical Activities.**   Most physical activities or exercises can be safely engaged in with moderation. The idea that almost any kind of activity increases the likelihood of an abortion or miscarriage has been largely abandoned. Walking, running, bicycling, dancing, tennis, golf, housework, and other activities of a similar nature can be participated in except that the activity should not be carried on to the point of overfatigue. At the first indication of the onset of fatigue, the activity should be discontinued. Plenty of rest and sleep are essential. Frequent rest periods during the day and occasional naps are recommended.

**Employment.**   Many women who are regularly employed wonder whether they should refrain from work during pregnancy. In general, if the work is not too strenuous, fatiguing, or stressful, there is no reason why it should not be continued throughout most of the gestation period. As a general rule, cessation of work during the final six weeks is recommended. Resumption of work following delivery depends largely on the physical condition of the mother. Generally work can be resumed five or six weeks after delivery, provided adequate care of the infant can be arranged for.

**Travel.**   With respect to travel, the following should be kept in mind. First, the possibility of a spontaneous abortion or labor occurring should always be considered and plans made to meet such an emergency. Second, trips are often long, uncomfortable, and fatiguing. If travel is necessary, the trip should be kept at reasonable length and duration. Travel during the last six weeks of pregnancy is to be avoided.

**Sexual Activity.**   Sexual intercourse, if desired, may be engaged in throughout pregnancy and orgasms may be experienced by a pregnant woman without possibility of harm to the fetus. In some women interest in sex is increased and response heightened as there is no need to employ contraceptive measures and there is no fear of an unwanted pregnancy. However many pregnant women exhibit a general loss of libido, a decline in interest in sex, and a disinclination to engage in coitus. Some prefer noncoital activities. In most cases following delivery, former feelings and desires return and usual patterns of sexual activity are resumed.

It is sometimes recommended that during the early months of pregnancy, intercourse be abstained from during the days when the menstrual cycle would normally have occurred when it is thought that coitus might trigger the uterine expulsive mechanism. At other times, intercourse is permissible except in cases of vaginal bleeding or if a woman has a history of miscarriages. It is generally desirable for the male to reduce the force of thrusting movements and depth of penetration. As pregnancy progresses a change in coital position may be necessary. The woman may assume a position astride the man or rear entry may be employed. Femoral intercourse may be substituted. Should any bleeding occur during or after intercourse or should there be excessive discomfort or pain, sexual relations should be stopped until after consultation with a physician.

Although it is generally recommended that sexual intercourse be avoided during the four or five weeks both preceding and following parturition, recent studies indicate that there is no necessity to place any restrictions on intercourse preceding childbirth. The possibility of harm or injury to the fetus is negligible. Following parturition, sexual relations may be resumed as soon as the vaginal discharge (lochia) ceases. Although it is generally believed that a woman cannot become pregnant while she is nursing her baby or until menstruation reoccurs, conception does occasionally occur hence contraceptive measures should be employed.

## THE EXPECTANT FATHER

While pregnancy and childbirth primarily involve the wife or prospective mother, the male member of the marital team, the prospective father, is also vitally involved. Sexual relations are almost with certainty due to be disrupted for a period of weeks preceding and following parturition. For some men, this may create a difficult period. However for most husbands, a thorough understanding of the physical and emotional problems which a woman encounters during pregnancy enables him to adjust to the situation and following the birth of the baby, normal marital relations are usually reestablished. The services of the wife's obstetrician should be available for the husband as well as the wife and the physician should take the time to discuss the various problems of pregnancy including the problem of sexual relationships and especially emotional problems that may develop. He should answer freely any questions that may arise concerning pregnancy and parturition. In this way, fears and anxieties can often be avoided or ameliorated.

Classes in prenatal planning or preparation for natural childbirth are often available and it is recommended that expectant fathers attend. Attending classes with the prospective mother does much to promote mutual interest and understanding and the recognition of problems likely to be encountered. A responsible, understanding, and sympathetic husband can do much to relieve the burdens of the mother who must experience the discomforts and bear the risks of pregnancy. He should make himself aware of what is involved in adequate maternity care and see that his wife receives the best. He will usually find it necessary to assume more of the responsibilities of the home and he should see that his wife is relieved of any heavy duties that are likely to be fatiguing.

The husband should be especially informed concerning signs of impending delivery and plans and preparations should be made for getting his wife to the hospital when birth is imminent. During early labor, the husband may be permitted as a visitor in the labor room, however in many hospitals, his presence in the delivery room may not be allowed. However for couples who choose the psychoprophylactic technique for natural childbirth, the presence of the husband is essential in both labor and delivery rooms and many hospitals make accommodation for and welcome the husband's presence. The advent of the father witnessing and assisting in the birth of his own child creates a feeling of togetherness and mutual understanding that may be lacking when the mother is alone throughout the process.

For many husbands the period of pregnancy and childbirth is one of considerable emotional tension and some exhibit symptoms similar to those of their wives as nausea, fatigue, and backache. Fears and apprehensions experienced by the mother may also be experienced by the father. Postpartum blues, a period characterized by depression and crying occurring a day or two after delivery are common in many women and may also be experienced by the father.

Finally, even though the period of pregnancy and childbirth may be difficult, most husbands survive. However, the addition of a third member to the household may introduce a disrupting factor especially if the wife devotes too much attention to the newcomer. The new baby may be regarded as a rival for his wife's attention or it may happen that the father pays too much attention to the new baby and the wife is neglected. In either case, the bond between husband and wife may be lessened instead of strengthened. Such cases are rare, however, and proper counseling usually can bring about mutual adjustment and understanding. Generally the addition of a new member to the family is one of life's great adventures and the father's role in the well-being of the family is greatly enhanced.

## MINOR COMPLAINTS AND DISCOMFORTS OF PREGNANCY
Women who are pregnant experience unpleasant situations and conditions to a greater extent than nonpregnant women. Some of these are caused by pregnancy; others are aggravated by it. Some of the conditions commonly encountered are the following.

### Nausea and Vomiting
Nausea or "morning sickness" during the early hours of the day is common. Usually it is limited to the first two months of pregnancy and then it

tends to disappear. Vomiting, if it occurs and persists, requires special medical attention as it may be a symptom of a more serious condition.

### Heartburn

This is a burning sensation in the chest accompanied by regurgitation of small quantities of a bitter, sour-tasting fluid. It is a common condition in pregnancy due to the upward pressure exerted on the stomach due to the increased size of the uterus. Antiacids such as milk of magnesia may bring relief and a reduction in the intake of fatty foods is recommended.

### Flatulence

This is a condition resulting from the accumulation of gas in the stomach and intestine. It produces a bloated condition and the desire to pass gas. The avoidance of gas-forming foods such as beans, swallowing small amounts of well-masticated foods, and the maintenance of regular habits of elimination sometimes relieves the situation.

### Constipation and Other Complaints

This is infrequent or difficult defecation or sluggish action of the bowels. The enlargement of the uterus often interferes with muscular activity of the intestine causing a slowing of peristaltic action. Important factors in the maintenance of proper bowel action are (1) the development of proper elimination habits by going to the toilet at a regular time each day; (2) adequate fluid intake, especially drinking a glass of water before breakfast; (3) adequate bulk in diet; (4) eating fruit, especially prunes or raisins before retiring, and; (5) avoiding the constant use of laxatives. The occasional use of laxatives may be justified but continuous use may cause constipation. In extreme cases suppositories or enemas may be resorted to but only under the physician's directions.

In addition to the conditions discussed, other complaints include:

*Varicose veins.* These are enlarged and swollen veins in the lower extremities, some with a knotted appearance resulting from stretching and distention of the weakened walls. Elastic leg bandages may be required. A standing position should be avoided.

*Hemorrhoids* or *piles.* These are enlarged and swollen veins situated near the anal opening. They are aggravated by hard bowel movements and excess roughage in the diet.

*Swelling in legs and ankles.* This is edema and results from an increase in the amount of fluid held in the tissues and its tendency to accumulate in the lower extremities. If persistent, it should be reported to the physician.

*Leg cramps.* These are painful, spasmodic contractions of muscles in the lower extremities especially in the calf muscles. Local massage frequently brings about relaxation and relief. Elevation of the legs and avoiding long periods of standing may bring relief. An increase in calcium intake is recommended.

*Pruritus.* This is itching of the skin, often relieved by application of a solution of sodium bicarbonate.


## Nosebleed and Nasal Congestion

The former is generally due to excessive dryness of the nasal mucosa; the latter to vasocongestion resulting in a swollen condition, thought to be an allergic reaction (allergic rhinitis) of unknown cause.


## Dyspnea

This is difficult breathing or shortness of breath, commonly the result of pressure of the uterus against the diaphragm.


## Vaginal Discharge or Leukorrhea

A thin, yellowish discharge from the cervical glands of the uterus is normal but if the discharge becomes thick, yellowish, and profuse, it is usually indicative of a vaginal infection of some kind. It should be brought immediately to the attention of the physician.


## Miscellaneous Conditions

Insomnia, backache, tendency to faint, increased frequency of urination, painful uterine contractions, periods of depression, and other more serious complications may be experienced. In all cases, if the condition is persistent, a physician should be consulted.


## CHILDBIRTH OR PARTURITION

The period of *pregnancy,* also called *gestation,* is normally 280 days or 40 weeks. At the end of this period called *full term, delivery* or *parturition* occurs (Fig. 24). Contractions of the uterus, usually accompanied by *labor pains* force the amniotic sac into the cervical canal bringing about dilatation of the cervix. This results in a shortening and widening of the cervical canal, a process called *effacement,* so that the edges are brought upward around the presenting part usually the head of the fetus. This is the *first stage* of delivery. During this stage, the amniotic sac usually bursts and there is the flow of waters, the *amniotic fluid.* During the *second stage* of labor (stage of expulsion), the body of the fetus is expelled through the birth canal and the actual birth occurs. This is accomplished by uterine contractions aided by bearing-down efforts (contractions of abdominal muscles) which increase intra-abdominal pressure. The *third* and *final stage* of labor consists of the separation of the placenta from the uterine wall and its expulsion. This usually occurs 20 to 30 minutes after the birth of the baby. The placenta and its attached membranes constitute the *afterbirth.*

The length of labor from the onset of the first labor pain to the conclusion of the third stage, that is, the expulsion of the afterbirth, may vary from 3 hours or less (very rapid labors) to 24 hours or longer. For women bearing their first child, the average length of labor is about 14 hours; for women

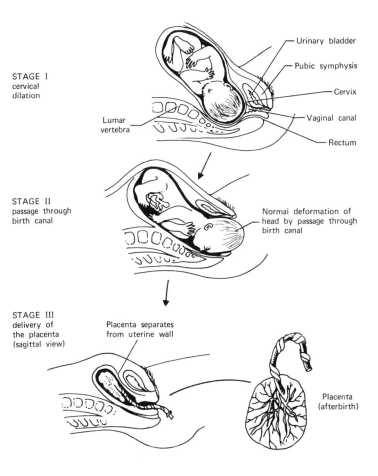

STAGE I
cervical
dilation

Urinary bladder

Pubic symphysis

Cervix

Vaginal canal

Rectum

Lumar
vertebra

STAGE II
passage through
birth canal

Normai deformation of
head by passage through
birth canal

STAGE III
delivery of
the placenta
(sagittal view)

Placenta separates
from uterine wall

Placenta
(afterbirth)

**Figure 24   Stages in normal delivery. (From J. Schifferes,** *Healthier Living,* **3rd ed., Wiley.)**

bearing their second and subsequent babies, the time is 4 to 6 hours less. Women aged 35 or over giving birth to their first baby usually have a longer period compared to women in their twenties.

Termination of pregnancy may occur before full term. The fetus at 28 weeks (seven months) is termed *viable,* that is, capable of living outside the uterus, although with mechanical aids such as incubators, babies much younger can be kept alive. Before this time, termination of pregnancy is called an *abortion* or *miscarriage;* after 20 weeks but before full term, a *premature birth.* However since the exact duration of pregnancy is difficult to determine, a baby weighing roughly less than 5½ pounds but more than 2¼ pounds is designated *premature;* between 2¼ pounds and 1 pound, *immature;* less than 1 pound, an *abortion.*

When normal delivery cannot occur, as in cases of mechanical obstruction of the birth canal or ectopic pregnancy, a *Cesarean section* may be resorted to. This is removal of the fetus through an incision made through the abdominal wall and wall of the uterus. Cesarean sections have become increasingly common and now comprise about 7 percent of all deliveries. Although a woman can experience an indefinite number of Cesarean sections, it is generally recommended that after the third section, measures be taken to prevent subsequent pregnancies.


*NATURAL CHILDBIRTH*

The delivery of a baby is usually accompanied with a variable amount of pain, sometimes slight, sometimes severe. Women have usually accepted this as a natural and inevitable consequence of childbirth but with the development of modern medicine and especially the use of analgesics and anesthetics for the relief of pain, efforts have been made to apply them to the process of parturition. However, the use of any type of agent for the relief of pain is accompanied by some danger and the effects are twofold since both mother and the newborn are involved. Anesthesia ranks fifth among the causes of maternal mortality and babies frequently suffer from anoxia with resultant respiratory difficulties and possible brain damage. Also, the failure of the mother to participate consciously in the act of labor sometimes may be a factor in postpartum depression.

As a consequence, special efforts have been made to bring about relief from pain through psychological training methods. A number of different techniques under various names have been developed. In general, they fall within four categories designated natural childbirth, autogene training, hypnosis, and psychoprophylaxis.

*Natural childbirth,* according to its proponents, is a normal, functional activity and should be painless. They believe that fear, apprehension and tension are the primary factors in producing a pathological condition that results in pain and that the elimination of these attitudes tends to bring about painless parturition. Physiotherapeutic exercises and psychological education employing these principles are now widely employed.

*Autogene training* utilizes psychological and exercise techniques but it also emphasizes the importance of relaxation and release from tension to reduce childbirth pain.

*Hypnosis* for the control of pain has been employed for many years and is effective during childbirth when employed by a competent hypnotist on a receptive patient.

*Psychoprophylaxis* is a method widely used in Europe and especially in France where it was promoted by Dr. Fernand Lamaze. It is based on the principle of developing conditioned reflexes or responses by which a patient blocks out painful sensations by substituting another stimulus at the appropriate time. Controlled respirations play an important role in this method. By concentrating on rapid, shallow breathing, parturition is facilitated and the sensations of pain are less noticeable.

A number of training programs have been developed, some of which utilize one of the above concepts while others may make use of some of the principles from all. Emphasis to a varying degree is given to educational and psychological preparation, body exercises, breathing and relaxation techniques, assistance during labor, and so on. Their effectiveness depends on a large number of factors.

In most communities, classes in "childbirth education" are now available. In these classes, instruction in the anatomy and physiology of the reproductive organs, changes that occur during pregnancy, the physiology or labor and delivery, alternate methods of delivery, and care of the newborn are covered in detail. Especial attention is given to preparation of the prospective mother (and father) psychologically and emotionally for the event of childbirth and parenthood. As members of the class form a "group" with common interests and concerns, learning from others and from women who have recently given birth to a baby does much to alleviate undue concerns and worries. Doctors and nurses often participate in these classes and contribute from their firsthand knowledge and experience.

### Delivery in the Home

Many prospective mothers wish to have their baby "in the home as nature intended." While there may be some advantages, because of the dangers involved, most obstetricians recommend against such. In a hospital, facilities are available for taking care of emergency situations that may involve extreme pain, convulsions, hemorrhage, trauma, mechanical difficulties in delivery, and other conditions. Postpartum care for both mother and infant are available, and the mother gets a few days rest she would not get in the home. Home delivery is not recommended except under emergency conditions.

## COMPLICATIONS OF PREGNANCY
Although in most women, the process of conception, development, and childbirth occur normally, among a certain number, various complications may develop that may lead to difficulties in delivery. The following are some of the difficulties sometimes encountered.

### Premature Delivery
This is the condition in which the baby at birth has a weight of less than 5½ pounds. Such babies are usually born prior to the thirty-sixth week of pregnancy. In these babies, the organs have not developed to their full extent hence their chance of survival is less than in a baby which develops to full term. Babies weighing as little as 2¼ pounds have a chance of survival; those weighing less may survive, their chances depending on their degree of development.

The primary cause of premature delivery is not known because the precipitating factor that brings about the initiation of uterine contractions is yet to be discovered. However a few conditions are known that tend to be associated with premature delivery. Twins are usually delivered early. If the first baby is premature, the chances are increased that the second and subsequent babies will be born premature. Structural disorders, infections of the uterus, systemic disorders such as diabetes and hypertension, infectious diseases as syphilis and tuberculosis, and endocrine disorders are causative factors. Disorders of pregnancy itself may be involved. Maldevelopment or malnutrition of the embryo, abnormalities in placental structure, and erythroblastosis fetalis resulting from Rh incompatibility of parents may be causative factors.

### Toxemias of Pregnancy
These comprise a group of disturbances characterized by persistent vomiting and a number of other signs and symptoms including elevated blood pressure, albuminuria, edema, and, in serious cases, convulsions and coma. In this group are *preeclampsia* and *eclampsia,* serious disorders of pregnancy. The possible occurrence of these conditions emphasizes the importance of prenatal care for the pregnant woman.

### Difficult Delivery or Dystocia
Delivery for some women may be easy, of short duration, and with little pain or discomfort. For other women, especially women experiencing their first pregnancy, it may be prolonged and difficult. Various causes include malposition of the fetus, structural abnormalities (especially those involving the pelvic girdle), and physiological disorders such as uterine inertia.

The baby is usually born with the head coming through the birth canal first with chin bent down on the chest. This is the termed a *head* or *cephalic presentation*. When a part of the body other than the head is presented first, such constitutes an unusual or abnormal presentation. Various types of presentation include a face, brow, breech, shoulder, or transverse presentation. These positions usually impede the process of labor and make for a difficult delivery. Sometimes special instruments including obstetrical forceps are used to assist in delivery. When vaginal delivery is impossible, a *cesarean section* in which the fetus is removed through an incision made through the abdominal wall and the uterine wall is usually required.

Structural disorders that may lead to difficult delivery include cysts or tumors of the uterus, a pelvis of abnormal size or shape, abnormal placement of the placenta, or abnormal implantation. Sometimes the placenta is attached to the lower portion the uterus where it covers wholly or in part the opening to the cervical canal. This condition, called *placenta previa* usually results in severe hemorrhaging and premature delivery. Hemorrhaging during the latter part of pregnancy may also be due to premature separation of the placenta, that organ becoming partly or wholly detached from the uterine wall before the baby is born. This condition is often fatal to the fetus, which is deprived of oxygen before its lungs are functional. Ectopic pregnancies, in which implantation is outside the uterus (in the uterine tube or body cavity), necessitate removal by cesarean section.

In a normal delivery, uterine contractions occur every two or three minutes, the contractions beginning at the top of the uterus and proceding downward toward the cervix in a wavelike fashion. In *uterine inertia,* the contractions are uncoordinated, tending to start in the middle and spread both ways. They also occur at irregular intervals and are irregular in intensity and duration. As a result, the cervix is not dilated and women tend to experience prolonged and painful labor. The condition is sometimes alleviated by an induced period of rest or the use of a uterine stimulant such as oxytocin, a pituitary hormone.

## MULTIPLE BIRTHS

Normally a single birth results from a pregnancy, however multiple births occur in approximately these ratios: twins (once in 80–90 births), triplets (once in 7000–8000 births), quadruplets (once in 700,000 births), quintuplets or more (rarely, although the incidence is increasing with the use of fertility drugs).

Twins are of two types, identical and fraternal. *Identical* or *true twins* develop from a single fertilized ovum, hence they are *monozygotic*. They have the same chromosomal makeup and are remarkably alike in physical, physiological, and mental traits. They are always of the same sex, and because they develop from the same blastocyst, they are enclosed in a single chorionic sac and have a common placenta. About one third of all twins are of this type.

*Fraternal* or *false twins* result from the simultaneous development of two ova fertilized by two different spermatozoa. They are *dizygotic twins*. Twins of this type each develop their own chorionic sac and placenta. They may be of the same or different sexes and they resemble each other no more than brothers and sisters of the same family.

The same explanation holds for the occurrence of triplets and quadruplets. They may all arise from the same fertilized egg, in which case they would be identical; or from a combination of both, as triplets arising from two eggs, one developing into identical twins and the other developing as a single individual; or all three may develop from separate eggs.

Multiple births, especially those arising from multiple ova, may be due to a number of factors. Heredity is of significance as indicated by the tendency of twins of the two-egg type to occur in certain families, or to occur successively in a single family. The age of the mother is another factor, twinning occurring more frequently as mothers get older. Finally twinning varies with country and race, blacks having more two-egg twins than whites and American whites more than Japanese. The incidence of one-egg twins is fairly constant in various races and groups and probably is a fortuitous occurrence.

Occasionally twins are born with bodies fused in varying degrees. These are *conjoined* or *Siamese twins*. They are the result of an incomplete division of the embryo at an early developmental stage probably by the end of the second week. Twins of this type may be approximately equal in size and development or one may be markedly smaller than the other in which case it is designated a *parasite*. The degree of fusion may be slight permitting surgical separation after birth, but often it is extensive involving shared organs which do not permit separation. This type of twinning is rare, occurring an estimated once in 100,000 births.


*ABNORMAL DEVELOPMENT*

A considerable amount of variation occurs in the development of the human body, however when the body as a whole or a specific part deviates from the generally accepted range of variation, such is referred to as a *malformation* or *anomaly*. Sometimes a fetus is so grossly malformed that it bears little resemblance to a normal baby. The study of abnormal development is called *teratology*.

Birth defects may be structural or physiological. *Structural abnormalities* may involve the absence of an organ, excessive or reduced size of a part, failure of an embryonic structure to atrophy, failure of an opening to develop, failure of an embryonic opening to close, the fusion of parts that are normally separate or failure of fusion when such should occur, the splitting of parts normally single, duplication of parts, stenosis (the abnormal narrowing of a duct or opening), abnormal migration of parts, or misplacement or reversal of organs.

*Physiological abnormalities* or defects include such conditions as abnormal protein metabolism, phenylketonuria (PKU), albinism, hemophilia, and many others. These conditions are usually of genetic origin.

*Causes of Maldevelopment*

Birth defects fall into two categories based on their causes. They are either inherited or acquired. *Inherited defects* are those that result from hereditary determiners or genes transmitted through the egg or sperm. These defects, if not lethal (terminating life or preventing reproduction), tend to recur generation after generation, however, sometimes skipping a generation. New characteristics may appear as a result of mutations or sudden changes in the genes. A change in the number of chromosomes, as an increase or decrease, usually results in a defect of some type.

*Acquired defects* are the result of some environmental factor or influence acting on the developing embryo or fetus that alters the normal processes. In development, each organ or structure passes through a *critical period* during which a series of rapid changes occur—for example, when the neural plate, a solid plate of cells, transforms into a tube from which the spinal cord and brain develop—or in the development of the face when three separate elements fuse to form the upper lip and the two halves of the palate unite to form the roof of the mouth. If a significant environmental change occurs during this critical period, as a decrease in oxygen supply or the presence of an adverse drug, hormone, or other chemical substance, development can be slowed or speeded up and normal processes fail to occur. This could account for brain and spinal cord defects and hare lip or cleft palate mentioned.

Other factors that can affect development adversely are mechanical factors (such as a constricted umbilical cord) radiations from X rays, or radium, chemical substances from the mother's blood (such as Rh-negative antibodies), or disease organisms involving either the mother or fetus or both.

The most critical period in the development of a human embryo is the period of the *third to eight weeks,* during which all the organ systems come into existence. During this period a prospective mother should take special care to avoid subjecting her body to physical, chemical, or emotional changes that might be of significance to the developing embryo. The ingestion of drugs and medicines should be guarded, exposure to disease—especially rubella or German measles—should be avoided, and adequate nutrition should be provided for the needs of the developing embryo and the changes that take place in the mother.

The age of the mother is also a factor in the incidence of certain abnormalities. Down's syndrome or *mongolism* is a condition characterized by mental retardation, broad face, flat nose, oblique eyes, and various malformations. It is due to an extra chromosome (specifically chromosome 21 or a part of it) in one of the sex cells. In women aged 31 to 35, its rate of incidence is 10.9 per 10,000 births, but for women age 41–45 its incidence is 185 and for women between the ages of 46 and 50, it is 869.


*Extrauterine Pregnancy*
Normal development occurs within the uterus but sometimes an ovum may develop outside of the uterine cavity. This results in *extrauterine* or *ectopic pregnancy*. The blastocyst may develop within the ovary (*ovarian pregnancy*), in the abdominal cavity (*abdominal pregnancy*), within the uterine tube (*tubal pregnancy*), or within that part of the tube that is in the wall of the uterus (*interstitial pregnancy*). Since normal development usually is impossible for a fetus developing in one of these locations, pregnancy is usually terminated and the fetus removed by Cesarean section. Only rarely do they go to full term.


*THE DETERMINATION OF SEX*
Neither the male nor female reproductive system is fully developed at birth. In early embryonic development, there is a single, indifferent system (Figs. 4, 5) and sex cannot be distinguished. At about the age of six weeks, the structures begin to differentiate into organs which distinguish the sexes. This is dependent upon the chromosomal makeup of an individual as determined at fertilization. There are two types of sperm. One contains an X chromosome; the other a Y chromosome. Every egg contains an X chromosome (Fig. 1). If an X sperm unites with an X egg, the resulting zygote or fertilized egg (XX) will develop into a *female*; if a Y sperm fertilizes an X egg, the resulting zygote (XY) will develop into a male. The difference between X and Y sperm cannot be detected by microscopic observation.

Since the reproductive systems of the two sexes have a common origin, there is a basic correspondence in certain of their parts (Figs. 4, 5). The *gonad* develops into a testis or ovary; the *phallus* into a penis or clitoris, the *labioscrotal folds* into the major labia or scrotum. Sometimes abnormalities in development occur and an individual of one sex may have structures that resemble those of the opposite sex. Such an individual is a *pseudohermaphrodite*. A true *hermaphrodite* possesses both a testis and ovary or an ovotestis. Such individuals are rare.

## SEX SELECTION OR PRESELECTION OF SEX

Man has always had the desire to be able to predict the sex of expected offspring or to select the sex of the newcomer. In various cultures, the value of the newborn baby was determined largely by its sex. Sometimes the male was more highly prized, sometimes the female. As a result, throughout the ages, many erroneous ideas and beliefs developed as to how the sex of a child *in utero* could be determined or how it would be possible for parents to produce a child of a desired sex.

With reference to the predetermination of the sex of an unborn child, it is now possible to determine such through the examination of the fluid within the amniotic sac. Through an operation known as *amniocentesis,* a hollow needle is introduced through the abdominal and uterine walls and some fluid withdrawn from the amniotic sac. By examination of the fetal cells contained in this fluid, the sex of an unborn child can be ascertained as the cells of females possess a small, dark body, the *Barr body,* located at the edge of the nucleus. This body is the condensed X-chromosome. This operation is not without risk and while it is valuable in some cases for clinical purposes, it is not recommended for general use.

With respect to the selection of sex of a prospective child, various techniques and procedures have been suggested and tried in an effort to assure that the child born would be of a particular sex. These have involved variations in coital position; times of coitus in relation to time of day, season, or menstrual period; type of food eaten; climate; age of individuals; acidity or alkalinity of the vagina; and other conditions. However, none of these have given consistent results in predicting the sex of offspring. The sex of a child is determined largely by chance, that is, whether the egg is fertilized by a sperm bearing an X or a Y chromosome.

Modern attempts of sex control have centered on efforts to separate the male-determining, Y-bearing sperm from the female-determining, X-bearing sperm. Although some investigators have claimed a measure of success, no sure method of separating these two types of sperm or of being assured that they would be capable of fertilizing an ovum after separation has been developed. Methods utilized in separation procedures include electrophoresis, various physical or mechanical methods, and chemical procedures. None to date have proven effective to the extent that definite control of the sex of offspring can be achieved.

One of the puzzles of reproductive physiology is the preponderance of male conceptions. X and Y sperm are produced in approximately equal numbers and theoretically equal numbers of males and females should be born. However, the ratio at birth is about 105 males to 100 females. Furthermore, the ratio at the time of conception is much higher, about 150 to 100. The viability of male embryos is much less than that of female embryos as indicated by the greater incidence of aborted and stillborn males.

One explanation of the greater number of male conceptions is the supposition that the male-determining, Y-bearing spermatozoa are lighter and more active than the female-determining, X-bearing sperm, hence are more likely to be the first to reach the fertilization site and fertilize the egg. This is merely conjecture and has no scientific support.

There have been a number of recent publications in which it is claimed that it is possible for a couple to choose the sex of their baby and specific instructions are given as to how to accomplish such. The activities recommended for producing a *male baby* are as follows: engaging in intercourse following ovulation, alkaline douching by the female before coitus, precoital abstinence, deep penetration by the male, and the experiencing of an orgasm by the female. For producing a *female baby*, the recommendations are engaging repeatedly in intercourse prior to ovulation with no intercourse following ovulation, douching with an acid solution before coitus, shallow penetration by the male with the female not experiencing an orgasm.

Evidence that the above measures, if followed, will result in a baby being of the desired sex is weak and questionable. Extensive studies have been carried on in the field of animal husbandry in an effort to develop breeding techniques that will provide the breeder with an animal of the desired sex but no effective method of sex selection has been developed. The same applies to human reproduction, however, if a child of a specific sex is especially desired, any or all of the above suggestions may be resorted to without possibility of harm; and the probability of success will be at least 50 percent.

# TEN
# INFERTILITY

Most couples who enter marriage plan to have children and usually take steps to see that the right number arrive at the right time (planned parenthood). This normally involves the purposeful use of contraceptive measures to prevent an unwanted pregnancy. However, some couples find that, even if no contraceptive measures are employed, conception does not occur. When this situation persists, a condition of *infertility* exists. It may be temporary and after a period of time, usually several weeks or months during which regular intercourse is engaged in, conception may finally occur. If failure of conception persists indefinitely, a condition of *sterility* exists, which may be permanent.

It is estimated that 10 percent of all married couples are completely sterile and unable to have children. An additional 15 percent are infertile to a marked degree and have fewer children than they desire, and 25 percent may experience infertility to a limited degree. Studies in the field of human fertility have revealed that sterility is usually not due to a single cause but is generally the result of a multiplicity of factors. Anatomical, physiological, pathological, and emotional conditions affecting either the husband or wife or both may be involved. As a result of modern developments in this field about 30 percent of barren marriages are rendered fertile and, with improvements in methods of diagnosis and treatment, the outlook is even more promising for childless couples.

In general, the female plays a more important role in infertility than the male. This is because, first, she normally produces only one egg every 28 days while the male produces millions of spermatozoa more or less continuously. Second, the egg has a very limited life span and capability of being fertilized. Third, the reproductive organs must not only serve for female functions but they must provide a path for the sperm to travel to reach the egg, and finally, the entire process of pregnancy and childbirth must occur before a live baby is brought into existence.

## CAUSES OF INFERTILITY IN THE FEMALE
Infertility in the female may be absolute or relative. *Absolute infertility* or *sterility* is that in which structural or physiological conditions make it

impossible for a woman to conceive. It may be the result of (a) developmental anomalies; (b) disease processes or surgical intervention, as in ovariectomy, hysterectomy, or tubal ligation; (c) physiological disorders as disturbances in the functioning of the endocrine glands; or (d) immune responses as the immobilization and death of spermatozoa or death of embryos resulting from immobilizing, agglutinating, or other antibodies present in female tissues.

*Relative infertility* is difficulty in conception that might yield to the possibility of pregnancy. Some possible causes that are susceptible to correction are malfunctioning of the ovaries or uterus, obstructed uterine tube, conditions within the genital tract affecting sperm transport, psychologic factors, and timing of intercourse.

### Malfunctioning of the Ovaries and Uterus

The normal, cyclic functioning of the ovaries and uterus are essential for conception. Failure of the ovaries to develop follicles or failure of the follicles to release their eggs is often a primary cause. This may be due to ovarian lesions (cysts, tumors), ovarian infections, or to disorders of the pituitary gland or hypothalamus whose hormones are essential for normal ovarian function. Hormones from the thyroid and adrenal glands are also essential.

Ovarian disorders lead to disorders of the uterus since the cyclic changes in the uterine endometrium during the menstrual cycle depend on the production of estrogens and progesterone by the ovary. Anatomical disorders as a prolapsed uterus, diseased conditions as endometriosis, or the development of cysts or tumors may prevent implantation of the blastocyst. Endocrine disorders may result in repeated abortions or miscarriage.

### Obstructed Uterine Tube

A patent (open) uterine tube is also essential for conception. Since sperm must pass into the tube to accomplish fertilization and the fertilized egg must pass to the uterus to continue development, any obstruction of the tube as may result from tubal infections or structural abnormalities could result in failure in conception. Spastic muscular constriction at the isthmus or faulty activity of oviductal musculature could also prevent conception.

### Conditions Affecting Sperm Transport

Conditions within the female genital tract may have a significant effect upon sperm deposited within the vagina and their transport to the fertilization site. Excessive acidity of the vagina or allergic responses in which a woman develops antibodies antagonistic to a male's sperm may exist. Excessive thickness of the cervical mucus or a spastic condition of the muscles surrounding the opening of the uterine tube into the uterus may prevent the sperm from entering the tube. Failure of the sperm to undergo capacitation may result in their inability to penetrate the membrane surrounding the egg. A substance, hyaluronidase, present in semen may be lacking. This substance is thought to play a role in the dispersal of follicle cells that surround the egg. These and other conditions make the course of sperm a hazardous one and often account for failure of fertilization to occur.

*Psychologic Factors*
These play an important role in sterility and often are the only factors involved. These are poorly understood, but fear of pregnancy or fear of the responsibility of parenthood may bring about reactions which prevent conception. In such conditions, adoption is often resorted to with the result that conception occurs shortly after the adoption of a baby.

*Timing of Intercourse*
The proper timing of intercourse so that sperm are deposited within the vagina at or near the time of ovulation is essential. Details as to its relationship to conception are given under the safe-period method of contraception. (pp. 155–157).

## CAUSES OF INFERTILITY IN THE MALE
The basic causes of infertility in the male are impaired production of spermatozoa, disorders in the secretory function of the accessory sex glands, and interference with the passage of spermatozoa through the genital ducts.

*Impaired Production of Spermatozoa*
This may result from (a) structural disorders of the testes, such as undescended testes or the existence of varicocele (enlargement of veins in the spermatic cord), hydrocele (accumulation of serous fluid about the testis), cysts, or tumors; (b) atrophy of sperm-forming tissues that may occur in mumps orchitis, in certain systemic conditions, following irradiation or resulting from elevation of scrotal temperature; (c) endocrine disorders involving the pituitary, adrenal, or thyroid glands; (d) nutritional deficiencies, such as lack of adequate proteins, vitamins, or minerals; (e)

autoimmune responses to antigenic factors in sperm or spermatogenic tissue; and (f) retrogressive changes as those that occur in aging.

As a result of these conditions, sperm may be absent (*aspermia*), or few in numbers (*oligospermia*), or there may be a high percentage of deformed or nonmotile sperm (*dysspermia*). Normal sperm count ranges from 40 million to 150 million per cubic milliliter of ejaculate. Of this number, 60 percent show active motility and abnormal forms should not exceed 20 percent.

### Disorders in Secretory Function of the Accessory Sex Glands

The accessory glands especially the seminal vesicles and prostate gland produce the major portion of the seminal plasma that comprises about 90 percent of the male ejaculate. Important constituents of their secretions are fructose and prostaglandins from the seminal vesicles and liquefying enzymes, zinc, and magnesium from the prostate. Normally semen coagulates immediately after ejaculation but liquefies normally within 20 minutes as a result of the action of liquefying enzymes. Inadequate prostatic secretion may prevent or delay liquefaction. Impaired secretion of fructose by the seminal vesicles impairs the motility and vitality of spermatozoa. Buffering agents secreted by the seminal vesicles are important in counteracting the adverse effects of acids on spermatozoa, consequently, disorders of the accessory glands especially infections may have a serious effect on fertility.

### Obstruction of Sperm Ducts

Any condition that might interfere with the passage of semen through the efferent passageways and its deposition within the vagina could be a factor in infertility. Structural alterations may involve any of the following: epididymis, vas deferens, ejaculatory duct, or urethra. Infections are a common cause of blockage of the narrow ducts. However, the most common cause of infertility in the male is *varicocele*, a form of varicose veins inside the scrotum. The enlarged veins press against the vas deferens, blocking the movement of sperm. Impairment of the nerve supply, stricture of the urethra, or pressure from an enlarged prostate are less common forms of infertility in the male.

### Treatment of Infertility

Infertility that continues over a period of several months should be treated by a gynecologist or andrologist trained in this particular field. An adrologist is a physician who specializes in diseases or disorders of men, especially those of the reproductive system. Most large cities have fertility clinics where special treatment can be secured. Fertility drugs are now available that are effective in some cases. Sometimes they are too effective resulting in multiple births. Usually a thorough examination of both partners will reveal the basic cause or causes of infertility and often times the condition can be corrected.

## ARTIFICIAL INSEMINATION
When natural conception is not possible, sometimes artificial insemination (AI) is resorted to. In this procedure, semen is deposited mechanically within the vagina or uterus of the female. The sperm may be obtained from the husband (AIH) or from another male or donor (AID). AIH is employed when the number of spermatozoa in the husband's ejaculate or the quantity of semen is inadequate. By collecting the semen from a number of ejaculations and preserving it until an adequate quantity is secured, AIH is sometimes effective.

If the donor is other than the husband, with the consent of both the husband and wife, sperm from a person of sound genetic background and resembling physically the husband is obtained and used for insemination.

Even though artificial insemination presents certain moral, religious, and legal problems, it is becoming a generally acceptable procedure. It is used extensively in the breeding of domestic animals.

### Test-Tube Babies
An important step has been accomplished in recent years in the fertilization of human eggs *in vitro* (in a glass vessel or container outside the body). Ova are obtained from the mature follicles of a woman or from the uterine tube following ovulation. These are then fertilized by sperm, either fresh or frozen, and the fertilized egg or zygote is then replaced in the uterus of a female, either the woman from whom the egg was obtained or possibly another woman whose uterus is in a receptive state. It has been reported that, in a limited number of cases, development has proceded to full term. Although the procedure is difficult to accomplish, it does provide a method by which a woman who cannot become pregnant by normal means might be able to give birth to a baby.

### SPERM BANKS
A sperm bank is a place where sperm are cryogenically stored, that is, stored at a very low or ultralow temperature. Spermatozoa in freshly ejaculated semen will retain their potency only about 48 hours, but when refrigerated properly, their ability to fertilize an ovum persists for an undetermined period of time. Cattle breeders were the first to successfully use refrigeration as a technique for preserving sperm and employing it in controlled fertility.

The procedure generally used in the storage of sperm is as follows. Semen is collected and mixed with a protective medium consisting of egg yolk, glycerol, and other substances. It is then put in plastic vials or straws and these are placed inside an aluminum, cigar-shaped container. The aluminum cylinder is then flash frozen and maintained in liquid nitrogen at $-196°C$ ($-320°F$). A waiting period of six weeks is customary before a patient may make a withdrawal. The customer may have his sperm destroyed at any time.

Over the past 30 years, thousands of pregnancies have been achieved through the use of frozen semen. The incidence of birth defects, abortion rates, and obstetrical complications roughly parallels that of typical pregnancies. The newborn appear to be both physically and mentally comparable to those born following regular pregnancies. Commercial semen banks are available in most of the large cities of the United States.

Sperm banking (cryobanking) is a method of controlling fertility but may also be looked on as fertility insurance. Some males who wish to undergo vasectomy consider sperm banks as a possible answer to the nonreversibility of vasectomy since prevasectomy specimens can be stored for a limited period of time. Semen specimens containing sperm with normal motility usually have a pregnancy rate of about 70 percent when used after thawing. However for unknown reasons, the semen of certain individuals does not freeze well. In these cases, if the sperm in a thawed specimen does not have a motility of 60 percent of the motility before freezing, the pregnancy rate drops to about 30 percent. Most sperm cryogenically stored begin to lose their motility after about four years of storage. The exact length of time that sperm retain their fertilizing capacity is not known.

Most depositors of semen are men who about to undergo a vasectomy. Others utilizing the process include men who work near radiation-producing equipment who fear the possibility of genetic damage from ionizing radiations and anonymous donors whose semen is used in artificial inseminations. The process is also used in the treatment of men with low sperm counts (oligospermia). Several specimens of semen are collected over a period of time and these later are pooled in an effort to increase the chances of fertilization.

A number of ethical and legal problems have arisen with the development of artificial insemination and sperm banks. The possibility of a popular figure selling his semen and fathering many offspring or the possibility of a woman having a child years after the death of her husband are examples. The need of laws to control commercial sperm banks to avoid possible abuses is obvious.

The use of sperm banks as a eugenic device by which the sperm of men of exceptional qualities could be preserved and utilized in the future would maximize the genetic potential of these men. The role of sperm banks in the preservation of germ cells in the event of a worldwide nuclear disaster might prove to be invaluable.

# ELEVEN
# BIRTH CONTROL;
# STERILIZATION; ABORTION

## BIRTH CONTROL

*Birth control* or *fertility control* includes any measures taken to limit the number of offspring produced. In a broad sense it includes *abstinence, contraception, contraimplantation, sterilization measures,* and *abortion procedures.* All of these in some way either prevent the production of germ cells and their union to form a fertilized egg or they prevent the development of a fertilized egg and the birth of a viable baby.

*Contraception* means literally the prevention of conception. However the term is now generally used as a synonym for birth control and includes any measures used to prevent conception and the development and birth of an embryo or fetus.

Birth control or contraceptive procedures are employed (a) when hereditary or congenital disease or the presence of genetic defects indicate the probability that the offspring would be afflicted; (b) when a pregnant woman has been subjected to illness, especially German measles, to drugs, such as thalidomide, or to excessive irradiation, which makes it probable that the offspring would be deformed; (c) for personal reasons, as avoidance of pregnancy out-of-wedlock, for the spacing of pregnancies, or limitation of family size for social, economic, health, or other reasons; or (d) as a means of alleviating worldwide population growth, which, if unchecked, will lead to overpopulation. Various organizations, some worldwide in scope, are promoting birth control under such terms as Planned Parenthood, Family Planning, Fertility Control, and Zero Population Growth.

Birth control or contraception includes the utilization of any method or procedure that (a) prevents the egg from developing and being liberated from the ovary or sperm from being produced and liberated from the testes; (b) prevents the sperm from being deposited within the vagina, or if deposited, prevents their passage to and union with the egg; (c) prevents the egg, if fertilized, from implanting in the lining of the uterus or, if it implants, prevents it from developing to full term. Some of the methods are simple and can be self-employed; others are more complicated and require

**149**

medical supervision. Methods such as sterilization and abortion techniques that involve alteration in body structures necessitate surgical intervention.

Birth control is a highly controversial subject and may be considered from various aspects: religious, moral, ethical, medical, and legal. Religious and moral attitudes concerning the use of contraceptive techniques vary widely. Some individuals or groups do not condone their use under any circumstances; others approve of their use only under limited conditions. Still others believe in their free use and that knowledge concerning them should be unrestricted and their use a matter of individual judgement. The widespread use and availability of contraceptives in the United States indicates that the latter view is the one most commonly held in this country today.

## METHODS OF BIRTH CONTROL OR CONTRACEPTION*

### Abstinence, Continence, or Celibacy

This is voluntary restraint or abstention from sexual activity and is the surest method of birth control. It is sometimes resorted to for religious or moral considerations but it is also an essential part of the rhythm method of contraception when abstention from sexual intercourse during the fertile period is indicated. For individuals with a weak sex drive, abstention is not difficult but for those with a strong sex drive or for individuals who have established regular habits for sexual satisfaction, abstinence may prove difficult. Abstinence for a limited period of time increases man's fertilizing capacity for a period of several days.

### Withdrawal or Coitus Interruptus

In this method, the penis is withdrawn just before ejaculation. It is one of the most primitive and also one of the most unsatisfactory methods. It requires a great deal of self-control as the sexual act must be terminated just prior to the moment of greatest excitement and pleasure. Both male and female are frustrated. Also, there is always the possibility that some semen may escape before ejaculation (it takes only a drop) or that sperm can be deposited at the external opening of the vagina. In either case, pregnancy is possible. The failure rate by this method is high. Its only advantage is that it is cheap and available when other methods are not. It depends on will power, and in a contest between will power and the sexual instinct, will power is usually the loser.

### Condom

The condom or rubber (Fig. 25) as it is commonly called, is a thin sheath of latex or other material that is placed over the erect penis prior to inter-

*See also the sections "The Contraceptive Sponge" and "The Contraceptive Pill for Men" in the Preface to the Dover Edition, p. viii.

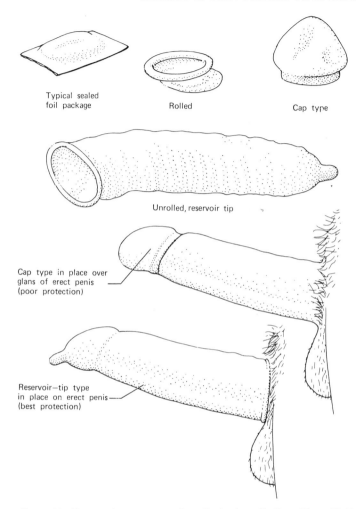

Typical sealed
foil package

Rolled

Cap type

Unrolled, reservoir tip

Cap type in place over
glans of erect penis
(poor protection)

Reservoir—tip type
in place on erect penis
(best protection)

**Figure 25   The condom—types and method of application. (From H. D. Swanson,** *Human Reproduction,* **Oxford University Press.)**

course. The semen is collected within the condom and consequently not deposited in the vagina. The condom also serves to protect the male against acquiring a venereal infection or transmitting one to a female. For this reason it is also known as a *prophylactic.*

**Advantages.**   Condoms are cheap and readily available; they are obtainable from drugstores or from dispensers commonly found in rest rooms of filling stations or motels. Furthermore, it is the only method other than vasectomy in which the responsibility for contraception is assumed by the male.

**Disadvantages.** A condom tends to dull sensitivity for both male and female partners and there is always the possibility of leakage or breakage from defective condoms but this is minimal with improved methods of manufacture.

**Effectiveness.** High when used properly and in conjunction with foam. Lubrication with a spermicidal cream or jelly is recommended. Prelubricated condoms are now available but the lubricating material is usually not spermicidal. The use of vaginal foam by the female for protection against possible breakage is recommended. Most condoms have a special reservoir at the tip for the accumulation of semen. If such is not present a space of about one-half inch should always be left at the tip of the condom. Care should be taken that the condom does not slip off during withdrawal especially if the penis is flaccid.

### Diaphragms and Cervical Caps

These are dish-shaped devices (Fig. 26) of soft rubber made to fit over the cervix of the uterus. Usually they have a semirigid rim for support. They come in various shapes and sizes, diaphragms averaging about three inches in diameter; caps are much smaller resembling a large thimble in appearance. They are inserted into the vagina in such a way as to cover the cervix thus preventing the entrance of sperm into the cervical canal. For many years their use combined with spermicidal jelly or cream was the procedure recommended by most birth control clinics. The diaphragm is widely used in the United States; the cap is more widely used in Europe.

**Advantages.** The use of a diaphragm or cap puts the control of pregnancy entirely in the hands of the female and the method is reasonably effective when used properly. It is a harmless method with no side effects.

**Disadvantages.** Diaphragms and caps are made in a wide variety of shapes and sizes, hence it is necessary that they be fitted by a physician or a specially trained nurse. The subject is instructed on how to use the device and later her ability to insert it properly should be checked. Changes in the shape of the vagina or cervix as occurs with growth, weight gain, or having a baby may necessitate refitting.

It is always necessary to use a spermicidal cream or jelly with a diaphragm because its main function is to hold these substances in place against the cervix. The cream or jelly should be applied to both sides of the diaphragm and around the rim. The appliance may be inserted up to two hours before intercourse and it should remain in place at least six hours or longer after intercourse. If a second intercourse follows the first, more spermicidal cream should be added before the diaphragm is removed.

**Effectiveness.** When used properly, diaphragms and cervical caps are fairly effective, the failure rate being from 5–10 percent.

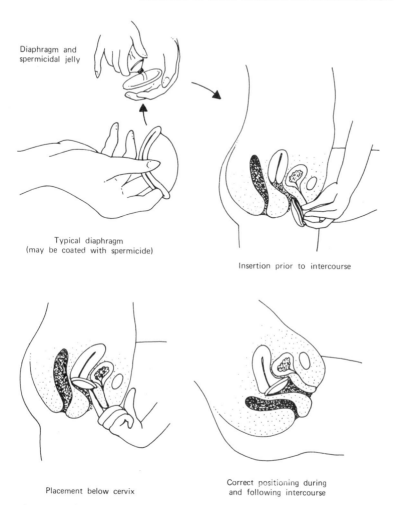

Diaphragm and
spermicidal jelly

Typical diaphragm
(may be coated with spermicide)

Insertion prior to intercourse

Placement below cervix

Correct positioning during
and following intercourse

Figure 26   The diaphragm and method of insertion. (From K. L. Jones, et al., *Sex*. Harper & Row.)

## Chemical Contraceptives

These are spermicial substances that, when placed in the vagina (Fig. 27), immobilize or kill the spermatozoa. They are usually combined with a gelatinous or oily base that acts to form a mechanical barrier preventing sperm from entering the uterus. Included are suppositories or pessaries, jellies, creams, foams, or aerosols.

Most effective of the group are the *foams* that are available in aerosol containers with special applicators. When introduced into the vagina, the effervescing foam forms a dense physical barrier blocking the opening to the uterus. The effervescence forces the spermicidal substance into the folds of the vagina which are not reached by creams and jellies.

*Suppositories* or *pessaries* are waxy pellets of soap, gelatin, or cocoa butter that are inserted into the vagina prior to intercourse. They melt at

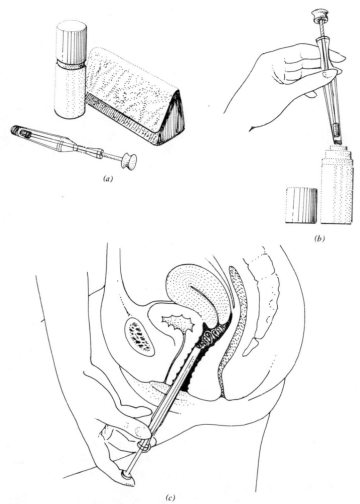

(a)

(b)

(c)

Figure 27   Application of spermicidal substance.

body temperature releasing their spermicidal agent and their oily base acts as an impediment to the movement of sperm. However, they tend not to be uniformly distributed throughout the vagina so their effectiveness is limited. Their chief advantage is their availability and ease of use.

*Jellies* are prepared with a gelatin base and are placed in the vagina by use of an applicator. *Creams* and *pastes* are prepared with a soap base. Jellies are more readily dispersed throughout the vagina but they have a tendency to leak out consequently are somewhat messy.

**Advantages.**   Chemical contraceptives are readily available at any drug store and may be obtained without prescription. They are easy to use and in general are harmless with no side effects.

**Disadvantages.**   To be effective, chemical contraceptives must be applied some time, at least 30 minutes, before intercourse and two applications should be made. If intercourse is repeated, the application should be repeated. In general, the creams and jellies are messy. Douches do not need to be resorted to but if they are used, a period of 6–8 hours should elapse since a douche tends to wash out or dilute the spermicidal substances. To a few individuals, specific chemical substances may be irritating or allergic reactions may result. Sensitive individuals should avoid their use.

**Effectiveness.**   When used alone, chemical contraceptives are only moderately effective, failure rate being about 10–20 percent. When used with a diaphragm or condom, effectiveness is increased, the failure rate being reduced to about 1 percent.


*The Safe-Period or Rhythm Method*

This method is based on the assumption that the ovary releases one ovum per menstrual cycle, about the fourteenth day, and that it is available for fertilization for only a limited period of time, about 24 hours. Spermatozoa within the uterus have a limited period of fertilizing power generally considered to be about 24 hours but there is evidence that it may extend for 72 hours. Consequently, there is a period of only four or five days, a *fertile period,* in each menstrual cycle that a woman is capable of conceiving. If intercourse is avoided during this period, pregnancy is unlikely to occur. Intercourse at other times during the cycle would not result in pregnancy, hence these would be "*safe periods.*"

For a woman having a 28-day menstrual cycle, the time sequence would be somewhat as follows:

| Menstruation | Ovulation | | Menstruation |
|---|---|---|---|
| 1 2 3 4 5 6 7 8 9 10 | 11 12 13 14 15 16 17 | 18 19 20 21 22 23 24 25 26 27 28 | 1 2 3 4 5 |
| Safe Period | Fertile Period | Safe Period | |

Represented in another way, the cycle may be presented as follows:

Menstruation

| 1 | 2 | 3 | 4 | 5 | 6 | 7 |
|---|---|---|---|---|---|---|
| 8 | 9 | 10 | (11) | (12) | (13) | (14) Ovulation |
| (15) | (16) | (17) | 18 | 19 | 20 | 21 |
| 22 | 23 | 24 | 25 | 26 | 27 | 28 |
| 1 | 2 | 3 | 4 | 5 |  |  |

Menstruation

In the above calendar, the days on which sexual intercourse should be avoided are encircled. These are the days during which conception is most likely to occur.

The effectiveness of this method depends on determination of the exact time of ovulation, which is usually about the fourteenth day *before* the next menstrual period. However, the onset of the next menstrual period can never be definitely known. Since the menstrual cycles vary in length, if they are fairly regular and average a fairly uniform length, adding a couple of days at the beginning and at the end of the estimated fertile period generally takes care of possible variations in the time of ovulation and possible time of conception. Using a calendar and keeping an exact record of the menstrual cycles is essential.

An adjunct is establishing the exact time of ovulation is to note temperature changes within the body. Following ovulation, there is a rise in body temperature of about 1°, which is maintained until a day or two preceding menstruation when it drops to the preceding normal (Fig. 28). *Basal body temperature* (BBT) is that taken the first thing in the morning after awaking and before engaging in any kind of physical activity. Temperature should be taken by rectum. A chart of daily body temperature over a period of months will sometimes show a fairly regular pattern of temperature change from which the time of ovulation can be calculated or determined.

Because of the variability in the length of menstrual cycles and the difficulty in determining the precise time of ovulation, the determination with accuracy of "safe" and unsafe or "fertile" periods is complicated. For a person or a couple who intends to use the rhythm method it is recom-

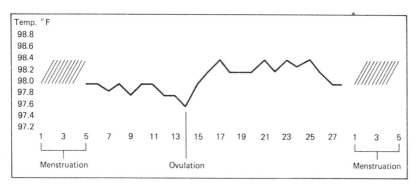

Figure 28   A chart showing representative changes in basal body temperature during a 28-day menstrual cycle. Time of ovulation is indicated by a slight drop followed by a rise in temperature for three successive days. Rise is due to secretion of progesterone indicating that a corpus luteum has been formed.

mended that they consult a physician or a family planning clinic for specific instructions.

**Evaluation.**   This method of contraception is, under certain circumstances, approved by the Roman Catholic Church, it being considered to be a "natural method" as it does not involve the use of artificial devices or chemicals, its effectiveness depending entirely on an understanding of the natural, physiological processes which occur within the body. It gives a limited degree of protection from an unwanted pregnancy but it is too uncertain to be depended on in all cases. The failure rate averages about 25 percent. It requires specific instructions in its use and a considerable amount of intelligence and self-control in its application so it is generally recommended only for married couples for whom an unwanted pregnancy might not be unwelcome or disastrous.

*Intrauterine Device*
Intrauterine contraceptive devices (IUD and IUCD) are small pieces of metal or plastic (Fig. 29) which are inserted into the uterus by a physician. They are usually pliable and are of various sizes and shapes (loop, spiral, coil, bow, shield, T-device). The device is placed within the uterus usually with the aid of an *introducer* or *inserter* by which it is passed through the cervical canal. It usually possesses a "tail" (IUD string or transcervical threads) consisting of one or several fine nylon threads that protrude through the cervix into the vagina. The presence of the tail noted by inserting a finger into the vagina enables the subject to know that the device is in place and has not been expelled.

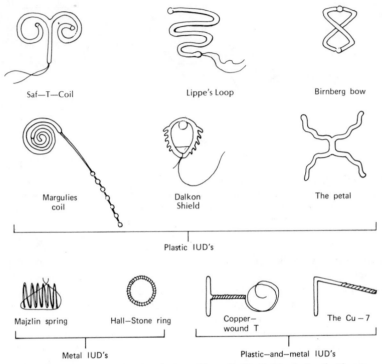

Saf—T—Coil            Lippe's Loop            Birnberg bow

Margulies            Dalkon            The petal
coil                 Shield

Plastic IUD's

Majzlin spring    Hall—Stone ring    Copper—    The Cu—7
                                     wound T

Metal IUD's                    Plastic—and—metal IUD's

**Figure 29  Types of intrauterine devices. (From C. W. Hubbard,** *Family Planning Education.*) **Mosby Company.**

The means by which an intrauterine device prevents pregnancy are not definitely known. Its presence does not prevent ovulation or fertilization of the egg if sperm are present. Nor does it prevent the passage of a fertilized egg through the uterine tube. Its action seems to be that of creating a traumatized condition within the uterus preventing the blastocyst from attaching to the uterine lining or, if it does attach, preventing its further development or bringing about its abortion. The IUD being a foreign body within the uterus is thought to cause protective cells (macrophages and white blood cells) of the uterus to react against the developing blastocyst or it may stimulate muscular movements that force the blastocyst from the uterus.

**Advantages.**   The chief advantage of the IUD is that once it is inserted no further action is required of the subject other than checking about once a week or before coitus to see that it is still in place. It is of special value to women for whom other methods of contraception are unavailable, undesirable, or disliked for any reason. An IUD can be removed by a physician at any time if pregnancy or a change in contraceptive technique is desired.

**Disadvantages.** In some individuals, following insertion of an IUD, excessive bleeding may occur. Some experience abdominal pain or pelvic discomfort. Excessive menstrual flow may occur and bleeding or irregular spotting may occur between periods. If such happens, the patient should check with her physician. Some women, especially nulliparous women (women who have borne no children), spontaneously expel the device such commonly occurring at the first menstrual period following insertion or sometime during the first year. For this reason, the cervical tail enables the patient or the doctor to know if it is still in place. If the device is expelled, the patient should not attempt to reinsert it as it is likely to be replaced in the vagina and not the uterus. There its effectiveness would be zero.

Occasionally other complications may arise from the use of an IUD. Infection of the uterus may develop and a number of cases of perforation of the uterus followed by serious pelvic inflammatory disease (PID) have been reported. In 1974, the widely used Dalkon Shield was withdrawn from the market at the request of the U.S. Food and Drug Administration because of a number of deaths (13–20) and over 200 cases of uterine infections associated with septic spontaneous abortions occurring in patients using the shield. It is thought that the infections were due to bacteria accumulating within the multifilament tail, which, during a pregnancy, is drawn into the uterus. Other intrauterine devices commonly used, for example, the Lippe's Loop (Fig. 30) and the Saf-T-Coil, each have a monofilament tail that is less favorable to bacterial contamination.

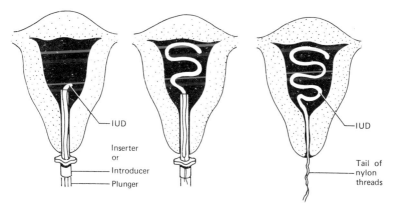

Figure 30   Insertion of an intrauterine device (Lippe's Loop). (From L. O. Crawley et al., *Reproduction, Sex, and Preparation for Marriage.* **Prentice-Hall, Inc.**)

**Effectiveness.** Overall effectiveness is fairly high but not as high as that of the Pill. Failure rate during the first year runs about 3–5 percent. It is most effective for women over 30 who have had several children. The most common cause of failure is an unnoticed expulsion of the device.

Some IUD's have included the addition of a copper wire about the stem. The first device of this type was called the Copper-7 (Cu-7); it was followed by the Copper-T (Cu-T). Their effectiveness was enhanced by the continuous release of copper ions into the uterine cavity. The exact way in which the metallic copper acts is not known.

Another IUD, the Progestasert, is similar to the Cu-7 except that it contains progesterone in its stem. The continuous release of small amounts of progesterone within the uterus creates a condition in which the endometrium is unreceptive to the blastocyst. A shortcoming of these devices is that they are effective for only a limited time. When the copper or progesterone is exhausted then the device must be replaced in two to four years or one year, respectively.

In September 1985 the Ortho Pharmaceutical Corporation ceased marketing its Lippe's Loop IUD, and in January 1986 G. D. Searle and Company ceased distributing its two IUD's (the Cu-T and the Cu-7) in the U.S. Thus, only one IUD remains available in the U.S., the hormone-releasing Progestasert, made by the Alza Corporation. However, the Lippe's Loop and the two copper IUD's represented 97 percent of all the IUD's sold in the U.S. in 1984.

Ortho and Searle did not stop marketing their IUD's because of newly discovered medical risks or because of pressure from consumer groups. Instead, the two companies stopped selling IUD's for economic reasons, the financial risks associated with potential lawsuits for alleged injuries by IUD's. Their decisions were based not only on the high cost of defending liability lawsuits but also on the fact that liability insurance for contraceptives is almost unobtainable in our current legal climate. This legal climate resulted from the many large damage awards and legal costs related to medical problems from the Dalkon Shield IUD, which eventually resulted in the manufacturer, the A. H. Robbins Company, declaring bankruptcy. This means that about 2.2 million women who are IUD users, 7 percent of all women who use a contraceptive, will now have to turn to some other form of contraception.

## Oral Contraceptives; the Pill
Birth control pills, commonly called the Pill, are composed of synthetic estrogens and progestins which resemble the natural hormones produced by the ovary. The natural hormones are *estrogens* produced by the developing follicle and *progesterone* produced by the corpus luteum. The

secretion of both of these hormones is induced by *gonadotrophic hormones* (FSH and LH) produced by the pituitary. These ovarian hormones are responsible for the normal changes occurring in the endometrium of the uterus preparatory to menstruation. When the hormones are produced naturally, they act by a feedback mechanism on the pituitary by way of the hypothalamus, causing the pituitary to stop producing gonadotrophins.

When these hormones, or synthetic hormones resembling them, are taken in pill form, the formation and release of gonadotrophins by the pituitary is inhibited. As a result, no follicles develop in the ovary, consequently no eggs are developed or released (Fig. 14). In the absence of an egg, no pregnancy can occur.

Pills are available in three types: designated combination, biphasic, and triphasic. In the *combination pill,* the estrogen and progesterone are combined in a set proportion in all the pills. All 21 pills contain the same substances in the same proportion. A *biphasic pill* was introduced in 1982. The biphasic contains the same level of estrogen for all 21 pills, but there are two phases in the level of progesterone, a low level for the first 10 days, followed by a slightly higher level for the last 11 days. This is an attempt to reduce total level of exposure to hormones and to provide a hormone cycle more like the natural cycle. In 1984 a new *triphasic pill* was introduced in an attempt to match even more closely the hormone levels of the natural cycle. Again, the level of estrogen is the same in all 21 pills, but there are three phases in the level of progesterone, with an increasing level of progesterone every 7 days. Another type of oral contraceptive called the *sequential pill* was the first available pill in the U.S. It consisted of 16 pills that contained only estrogen, followed by 5 pills that contained both estrogen and progesterone (Fig. 31). The sequential pill is no longer utilized in the U.S.

Pills are usually packaged in containers of 21 or 28 and are taken, one a day, for a period of 21 or 28 days starting with the first day of the menstrual cycle. Day 1 is the day bleeding begins. The 28-day pill pack is an attempt to reduce the effort involved in keeping track of the days on which pills are to be taken, since the last 7 pills are inert or inactive. Thus, the woman begins the next pill pack the next day after her last pill of the 28-day pill pack rather than waiting 7 days.

Pills are available only by prescription. A medical examination is required and a physician determines if the pills are safe for the subject to take. There are certain conditions in which the use of the Pill is contraindicated. These include a history of thrombophlebitis, liver dysfunction, diabetes, known or suspected malignancy of the breast or genital organs, and mental depression.

There are a number of contraceptive Pills on the market differing in the specific hormones contained and in the amount of hormones in each pill. The doctor will prescribe the one best suited for each particular case.

Oral contraceptive pills

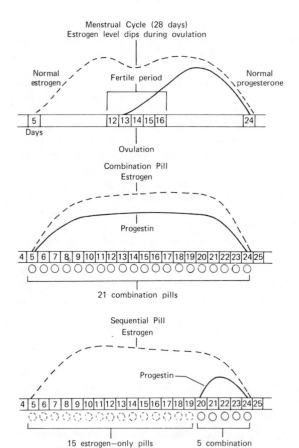

Menstrual Cycle (28 days)
Estrogen level dips during ovulation

Normal estrogen

Fertile period

Normal progesterone

5          12 13 14 15 16          24
Days

Ovulation

Combination Pill
Estrogen

Progestin

4 5 6 7 8 9 10 11 12 13 14 15 16 17 18 19 20 21 22 23 24 25

21 combination pills

Sequential Pill
Estrogen

Progestin

4 5 6 7 8 9 10 11 12 13 14 15 16 17 18 19 20 21 22 23 24 25

15 estrogen–only pills          5 combination pills

**Figure 31  The menstrual cycle and the oral contraceptive pill. (Diagram by R. M. Chapin from** *Time,* **April 7, 1967. Copyright Time, Inc. 1967.)**

The Pills should be taken *exactly as directed* on the days specified. It is desirable to take the Pill at a regular time each day so that a habit can be established. If a pill is missed on one day, two should be taken on the following day. If the Pill is missed on two days in succession, a double dose should be taken for the next two days, and if intercourse is engaged in, supplementary contraceptives (condom or/and foam) should be employed.

**Effectiveness and Advantages.**    The Pill is the most effective method of contraception, the pregnancy rate for women who take the Pill regularly being less than 1 percent. It is effective at all times and independent of the amount of sexual activity. It is relatively safe to use and its effects are reversible. In addition, it often has beneficial side effects. Among these are more regular menstrual periods, relief from menstrual cramps and premenstrual tension, stimulation of breast development, relief of acne, general improved well-being, and improvement in sexual pleasure resulting from freedom of fear of possible pregnancy.

**Disadvantages.**    As the Pill can only be secured by prescription, a physical examination is required, including a pelvic examination. Sometimes minor side effects occur such as a gain in weight, breast discomfort, nausea, and irregular bleeding or spotting. Use of the Pill also causes changes in the acidity of the vagina, which increases the likelihood of certain infections such as moniliasis. The use of the Pill also increases the likelihood of acquiring a venereal disease, which results from increased susceptibility of the vagina to infection and the decreased use of the condom as a preventative measure.

Occasionally major disturbances or complications may occur but these are rare. One is thromboembolism (blood clotting and blood vessel obstruction), the incidence of which is reported to be higher in women taking the Pill. The mortality rate for thromboembolism among pill takers is about 3 per 100,000 per year as compared with 1 per 100,000 among nonpill users. However, the mortality rate among women who became pregnant in 1983 was 8 per 100,000 per year; consequently, there is much less risk in taking the Pill than in becoming pregnant.

The use of oral contraceptives has also been associated with an increased risk of heart attack especially in women over 40 who are smokers. Sugar metabolism is altered and the unmasking of diabetes may occur. The incidence of depression with suicidal tendencies seems to be increasing among pill users, depressive personality changes occurring approximately in one of every three pill users. It has been suggested that the Pill might cause cancer or increase the incidence of fetal abnormalities but a causal relationship between the Pill and these conditions has not been proven.

An unanswered question and one about which there is considerable controversy is "How long should a woman take oral contraceptives?" Because the Pill interferes with the normal functioning of the ovary and the pituitary gland that controls it, some doctors suggest that after a period of three or four years on the Pill, there should be a period of abstinence of three or four months to allow these organs to resume their normal activity. Other doctors say that this is not necessary. Also there is a question of the long-term effects of the pill. At the present time, these are not known.

## OTHER CONTRACEPTIVE PROCEDURES

### Douching
This is the process of washing out the vagina by injection of a solution usually containing a spermicidal substance. It must be done immediately after intercourse and there is always the possibility that many sperm may have entered the uterus where no douche can reach them. It is the least effective method of contraception and, in fact, may increase the possibility of pregnancy by washing the sperm into the uterus.

Douching is highly advertised as essential for "feminine hygiene" or cleansing the vagina. Such, however, is not necessary as the secretions of the vagina usually are adequate for keeping it moist and clean. The introduction of antiseptic solutions can alter the normal bacterial flora of the vagina and if strong solutions are used, irritation of the vaginal mucosa may result. If there should be continuous itching or an unusual vaginal discharge, a physician should be consulted.

### The Morning-after or Postconception Pill
If intercourse occurs and no contraceptive measures have been taken, pregnancy can still be avoided by a medication, diethylstilbestrol (DES), a synthetic estrogen. DES is given as a pill or shot beginning two or three days after exposure and continuing for five days. This prevents implantation of the fertilized ovum. The medication must be taken under medical supervision and, since side effects are usually severe, it is not recommended except in extreme circumstances such as rape, incest, or when conventional methods fail. Its use as a contraceptive has only recently been authorized by the Food and Drug Administration; however, some physicians state that there is a definite cancer risk to daughters born by pregnant mothers who take DES.

### The "Minipill"
This is a pill containing only a minute dose of a progestogen. It is taken daily even during menstruation and its use eliminates many of the side effects that are caused by synthetic estrogens present in the Pill. It is been widely used in Europe and is now available in the United States. It acts prin-

cipally on the cervical mucus increasing its viscosity, thus impeding sperm migration. It also alters the endometrium making it less receptive. It does not prevent ovulation.

*Contraceptives of the Future*

Research is going on continuously with the aim of developing more effective methods of contraception. Some of the recent developments are the following; however, most have not been tested adequately to insure their safety for the general public.

**Prostaglandins.** One of the most recent and fascinating developments in the field of reproductive physiology has resulted from extensive research on *prostaglandins* (PG). These are fatty acid derivatives that have been found in many tissues (brain, lung, liver, kidney, uterus, seminal vesicles, and placenta) and in various fluids (seminal, menstrual, amniotic). They are biologically active in a number of ways. Because of their action in the stimulation of smooth muscles, they have been found to be effective in the induction of labor at term and in the induction of abortions. They are also thought to play a role in the regression of the corpus luteum, in the production of ovarian hormones, and a possible factor in IUD action. There is evidence that low prostaglandin content of the semen may be a factor in male infertility although its exact mode of action is unknown. There are a number of prostaglandins, grouped into four primary groups, designated *prostaglandin* A, B, E, and F. Of these groups, the E and F prostaglandins and their subtypes seem to be the most effective in their labor-inducing and abortifacient effects.

The possible use of prostaglandins as contraceptives is largely based on their action as a *luteolysin,* a substance that exerts its effect on the corpus luteum causing it to regress and cease progesterone and estrogen production. This is a primary factor in inducing the onset of menstruation. If an effective way of administration with minimal side effects could be developed, it is postulated that a once-a-month method of birth control might be made effective even though conception might have occurred. However, despite considerable publicity on the matter and enthusiastic expectations, much more research on the role of prostaglandins in birth control must be conducted before safe and practical methods of administration can be employed. Furthermore, more knowledge on possible ill effects as the induction of maldevelopment or adverse effects on bodily metabolism are essential before prostaglandins can be made available for general use.

Recently, one of the PGF's (dinoprost) has become available in the United States for use as an intraamniotic abortifacient during the second trimester of pregnancy. Intraamniotic instillation is accomplished by inserting a long needle through the abdominal wall into the amniotic sac. The average treatment-to-abortion interval is between 10 and 20 hours. Because of the shorter treatment-to-abortion time, lessened side effects, and greater safety of dinoprost as compared with the use of hypertonic saline injection, dinoprost is the present treatment of choice for an abortion during the second trimester of pregnancy.

**The Once-a-Month Pill.** This is a contraceptive pill that releases its agents, principally estrogens, slowly over a long period of time. Its effectiveness has not been definitely established.

**Injectable Contraceptives.** A new drug, Depo-Provera (medroxyprogesterone) has recently been approved by the Food and Drug Administration for limited use as a contraceptive. It is the first and, at present, the only injectable, long-acting contraceptive available. It needs to be injected only once every three months and it is effective in preventing pregnancies. Side effects include the possible continuance of infertility for a period of months after use has been discontinued. There is also a possible relationship to breast tumors and blood clotting.

**The Intravaginal Ring.** This is a silastic (rubberlike) circular device containing a progestogen made to fit around the cervix. It is inserted into the vagina on the first day of menstruation and left there for 21 days. Upon removal, menstruation occurs within a few days. Early reports indicate that this is a highly promising procedure.

**Silastic Implants.** These are small, plastic rods containing progestogens that are implanted under the skin, usually on the inner side of the arm. They release their progestogens slowly over a period of 5 years. The failure rate is less than 1 percent. Fertility returns rapidly after the implants are removed. Minor surgery under local anesthetic is required for insertion and removal. A two-rod system is expected to be available in the U.S. in 1987.

**Immunization Techniques.** Studies have been in progress for a number of years toward applying the principles of immunity to the field of reproduction and specifically to the area of contraception. The following procedures are being tested: (1) the injection of constituents of semen (antigens) into a female, whereupon the female's body produces antibodies that react against the particular substance injected; (2) the female is given injections of her own follicle tissue (an antigen) and her body reacts forming antibodies against her own ova; (3) the male is given injections of his own semen constituents (antigens) and his body reacts to form antibodies against his own sperm. Success with these techniques has been limited. A major problem is that of reversing the process once it has been established.

*Comparison of Contraceptive Methods*
It is difficult to compare various contraceptive methods because of the many variables involved. They vary in effectiveness or reliability, availability and ease of use, preparation required, possible side effects, and cost. Various personal factors are involved in their use so attempts to evaluate them by giving specific figures as to failure rates may be misleading. In general, the methods can be rated as follows.

| (Those Highly Effective) | (Estimated Effectiveness) (percent) |
|---|---|
| Sterilization | Nearly 100 |
| The Pill | 99+ |
| Intrauterine devices (IUD) | 95+ |
| Minipill | 95+ |
| Condom | 90+ |
| Diaphragm or cervical cap (with foam) | 90+ |
| (Those Moderately Effective) | |
| Chemical barriers alone | 80+ |
| Rhythm method | |
| For regular women only | 85 |
| For all women | 70 |
| (Those of Low Effectiveness) | |
| Withdrawal | 40–50 |
| Douche | 30–40 |
| None | 10–20 |

## MALE CONTRACEPTIVE TECHNIQUES

Most of the contraceptive techniques, with the exception of the condom and withdrawal, depend on their action within the female and their effectiveness depends on how, when, and to what extent they are used. As a result, special attention has been given in recent years to the development of procedures that would be effective in the male and be under his control. The structure of the male reproductive system, however, presents only a limited number of areas that are susceptible to modification or control. These are the *testes,* involved in the processes of spermatogenesis and androgen production, the *epididymis,* in which sperm are stored and maturation takes place, the *accessory glands* (seminal vesicles and prostate) in which seminal plasma is produced, and the *conducting ducts* through which semen passes to the female.

Attempts to develop a male contraceptive comparable to the Pill have generally been unsuccessful. Procedures that have been moderately successful in preventing sperm formation in lower animals have had limited success in humans. Among these are the injection of androgens or the placing of androgen implants beneath the skin that would theoretically suppress spermatogenesis through their inhibiting effect on the production of testis-stimulating hormones by the pituitary. The spermatogenic suppression effects have been accompanied by a reduction in the production of androgens resulting in the reduction of libido, an undesirable effect. Several drugs have been found that have an adverse effect on spermatogenesis but their use is contraindicated because of deleterious side effects.

Sperm, on leaving the testes, accumulate in the epididymis and there undergo maturational development involving the acquisition of the capacity for motility and fertilizing ability. An agent that would prevent this process from occurring would be an effective contraceptive. Progesterone implanted subdermally has this effect in some experimental animals but it has not been proven to be effective in man.

Measures that would alter the chemical nature of the secretions of the accessory glands (seminal vesicles and prostate), secretions essential for the transport, nourishment, and proper environment for the sperm have not been successful. Methods to induce the development of immunologic factors in the male that would deactivate or agglutinate sperm or destroy their fertilizing power have been studied but no practical way of accomplishing such has been developed.

Efforts have been made to develop mechanical devices by which sperm transport through the reproductive ducts especially the vas deferens can be regulated or blocked. These include a clip that is fastened on the vas deferens, which blocks the passage of sperm; an intravas device that permits the passage of sperm but renders them incapable of fertilization; a plastic or silicone plug that can be injected or placed in the vas deferens to block sperm transport; and a microvalve that can be manually operated to regulate the passage of sperm. Most of the mechanical devices present the problem of requiring sufficient pressure to close the lumen of the duct and yet not injure the tube (pressure necrosis) to the extent that reversibility is impossible. The success of these devices to date has been limited.

A promising development in the field of controlling spermatogenesis has been in the local application of heat to the testes. The testes within the scrotum have a temperature slightly lower (2–3°C) than body temperature. If the temperature is raised only slightly by experimental means or as occurs naturally in undescended testes (cryptorchism), the production of sperm by the seminiferous tubules ceases although the ability of the interstitial cells to produce androgens is not impaired.

As a consequence, it has been claimed that the wearing of a tightly fitting, closely knitted jockstrap by a male for a period of four weeks will generally result in partial or total infertility. Normal sperm production is resumed in about three weeks after discontinuance of its use. Subjecting the testes to hot water (120–130°F) for 30 minutes or more on several successive days leads to a significant reduction in sperm production.

On the basis of these results, the use of infrared radiation, microwaves and, most recently, ultrasound as a means of increasing the temperature of the testes and suppressing sperm formation is being studied extensively. Present indications are that the use of ultrasound is the most effective of the methods proposed, the desired effects being produced more quickly and at a lower temperature. When used at the proper intensity, hormone production is not impaired hence libido is not affected.

## STERILIZATION

Sterilization is the process by which a person is rendered incapable of reproducing. It is resorted to in the female when anatomical or physiological disorders make the bearing of offspring difficult or impossible, when the reproductive organs are diseased or in certain pathological conditions involving the entire body, when genetically defective offspring are likely to be produced, or for various psychological, economic, or other reasons. In both sexes it is sometimes resorted to as a method of birth control, which is 100 percent effective at all times, requiring no thought or action of any kind, and with a minimum of side effects. In some states, sterilization may be legally authorized as a means of preventing the propagation of undesirable or dysgenic types or for parents who are physically or mentally unable to take care of their offspring.

Sterilization may be accomplished by surgical removal of the gonads, by sectioning or ligating the ducts involved in transporting the germ cells, or, in the female, by removal of the uterus. It may also be accomplished by subjecting the gonads to radiation, using either radium or X rays. Sometimes sterilization occurs naturally as a result of pathological processes as in atrophy of the testes following orchitis occurring as a complication of mumps. Injury, the development of tumors, or the occurrence of a varicocele may be factors in testicular failure. Cysts or tumors are common causes of ovarian failure.

### For the Male

**Castration or Orchiectomy.** This operation, the removal of the testes, is rarely performed as it removes not only the source of male germ cells but also the primary source of male hormones. If the testes are removed before puberty, a practice followed in certain cultures as a religious rite or to produce male singers with high-pitched voices, the sex organs remain undeveloped and secondary sex characters fail to develop; interest in sex and sex drive is lacking. However if the testes are lost or inactivated after maturity has been attained as may occur as a result of injury, radiation, or disease, the ability to perform sexually may be retained. This is due to the fact that the adrenal gland is capable of producing androgens in amounts adequate to maintain the sex organs in a functional state thus enabling sexual activity to continue. Injections of testosterone may aid. However, in all cases, the male is sterile. Castrated males are called *eunuchs*.

**Vasectomy.** This is an operation (Fig. 32) involving the removal of a section of, or the tying off (ligation) of the two vasa (ductus) deferentia or

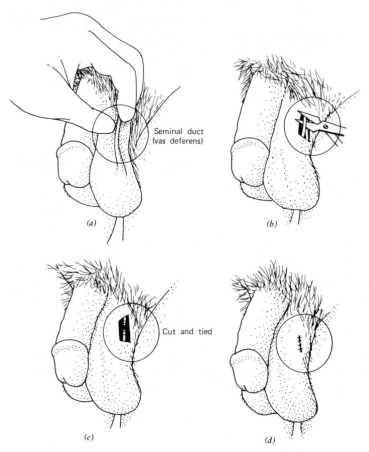

**Figure 32   Vasectomy, or sterilization by severing the vas deferens. (From K. L. Jones, et al.,** *Sex,* **Harper & Row.)**

sperm ducts. When this is done, the sperm, which continue to be produced, pass into the epididymis, which becomes turgid. There they die and are destroyed by macrophages. Since the sperm cannot pass to the urethra, the semen on ejaculation is free of sperm and impregnation is impossible. Since the semen consists principally of seminal plasma (about 90 percent), the amount of ejaculate is only slightly reduced. The operation does not affect a man's ability to perform the sexual act or alter his sexual desire. The ability to achieve an erection is not impaired.

The operation can usually be performed in a doctor's office under a local anesthetic. The male is not immediately sterile after the operation for sperm may remain in portions of the sperm ducts for a period of several weeks. A man should have intercourse several times using contraceptive methods and then have his semen examined after three or four weeks to be sure that spermatozoa are not present.

The principal objection to vasectomy is that it is more or less permanent and sometimes the vasectomized male changes his mind and wishes to become fertile again. Although reconnection (reanastomosis or vasovasostomy) of the cut ends of the vas deferens is possible in some cases, it is a difficult procedure and success can not be counted on with certainty. Also when such has been accomplished, although motile sperm appear in the ejaculate, pregnancy rates are lower than normally expected. This may be the result of sperm escaping from the epididymis into the tissues and causing the development of autoimmune bodies that act against the sperm.

### For the Female

**Castration or Ovariectomy.**  This is removal of the ovaries and is rarely done except when the ovaries are diseased as it results in complete and permanent sterilization and induction of a premature menopause. Hormonal disturbances usually result but these can be alleviated by the administration of ovarian hormones. If the ovaries are removed after maturity, sexual response is not altered significantly, in fact it may be improved as the fear of pregnancy is removed.

**Hysterectomy.**  This is removal of the uterus, usually done for correction of pathological disorders such as uterine bleeding, neoplasms

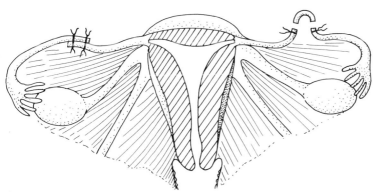

Figure 33  Female sterilization by tubal ligation. On the left side, a portion of the tube is removed and the two ends ligated; on the right side, the cut portions of the tubes are overlapped and doubly ligated.

(tumors), or infections. This results in complete and permanent sterilization, however sexual response is not significantly altered; sometimes it is
improved.

**Salpingectomy.** This is the removal of the uterine tubes. It is now
rarely done for purposes of sterilization as a simpler technique known as
tubal ligation is usually employed.

**Tubal Ligation.** This is a tying off (Fig. 33), fusion by electric heat, or
removal of a portion of each of the uterine tubes. It is a relatively simple
operation and may be accomplished by going through the abdominal wall
(*laparotomy*) or through the vaginal wall. When the tubes are ligated, the
fertilized egg cannot enter the uterus or the sperm enter the tube to fertilize the eggs. The operation results in permanent sterility and in most cases
is not reversible.

All of the female operations listed above fall in the category of major
surgery because all involve the opening of the body cavity and are usually
performed under general anesthesia. There is always the danger of hemorrhage or infection so these operations should be performed only by
qualified surgeons.


## ABORTION

If sexual intercourse is engaged in, unless contraceptive measures are
employed, pregnancy is likely to occur. If a woman is married, the pregnancy may be welcomed. On the contrary, the pregnancy may not be
desired for any of a number of reasons: physiological, psychological, social,
or economic. If a woman is not married, pregnancy usually results in
serious adjustment problems. If it proceeds to full term, the baby is born an
illegitimate child and the problems of a single woman bringing a child to
maturity without the assistance of a father can be overwhelming. Usually
the child is given up for adoption.

What is the other alternative to an unwanted pregnancy? It is the
termination of pregnancy accomplished by having an abortion.

Termination of the development of an embryo or fetus before the
twenty-eighth week (age of viability) is called an *abortion*. Abortions may
be natural or they may be induced. *Natural or spontaneous abortions* are
common occurring in 10–15 percent of all pregnancies. Most of these abortions are preceded by the death of the embryo or fetus hence the abortion
is Nature's way of eliminating malformed or diseased embryos or preventing pregnancy from proceeding in an individual incapable of bearing
young. Abnormalities are usually of genetic origin or they may be the result
of endocrine imbalance or of toxic or pathologic conditions within the
mother's body.

An *induced abortion* is one that is brought about artificially. In most
states, until recent years, an abortion other than to save the life of or

preserve the health of the mother was considered to be a criminal act. However, views pertaining to abortion and laws governing abortion are changing as evidenced by the action of the United States Supreme Court which, in a recent decision (January 22, 1973) struck down most state laws against abortion by ruling that during the first three months of a pregnancy, the decision to have an abortion lies solely with the woman and her doctor and that a state cannot interfere with that decision. With regard to an abortion during the next three months, the state may make laws with regard to protecting the health of the mother but it cannot forbid an abortion. In the late stages of pregnancy, after the fetus has reached the age of viability, the state may forbid an abortion except when it is deemed necessary for the preservation of the life or health of the mother. This, called a *therapeutic abortion,* is resorted to in cases of pernicious vomiting of pregnancy, eclampsia, and certain other pathological conditions in which termination of pregnancy may be imperative.

As a result of these rulings, any woman who has an unwanted pregnancy can legally have an abortion if she so desires. Over 1.5 million abortion procedures were performed in 1985. Having an abortion involves serious legal, religious, and moral considerations; various counseling agencies have been established to assist a pregnant woman in deciding on the proper course of action. Above all, no woman should allow a criminal abortion to be performed. Such abortions are usually performed by unqualified persons and often under conditions that may lead to infection and possibly death. Usually no anesthetic is used and postoperative care is lacking. Previous to the Supreme Court decision, legal abortions were available only in a few states such as New York, Hawaii, and Alaska.

Sometimes self-induced abortions are attempted. Sharp pointed objects are inserted into the vagina and attempts made to penetrate the uterus. Perforation of the vagina or uterus is the usual result. The injection of soap or detergent solutions under pressure is especially hazardous. The use of uterine pastes containing caustic substances may result in ulceration when placed in the uterus or vagina. Various drugs are used with the idea that they will produce an abortion and induce menstruation. Claims for such drugs are usually extravagant. There are a few drugs that act on the uterus and tend to bring about an abortion but their action is unreliable and their effects such that they must be given under medical supervision. Many of these drugs are poisons and can kill a pregnant woman or result in a grossly deformed baby. Death often results from illegal or self-induced abortions. Uterine perforation can occur. Infection of the uterus and uterine tubes is common with subsequent sterility or death. Hemorrhage and air or fluid embolism may occur, especially when substances as soap solutions are forced into the uterus.

A *legal abortion* when performed by an experienced physician during the first three months of pregnancy is relatively safe. A commonly used method is *dilatation* and *curettage* (D&C). In this operation, the opening into the cervix is stretched (dilatation) and the lining of the uterus scraped (curettage) by a spoonlike instrument (*curette*). The operation is brief, lasting only about half an hour and is usually done under a local anesthetic. It involves only a day or two stay in a hospital or clinic and sometimes is done on an out-patient basis.

Another method employed is that of *uterine aspiration* or *vacuum curettage*. A suction curette (*vacurette*) is used to dislodge the embryo and by vacuum pressure the *conceptus* (embryonic and placental material) is drawn into a container. The operation is done under a local anesthetic and requires only about 10 or 15 minutes. This procedure is sometimes employed in the absence of positive proof of pregnancy. It is then referred to as *menstrual extraction*.

Another method used is that of *intraamniotic injection* employed principally after the fifteenth week of pregnancy. This involves the puncture of the amniotic sac (*amniocentesis*) and withdrawal of the fluid contained in it followed by injection of a strong saline or glucose solution. This solution stops development of the fetus and hormone production by the placenta. Premature labor with expulsion of the fetus follows within a day or two. This operation is the most dangerous of the abortion techniques but its mortality rate is less than that for normal pregnancy.

A new method by which an abortion may be safely performed during the second trimester has recently been developed. It involves the insertion of sticks of laminaria, a Japanese seaweed, into the cervix of the uterus the day before the operation is to be performed. The laminaria causes the cervix to soften and dilate to the extent that instruments can readily be accommodated. On the day of the operation, the laminaria is removed and the region anesthetized. Then by the use of a specially designed curette, forceps, and suction, the fetus and placental material is removed. The complication rate for this method is said to be significantly less and after effects less severe than in fluid exchange procedures.

## THE ABORTION PROBLEM

Abortion is a subject on which there are many and varied opinions. It is a problem of many facets and one for which there is no simple solution. It is first and primarily a personal problem for the person involved but every abortion may raise moral, medical, social, demographic, and psychological considerations on which there are far-ranging and divergent views.

From the moral standpoint, the nature and value of human life, when it begins, who should have control over it, and so on, are of primary concern. Most religions oppose abortion. In our Judeo-Christian society, Catholics and conservatives of most faiths oppose it; moderates condone it under certain circumstances; liberals and nonbelievers lean toward the extreme liberalization of abortion laws.

From a medical standpoint, a doctor who must perform an abortion often must violate certain principles of medical ethics that dictate that he seek to save and preserve life and do no harm. Some doctors refuse to perform an abortion under any circumstances and some hospitals refuse to permit a doctor to perform an abortion under their jurisdiction.

From a legal standpoint, to what extent is an embryo or fetus a human individual entitled to the protection of law? To what extent should society seek to dictate medical practices or control what a person does with his or her body?

From a social standpoint, to what extent should society seek to dictate the role a woman should play in our social system? To what extent should society act to prevent an increase in unwanted children? Also, how much should any group of persons, large or small, be permitted to impose its moral values on an entire society?

From a demographic standpoint, to what extent should society seek to limit population growth utilizing abortion as a means of population control?

From a psychological standpoint, the emotional effects of a pregnant woman not having an abortion or having an abortion must always be considered. Psychosomatically, the interrelationships between physical and emotional health are inseparable.

The psychological effects of having an abortion may be severe especially if complications occur. Guilt feelings may result, negative feelings toward sexual relationships may develop, or moderate to severe depression may be experienced. Generally speaking, however, among the majority of women who have abortions, psychiatric complications are short lived, their severity depending largely on religious background and training, difficulties encountered in having the abortion, attitudes of peer group and family, and the effectiveness of sympathetic counseling.

In general, the trend today is toward the liberalization of abortion laws and granting to a pregnant woman the right to decide, especially in the early stages of pregnancy, whether or not she wishes for the pregnancy to continue. Generally, it is difficult to conclude that all abortions are right, or all abortions are wrong; that all abortions are harmful, or all abortions beneficial; that all abortions are immoral, or all abortions moral. Each abortion needs to be judged on an individual basis, and a woman should be free (with some limitations) to have an abortion if she desires, and a doctor should be free to perform one.

Finally, there would seldom be an abortion problem if, when the sex act is engaged in, proper contraceptive measures are employed. Responsible couples owe it to themselves, to their families, and to society to take steps to prevent conception unless they are prepared to accept the responsibilities of parenthood. If this policy were followed, unwanted pregnancies would be few in numbers and the problem of abortion would seldom arise.

# TWELVE
# VARIATIONS IN SEXUAL
# BEHAVIOR

*DEVIANT OR VARIANT SEXUAL BEHAVIOR*

Deviant or variant sexual behavior violates the norms of a particular society. Sexual behavior is manifested in many ways. In all human societies, some activities have been regarded as normal, others abnormal; some acceptable, others not acceptable; some natural, others unnatural; some moral, others immoral. What is to be considered "normal" or "natural" or "moral" sexual activity? Much of the American attitude toward sex has been the result of the Judeo-Christian view on the nature of sex: that the sex act should lead to procreation and that any activity that did not lead to or result in production of offspring was contrary to natural law, hence regarded as unnatural, unclean, immoral, or sinful. The term "perversion" or "crime against nature" has been applied to many activities and severe legal penalties established for violations.

Normal or abnormal sexual activity may mean different things to different persons. To the theologian, normal activity is that which is moral and right according to a religious code; to the sociologist, any activity that is commonly engaged in by its members and that is beneficial to a society is normal; to the psychologist, any activity that is conducive to mental health and does not result in emotional disturbances could be considered normal; to the physician, any activity that is not injurious to physical or mental health may be considered acceptable. Because one's sexual behavior depends on biologic makeup, training and experiences, and social conditions and customs, sex activity can range all the way from total abstinence to involvement in one or a few or several types of activities, some of which might be classified as deviant.

One of the driving forces behind the accomplishment of the sexual act is the desire to achieve an orgasm. There are several ways by which an orgasm can be induced other than through heterosexual intercourse. Are these methods to be considered as sexually deviant? To what extent do individuals participate in these activities? In general, little is known about the sexual behavior of most people. Most sexual activities are performed in private and even investigations by sex researchers have only to a limited extent revealed the types, frequency of, and distribution of the various forms of sexual behavior. However, through the studies of Kinsey, Masters and Johnson, Money, and many other researchers, much new information concerning various activities and practices has been acquired. This, accompanied by a marked change in attitudes toward sexual behavior with emphasis on greater freedom in sexual expression, has increased our knowledge of and understanding of the various types of sexual response.

Strictly speaking, there is no form of sexual behavior that is intrinsically deviant. Behavior, both sexual and nonsexual, is the result of many factors acting on an individual. Genetic factors determine one's basic biological and physical makeup and, in general, the nature of one's sexual drive. In addition, training, education, and experiences occurring during development have a molding action  and finally, sociological factors as determined by the group in which a person develops exert their effects.

Sexual behavior becomes deviant when it violates one of the norms of our collective society. Norms fall within three categories: legal, societal, and behavioral. Laws, in general, seek to prevent, by making punishable, behavior that is not acceptable to the majority of a populace. Societal norms or mores are the accepted traditional customs and practices of a particular society generally considered desirable for its welfare. Behavioral norms are those actually practiced by the members of a society. Medical, religious, and ethical considerations are also utilized in evaluating sexual behavior. If a practice is injurious to the bodies of other individuals or to the health of the individual involved or if it offends by violating basic moral and religious teachings, it may be considered as deviant behavior.

There is often a lack of correlation between laws, mores, and behavior. Masturbation, for example, is almost universally condemned by religious groups and by parents, especially in their children, yet it is engaged in either occasionally or as a regular practice by nearly all males and most females. Premarital intercourse is illegal in most states and generally disapproved of, especially by parents, yet it has become increasingly acceptable and widespread in both sexes. Oral sex is now commonly practiced. Because of the large number of individuals involved in these types of behavior and because these activities are closely associated with and intimately related to traditional sexual activities, they are sometimes regarded as *acceptable deviations* or *variations*.

A second type of deviance or variant behavior includes offenses against individuals such as incest, sexual molestation of children, exhibitionism, voyeurism, and aggressive and assaultive offenses such as rape. These types of offenses are generally disapproved of by society and most are subject to legal penalties. Because these offenses usually involve an individual with a disordered personality, this type of deviance is considered as pathological.

A third type of deviance or variant behavior is that which involves the development of a social structure. This type includes nudism, prostitution, and homosexuality. Individuals involved in these activities become a part of a subculture constituting a group somewhat separated from and excluded from the main body of society. Homosexuals and prostitutes each function in a world requiring a specific knowledge of its ways and having a specific culture of its own.

The following includes a brief discussion of the principal types of sexual deviations mentioned.

## ACCEPTABLE VARIATIONS

### Masturbation
This is the process of inducing an orgasm by any means other than sexual intercourse. In both sexes it is usually accomplished by self-stimulation of the genital organs (automanipulation). In the male, grasping the shaft of the penis and applying an up and down milking action usually quickly brings about an ejaculation. Gently touching the moistened glans increases the pleasurable feeling. In females, gently stroking and rubbing the pubic area will usually bring about erection of the clitoris and swelling of the labia minora. Gentle pressure applied to the shaft of the clitoris accompanied by rhythmical pressure movements will usually induce a response. Manipulation of the mons area and, in some women, breast stimulation will lead to an orgasm.

Masturbation may also be accomplished by other forms of friction. In the male, rubbing the penis tightly against a hard mattress or holding it tightly between the thighs and performing coital movements will usually induce an orgasm without manual manipulation. Some males employ devices that serve as a substitute vagina. Sexual fantasies are usually a part of the act. In the female, riding astride an object as a bicycle seat or saddle, or crossing the legs and applying rhythmic pressure, or drawing the thighs up against the abdomen and performing coital movements usually are effective methods.

Some females may insert an object into the vagina when they masturbate that may provide additional pleasure. Commonly it is the finger but other objects having the shape of a penis, as a candle, may be used. An artificial, erect penis called a *dildo* or specialized devices as the Japanese benwa or Chinese tickler are sometimes employed. Electric vibrators with specialized attachments are available and utilized in female masturbation. Direction of a stream of water to the clitoral area is highly stimulating. For this purpose, whirlpool baths, hand showers, and certain types of bidets are employed.

Masturbation is a normal activity of childhood and nearly all children manipulate their genital organs and engage in sexual play. It is common in early puberty and continues throughout adolescence when normal sexual outlets are denied or are unavailable. With the development of adulthood and the assumption of normal sexual activity, the desire or need for masturbation generally recedes.

Masturbation is much more common than is generally assumed, studies showing that over 90 percent of all men and over 60 percent of all females masturbate at some time in their lives. Many erroneous ideas have developed as to its possible ill effects on health. It has been alleged to cause various kinds of ailments including mental disturbances, even feeble-mindedness and insanity, skin disorders especially pimples, menstrual disorders, exhaustion of sex organs through loss of "vital fluid," and with it the loss of manhood and decreased sexual vitality.

What are the facts? There is no truth in any of these claims. There is no evidence that any physical or physiological harm results from masturbation. It may even be desirable under certain circumstances as when sexual relations are difficult to achieve or are denied. From a religious standpoint, masturbation is generally condemned and consequently feelings of guilt may be associated with it. Masturbation is considered undesirable if it is used as a substitute for normal sexual relations or if excessive feelings of guilt, shame, or disgust result from its practice.

*Premarital Sex*
One of the most significant changes that has taken place as a result of the sex liberation movement of recent years is the increased frequency and more widespread acceptance of premarital intercourse. Studies indicate that about 75 percent of single women have engaged in intercourse as compared with about 33 percent a generation ago. Among men, there was only a slight increase indicating that there is a trend toward abandoning the double standard that permits sexual freedom for the male but not for the female.

Premarital intercourse may be of the promiscous type or it may be selective and restrictive. Males tend to have intercourse more frequently and with more partners, females not so frequently and more often restricted to a single male. Often it is the result of a close association usually involving petting and the development of an intimate relationship with emotional involvement that may lead to marriage. Such is the case for about half of the females who engage in premarital intercourse. However this type of relationship does not always lead to marriage for while there may be a commitment to each other, still the legal bond is lacking and each is free to associate with others with whom they may wish to try a similar relationship.

Whether an individual should engage in premarital intercourse is sometimes a difficult question to answer. For young individuals, especially girls, society assumes a protective attitude and most states have age-of-consent laws that establish a minimum age, usually 18, below which sexual relations are prohibited. However considering the large number of violations, these laws are rarely enforced.

A number of factors may affect one's decision concerning premarital intercourse. One's attitude concerning sexual activity as molded by his or her family life and instruction, religious training and beliefs, and contact with peers, all play a role in decision making. Furthermore, the possible benefits or penalties of such an action must or should be considered.

**Teenage Pregnancy.** A problem of national importance is the continued high level of teenage pregnancies. It is not known with certainty how many teenagers become pregnant each year but in 1981 approximately 13 percent of the young women ages 15 to 19 were reported pregnant, resulting in about 527,000 births. Women younger than 15 years of age had approximately 12,000 births. While young black women represented only 12 percent of the 15 to 19-year-old age group, they accounted for one-third of all births for this group. The proportion of births that were out of wedlock was 35 percent of the white teenage births and 86 percent of the black teenage births. The exact number of teenage pregnancies is difficult to arrive at because of the questionable records of the number of miscarriages and the number of abortions obtained by pregnant teens. It is estimated that in 1981 about half of all pregnancies of teenagers ended in a live birth, about 35 percent ended in abortion, and approximately 15 percent ended in miscarriage or stillbirth. White teenagers were more likely to terminate their pregnancy by abortion (45 percent) than were black teens (20 percent). The highest rate of abortions for any age group was for the 18 to 19-year-olds, of whom 6 percent had abortions.

Figures indicate that 1 out of every 10 girls now in school will become pregnant out of wedlock before they reach the age of 18. Of those who give birth to babies, about 85 percent will attempt to mother their child; the remaining 15 percent will give their babies up for adoption. Most of the girls will remain at home during their pregnancy only about 5 percent being served by maternity homes.

Among the factors that account for the increase in teenage pregnancy are the following: an increased awakening in sexual interest among teenagers accompanied by a decline in parental authority and a breakdown in traditional controls, a breakdown in the double standard whereby sexual relations are now as permissible for girls as for boys, a general disinclination of girls to make use of contraceptive techniques; a lessening or decline in religious restraints, a lessening in the stigma attached to out-of-wedlock pregnancy, and the constant subjection of young people to sexual stimuli through song and dance, the radio, movies, and television.

Teenage pregnancy creates major problems in various fields. Schoolage mothers, whether married or unmarried constitute high risks in three areas: educationally, medically, and socially.

From the *educational* standpoint, pregnancy is one of the most common causes for female high school dropouts in the United States. School officials usually force pregnant girls to cease attendance and frequently will not allow a mother to reenter school until adequate plans for care of the newborn are effected either through adoption or utilization of day care centers. As a result, a girl's education is usually interrupted for a period of a year or more and when she does return, she is often subjected to humiliation and penalties that are nothing more than punishment for her indiscretion. Attitudes are slowly changing and pregnant girls are now being encouraged to remain in school as long as possible. Steps are being taken to provide special educational programs that will enable girls who become pregnant to complete their education to the extent that they may become productive adults.

From a *medical* standpoint, pregnant teenagers constitute a high-risk group. Because of physical immaturity and lack of prenatal medical care, complications of pregnancy as excessive weight gain, hypertension, and toxemia occur with greater frequency. Complications are more common in unmarried mothers than married mothers. Premature deliveries occur more frequently and with prematurity, low birth weight. This results in an increase in the incidence of subnormal brain development, a factor in mental retardation. Infant morbidity and mortality rates are much higher in this group.

From a *social* standpoint, because of age, race, and low socioeconomic status, most pregnant teenagers have little available in the form of social services as counseling, care, and assistance. What services are available are utilized only to a limited extent. Maternity homes and volunteer agencies take care of only a minority of unmarried pregnant girls. Adoption facilities for nonwhite babies are almost non-existent. Day care centers for teenage mothers are extremely limited, contributing greatly to the underemployment of this group.

Because of the high number of divorces among those who marry young and the low economic status of unmarried mothers, most pregnant girls eventually end up on welfare roles. The suicide rate for this group is seven times the suicide rate for their peers.

While the above problems are serious for any unmarried woman, they are compounded for school-age, pregnant girls under the age of 16. Girls in this group have a greater proportion of premature babies and babies with low birth weights; they are less likely to finish high school; they are less likely to marry and if they do marry, they are more likely to become divorced. They are much more likely to have repeated pregnancies out of wedlock. This is supported by studies such as that done on 100 teenage, unmarried mothers at Grace New Haven Hospital, Yale University, in 1966. Of 100 girls who have birth before the age of 18, as many as 95 had repeat pregnancies and they produced 340 children within five years. A similar study at Grady Memorial Hospital at Emory University in 1967 showed that 122 patients aged 11 through 16 delivered either their second, third, or fourth babies. The year before a similar study at Emory identified 364 patients who had delivered a baby every year since the beginning of adolescence. Four of the patients age 25 had each delivered nine babies, six age 24 had delivered eight, and 15 patients age 23 had delivered seven babies.

Many factors are involved in the high incidence of teenage pregnancy. The lack of or limited use of contraceptive measures plays an important role. This may be due to nonvolitional factors as mental retardation, forced intercourse, unavailability of contraceptives, or lack of knowledge of their use. Contraceptive methods when used include the use of the Pill, the condom, and withdrawal. Rarely or never used are the diaphragm, contraceptive jelly or foam, and the intrauterine device. As the most effective methods of contraception require a visit to a physician, necessitating parental consent, most teenage girls are denied the opportunity of using them.

However many teenage girls purposely avoid the use of contraceptives. The reasons for such are numerous and varied. Among them are the belief that pregnancy will not occur or if it does, an abortion is readily available. Some believe that the use of contraceptives is abnormal and possibly harmful, that their use involves preparation for intercourse with resulting feelings of guilt, that their use is too much trouble, that the possession of contraceptive materials would result in embarassment especially if found by parents, that the spontaneity of the sex act is interfered with, that the thrill of the sex act is enhanced by the risk of pregnancy, and that assuming the risk is a demonstration or proof of love. Boys generally show little concern as to whether the girl becomes pregnant of not leaving the matter of contraception largely up to the girl.

The problem of teenage pregnancies is a serious one because babies that are born are usually unwanted, unplanned for, and usually inadequately cared for. The mothers suffer physically and mentally; they are deprived of adequate education and denied the opportunity of playing a meaningful role in our society. Pregnancy out-of-wedlock is generally condemned by society, yet our society fails to provide the basic fundamentals of sex education, information on contraception, and free access to contraceptive materials by which pregnancy can be prevented. Adequate medical care is not provided and counseling and guidance are minimal.

Who is responsible for this situation? The family, through its abdication of teaching about sex and its failure to provide two of the most important necessities of life, that is, love and understanding, is an important contributor to the problem. The church, for remaining silent, or teaching patterns of conduct unrelated to the realities of everyday life; the school, for its failure to provide courses in sex education, marriage, and family life; and society, which penalizes a girl for becoming pregnant and yet fails to provide the necessary information concerning materials and techniques for preventing such—all must share the blame. It is hoped that society in the future will face the problem squarely and take steps to alleviate a condition that is so costly in human resources.

## Oral-Genital Sex

Oral-genital sex, or simply oral sex, includes cunnilingus and fellatio. *Cunnilingus* is the process of orally stimulating the female genital organs by the application of the tongue to the clitoris and labia. *Fellatio* is the application of the lips and tongue to the penis. Both are extremely sexually stimulating tending to heighten sexual feelings and to bring on an orgasm.

These procedures, together with other forms of noncoital sexual behavior, have in the past been considered "perversions" and their practice regarded as "abnormal," however studies in recent years have revealed that a majority of males and females, both in and out of marriage, have participated in oral sex encounters at some time in their lives. To some persons these practices are considered unhygenic or esthetically undesirable, in which case they should be avoided. Also because strong taboos sometimes exist concerning these practices, feelings of guilt may result from participation in them. However among couples where true understanding exists, and if the practice is acceptable to both partners involved, the application of the tongue or lips to any part of the body may enhance lovemaking and greatly accentuate sexual feelings and pleasure.

## SEXUAL VARIATIONS INVOLVING DISORDERED PERSONALITY*

### Exhibitionism

Exhibitionism is a form of sexual behavior in which an adult male obtains sexual satisfaction by exposing his genital organs to women or children, often in a public place. He may or may not have an erection; he may engage in masturbation, if not during the act, immediately following. Satis-

*See also the section "Sexual Variations" in the Preface to the Dover Edition, pp. viii–ix.

faction comes principally from observing the reaction of the victim, which is usually surprise, disgust, and fright. Fear is usually unwarranted as the exhibitionist rarely attacks or molests the victim.

The exhibitionist is usually middle aged and may be married or unmarried. In most cases he is sexually suppressed or inadequate and uses this technique to demonstrate his masculinity. They are considered to be mildly neurotic and in need of psychiatric attention. Sometimes exhibitionism is practiced by older men in whom it is considered as retrogression to a childhood form of sexuality and a compensation for sexual impotence.

Closely associated with exhibitionism is the person who makes obscene telephone calls to women. Usually the victim is unknown and gratification is achieved from the embarassment or emotional response of the person called. The proper way to handle such calls is to hang up immediately. If the call is repeated, notify the telephone company, which will attempt to trace the call to its point of origin.

### Voyeurism

Voyeurism is the practice of obtaining sexual satisfaction by surreptitiously viewing the sex acts of others, viewing a naked woman, or watching one disrobe. Such a person is the so-called "Peeping Tom." Most men derive pleasure and usually become sexually aroused when viewing a nude female as manifested by their interest in viewing topless waitresses, burlesque shows with their striptease acts, and the popularity of sex magazines featuring the nude female. The voyeur, however, usually prefers that the viewing be done in private and the victim not know that she is being watched. This involves peeping through windows, an activity that involves considerable risk but, at the same time, provides an element of excitement. Dangers encountered are the risk of being caught, which frequently happens as a result of being observed by neighbors or passersby. Offenders, when arrested, generally turn out to be young men whose heterosexual activities are inadequate. Peeping, accompanied by masturbation, provides an outlet for such individuals and, when repeated, often becomes a compulsory activity.

### Sadomasochism

Sadomasochism (S-M) includes sadism and masochism, which are mirror images of the same behavior. Sadism is the achievement of sexual satisfaction through the infliction of physical pain and suffering on the recipient. This may be by beating, slapping, whipping, pinching, biting, burning, or other cruel actions. Such actions are essential in order for a sadist to achieve an orgasm.

Sadism is a distinctly aggressive activity and little is understood as to its basic causes. Every sex act involves certain aggressive activities and biting gently or nipping the skin is often a normal part of sexual foreplay. Even painful acts may be performed in the heat of passion but at that time, pain perception is dulled and the pain generally unnoticed. Sadism is sometimes manifested in sex crimes in which the victim is attacked, then afterwards murdered and her body mutilated.

Masochism is the submission to pain and humiliation in order to achieve sexual satisfaction. Sadism and masochism may be engaged in mutually to the satisfaction of both participants. It is resorted to in both heterosexual and homosexual activities. Masochism is sometimes associated with a guilt complex and the desire to be punished for sexual misbehavior.

## Sodomy

Sodomy is the act of anal intercourse, usually performed between males although it sometimes occurs in heterosexual relations. It is commonly practiced by homosexuals. When anal intercourse is between an adult and a child, it is called *pederasty*. Sometimes the meaning of the term is broadened to include oral-genital contacts (cunnilingus and fellatio) and intercourse with animals (bestiality), all of which are included under the sodomy laws that consider these activities as "crimes against nature."

## Nymphomania and Satyriasis

Nymphomania and satyriasis are terms that refer to exaggerated and compulsive sexual activity, *nymphomania* being applied to the female, *satyriasis* to the male. There appear to be various degrees to which the sex drive is expressed, all the way from total indifference and lack of interest in sex to excessive preoccupation with sexual activities as represented by individuals belonging to the groups mentioned. There are individuals who are hyperactive and capable of experiencing multiple orgasms daily for long periods of time but these comprise only a small portion of the population. This is thought to result from the excessive desire on the part of an individual to manifest her femininity or his masculinity, a form of neurotic behavior that has its origin usually in emotional conflicts which develop early in life.

## Fetishism

Fetishism is the condition in which sexual satisfaction is obtained from a *fetish,* an inaminate object or a part of the body not usually considered to be sexually stimulating. The attachment may be toward objects of clothing (bras, panties, girdles, a shoe, or handkerchief) or toward parts of the body (the limbs, feet, hands, hair, breasts). It may also be directed toward charcteristics of the body, such as odor or voice.

Fetishism is largely limited to men and the object is usually essential for obtaining sexual satisfaction. For some it is necessary in order to bring about an erection preliminary to intercourse. Often times the selection of a sex partner is limited to those possessing the fetish. It is commonly used as an aid in masturbation.

The causes of fetishism are uncertain since a certain amount of fetishism is encountered in most individuals, sexual feelings being aroused by the sight of items of clothing, especially those associated with sex appeal. One view is that the fetish is something that a person has learned to associate with sexual activity and he fanticizes to the extent that he tends to let the fetish substitute for the real thing. A second view is that the fetish exists in the unconscious and symbolizes and serves as a substitute for a woman. In either case fetishism is a symptom of sexual immaturity.

*Incest*

Incest is sexual intercourse between closely related persons who, by law, are prohibited from marrying. It may be between father and daughter, mother and son, uncle and niece, aunt and nephew, or between cousins. Sometimes other relationships are included. Incest is the only sexual offense that is universally condemned in all societies although in certain ancient civilizations, as among the Incas and Egyptians, brother-sister marriages were permitted in royal families.

In the United States, incest occurs commonly in families of the lower socioeconomic groups and generally involves individuals of substandard intelligence. Cases of incest often involve neurotic individuals living under crowded conditions as in city ghettos. Drunkenness is frequently a factor.

However, incest among the general population occurs more frequently than is generally supposed, occurring commonly in sparsely populated areas. It may involve individuals in families at all socioeconomic levels, and strangely, it often involves highly moralistic and religious persons. Fathers may take advantage of their daughters, brothers their younger sisters, uncles their nieces. Mothers may entice their sons into sexual activity. Often, the sexual activity is mutually agreeable. Since most cases never come to light, it is difficult to know the exact incidence of incest but it is definitely higher than statistics indicate. In families where the father shows an obvious erotic interest in his daughter, especially when she is at an age when she is becoming sexually attractive, special counseling or even psychiatric treatment should be insisted on. The trauma inflicted upon a young girl subjected to sexual attack by one of her own family, especially her father, is often more severe than that resulting from rape by a stranger. Any girl who is subjected to sexual advances from one of her own family should not hesitate to tell her mother (or her father, if he is not the one involved) and if no action is taken, a concerned counselor should be consulted and in extreme cases, the law should be requested to step in.

## Pedophilia

Pedophilia is the use of children for sexual gratification, usually considered by law as "child molestation." It most often involves contact with the child by a relative, a family friend, a neighbor, or a nonrelative living or working in the home. Usually sexual intercourse is not involved, the act consisting of fondling the genitals of the child and having the child reciprocate. It is commonly thought that homosexuals are the principal offenders but such is not the case since homosexuals tend to restrict their activities to other adults.

Heterosexual offenses involving adult men and children are generally looked on with great aversion. Offenders are usually individuals with impaired judgment, such as mentally retarded individuals, psychosexually immature persons, and senile men with deteriorating mental faculties.

## Bestiality

Bestiality is the practice of engaging in sexual relations with an animal other than a human. The practice is resorted to by both sexes, males copulating with female animals and females receiving the penis of male animals. The idea that such contacts made by females would lead to conception resulting in the birth of a monster was an erroneous concept held during the Middle Ages. Sometimes animals are trained for oral-genital contacts or used as an aid in masturbation.

Bestiality is generally condemned and individuals caught in the act are subjected to ridicule. However it occurs more frequently than is generally suspected. It is common among boys in rural areas where farm animals are the usual objects involved. Men living in isolated regions may engage in this activity. Sexual relations with animals are usually utilized as a sexual outlet when normal heterosexual outlets are not available.

## Necrophilia

Necrophilia is an erotic interest in and the use of a dead body for sexual satisfaction. It may involve erotic stimulation in the presence of or viewing a corpse, called *necromania,* or the desire for sexual contact including coitus. It is a rare phenomenon usually involving a severely deranged, psychotic male, so maladjusted that if a corpse is not available, he may kill a woman to obtain a body. Often mutilation of the corpse follows sexual relations. Necrophilia is considered one of the most extreme sexual deviations.

## Transvestism

Transvestism is the practice of wearing clothing of the opposite sex, the transvestite, usually a male, achieving sexual satisfaction in so doing.

Homosexuals, both male and female, may dress in clothing of the opposite sex to identify themselves as homosexuals or as a sign of the role he or she prefers to play in homosexual activity. However, a true transvestite is a heterosexual individual who has a compulsive desire to dress and act as one of the opposite sex. He will dress in female clothes, especially underwear, wear a wig, and apply cosmetics. He may remove excess hair from the face and body and take estrogen to develop the breasts and hips. In addition to impersonating the appearance of a female, he will attempt to duplicate their actions and movements. Some capitalize professionally by being female impersonators.

Transvestites usually grow up in families in which the female plays a dominant role. Typically, the mother is a cold, stern, and abusive woman; the father weak, ineffective, and frequently absent. Cross dressing usually begins in childhood with the child dressing in female garb and showing a fetish interest in female clothes. Male aggressive activities are usually discouraged. Transvestites obtain emotional satisfaction by dressing as a female, however one of the problems is that of being discovered. Most of their activities are carried on in secret and they usually live isolated lives often with feelings of guilt. They may marry, the wife sometimes accepting and even encouraging his transvestism. His family life may be difficult, especially keeping his activities from his children.

*Transsexualism; Gender Dysphoria*
Transsexualism is the condition in which a person, usually a male, wishes to be changed into a person of the opposite sex. He is a person with normal male physical characteristics with respect to gonads and secondary sexual characteristics, however he believes that he is truly a female and wishes to get rid of his male organs and become a female. He may engage in homosexual as well as heterosexual relations but since such individuals regard themselves as females, they do not consider their relations with other males as being homosexual.

The desire to become a female is sometimes so strong that the male transsexual will go to great expense and trouble and endure pain and suffering in order to become a "woman." This involves *sex-change* surgery that is difficult to obtain in the United States but is available in certain foreign countries, especially Denmark. For males, this necessitates the removal of the penis and testes and the creation of an artificial vagina accomplished by reconstruction of the pelvic tissues. Supplemental operations to alter the shape of the larynx, breasts, and hips supplemented by hormone therapy to bring about the development of secondary sexual characteristics are resorted to. Such operations, however, never bring about the production of a normal person of the opposite sex capable of reproduction.

For the female transsexual, transformation into a "male" is a more difficult procedure. It involves removal of the breasts and internal sex organs, creation of a scrotum from labial tissues, and the formation of an artificial penis from the clitoris. The "penis" formed from skin grafts is never capable of erection and is incapable of penetrating a vagina unless supported artificially.

Transsexuals may obtain legal status, change their name, and they may marry. Some adopt children. The causes for transsexualism are obscure. It is often related to overpossessive mothers who raise their boys as girls, encouraging them to talk, act, dress, and play in a feminine manner, or to fathers who, in the absence or illness of the mother, encourage their daughters to identify with them in masculine activities.

### Rape

Rape is the crime of forcing a female by the use of threats or violence to submit to sexual intercourse without her consent or against her will. Sometimes the meaning of sexual intercourse is broadened to include fellatio and acts of sodomy. A woman who consents through fear or if she is mentally deficient or unconscious from sleep, alcohol, or narcotics is also considered to be a victim of rape.

Statutory rape is intercourse between a man and a girl below the age of consent which in most states is 18. Even though the girl may lie about her age and give her consent, the man may be found guilty of rape and be imprisoned. In all states rape is considered a felony.

Group rape is that which involves two or more rapists. The initiation of the attack is more often not by the group leader but by one of the members who needs ego-enhancement from the group. Group rapes constitute 10 to 20 percent of all rapes. Most group rapes are by black males who rape black females.

The incidence of rape in all categories is on the increase. The chance that a woman will be raped sometime in her lifetime has been estimated to be from 1 in 10 to 1 in 4. In 1983, over 78,000 cases of rape were reported. However, the actual number of rapes is considered to be from 4 to 10 times higher, the discrepancy being due to lack of reporting because of the fear or humiliation and embarrassment often inflicted upon the victim. The pregnancy rate from rape has been estimated to be about 2 percent.

The majority of rapists are young males between the ages of 18 and 24. Most rapists are unmarried but a sizable minority includes older married men with families. Most rapists come from families in the lower socioeconomic strata. Many come from broken homes or homes in which they had been subjected to neglect and brutality. Most are of subnormal intelligence and work at unskilled jobs. Many are repeaters who commit the offense over and over again often following a particular pattern.

In a study of convicted rapists conducted by the Institute for Sex Research founded by Dr. Alfred C. Kinsey, rapists were found to fall into five general groups. The most common type was the *assaultive type*, comprising about 30 percent of the rapists studied. Men in this group were sadists who felt impelled to inflict physical pain on their victims, their sexual satisfaction being achieved principally through the use of force. Often, unnecessary violence or threats of physical harm were used to complete the rape. Most men of this group exhibited a pronounced hostility to women. The term "sex maniac" is often applied to rapists in this category.

The second type comprising about 15 percent of those studied were considered *moral delinquents*. These men were not sadists and, in general, were not hostile to females. They were primarily interested in one thing, coitus, and regarded women primarily as sex objects existing to satisfy their sexual needs.

A third type, comprising about 15 percent, included the *drunken variety*. Men who committed rape under the influence of alcohol. Individuals in this group were frequently of subnormal intelligence. Alcohol seemed to act principally in removing inhibitions that permitted the release of aggressive and often vicious behavior toward their victim.

A fourth type comprising 10-15 percent was the *explosive type*. Many rapists in this group were average, law-abiding persons whose aggression occurred suddenly and often without reason. Most had no history of sexual aggression and their actions were unexpected. Sometimes their behavior was vicious and brutal. Most men in this group exhibited personality defects, most of which had their origin in early development. Abuse and brutality in childhood were a common characteristic. Stress and strain ultimately led to an eruption in the form of sexual aggression.

A fifth type, comprising 10-15 percent was the *double standard* variety. Men of this group tended to divide women into two types, good and bad. Good women were those whom they respected and would not think of mistreating; bad women were those whose actions were interpreted as indicating that coitus was agreeable to them. These women were considered promiscuous and fair game for sexual advances. Moderate force or threats were considered justifiable by members of this group.

The above groups included about two thirds of the rapists studied. The remainder included rapists who were mental defectives, psychotics, and severely neurotic individuals. Characteristic of most attacks on women was a basic hatred of females, the sex act being committed not primarily for sexual satisfaction but more for the release of hostile feelings toward women and to belittle and degrade the female sex.

The attacks by rapists are seldom spontaneous; most are premeditated. The victim may be a particular female for whom they lie in wait or she may be any female who happens to become available. The latter is especially true in gang rape.

The victims of rapists present a greater range of variability than the rapists. Most are between the ages of 18 to 25, although they may vary in age from young children to women in their eighties. They may be ugly or attractive; they may be well or poorly dressed; they may appear sexually seductive or they may be of the prim, prudish type. The only characteristic common to all is that they are female and unfortunate in being in the wrong place at the wrong time.

Where do rapists make their attacks? They occur in back halls, basements, elevators, and on roofs of buildings, in abandoned buildings, in alleys and deserted streets, in bushy areas in parks, in cars, sometimes in city streets but often along little-traveled country roads. They may occur in the victim's home or apartment. More attacks occur in summer than in winter and on weekends than during the week.

A myth commonly held by rapists and by men in general is that most women secretly wish to be dominated by males and welcome forceful and sometimes brutal sexual attacks, that the resistance offered is only a token resistance, and that actual penetration could not occur without cooperation on the part of the female. It is also commonly thought that sexual advances may be encouraged by the female through her actions or dress. Studies indicate that all of these ideas are false; that most rapes are humiliating and degrading experiences that leave the victim emotionally scarred, often for life. Cooperation by rape victims is usually only granted under serious threat of bodily harm.

The laws governing rape differ in various states but often times they favor the offender rather than the person raped. Corroborating evidence that the act actually occurred is usually required together with evidence that force was exerted and that resistance was offered. A woman's sexual history is often examined and any activity of a questionable nature can sometimes be used in defense of the rapist. Because of the unsympathetic attention a woman often receives from the police and the necessity of repeatedly going over the details of an experience one wants to forget plus the hostile cross-examination by lawyers defending the rapist, women often prefer to remain silent and not prosecute.

With the development of the feminist movement, certain women's organizations are taking aggressive action toward having the laws pertaining to rape changed to give greater credibility to the victim and remove provisions that make the conviction of a rapist difficult. A few states, among them California and Michigan, have recently changed their laws regarding the introduction of sexual histories of the victims and no longer permit such to be entered in a trial. Centers for victims of rape are being established in various cities and on college campuses. These centers provide the victims with counseling service, legal and medical advice, and a sympathetic understanding. Educational campaigns are also promoted to alert the public to the problem of rape and how to deal with it.

**Prevention of Rape.**   A woman, to reduce her chances of becoming a victim of rape, should avoid those situations that often lead to forcible rape. Among the common potential rape situations which should be avoided are the following:

**Hitchhiking.**  Don't hitchhike. If you must, do so in company with another person. Never hitchhike in a car with two men.

**In Your Home or Apartment.**  See that all doors have safety chains. Do not open the door and admit any person unless you know the caller. Never hide a key outside the door. Do not use laundry rooms unless in the company of another person, especially late at night. Do not enter or remain in an elevator with a man you do not know.

**In Your Car.**  Do not get into your car until you have checked the back seat with the overhead light on; keep doors locked at all times; keep gas tank full to avoid running out of gas; never pick up a hitchhiker.

**On the Street.**  Avoid being alone if possible, especially at night; avoid dimly lighted streets and deserted areas.

**For Protection.**  Carry a police whistle. Do not carry a gun, knife, or tear-gas pen, unless you are adept at using them. Weapons can often be taken from you and used against you. It is also illegal to carry and use a concealed weapon.

Learn basic techniques of self defense and use them if possible. If in an area where help might be available, yell FIRE, not HELP. If attacked, try not to panic. Try to talk to the threatening rapist, maybe he can be talked out of it. Don't accept willingly a victim's role. Some degree of resistance must be evident to secure a rapists conviction.

**Action Following a Rape.**  The police should be called immediately. Many police departments now have women detectives who interview rape victims. Do not touch anything; do not bathe or change clothes as evidence may be destroyed. Present any evidence of force as torn clothing or bruises. You will then be required to go to a hospital where a physician will examine you. He will take a case history of the rape, collect evidence of penetration or attempted penetration, assess the amount of physical injury, take steps to prevent possible venereal disease infection, give antipregnancy shots or pills, and assess the general overall effects. Then you will be encouraged to consult your personal physician.

Although there is a tendency for most women who have been raped to remain quiet and not report the crime, such an action merely means that the rapist remains free to prey on other women. Prompt action in reporting the attack increases the likelihood that the rapist will be caught and hopefully will act as a detering factor for other rapists.

A problem encountered in dealing with cases of rape is the possibility of an emotionally disturbed woman accusing a man of rape or attempted rape when no offense actually occurred or of a consenting woman, for revenge or other reasons, charging after the act, that force had been used and the act was committed against her will. Cases like this are rare but there is always the possibility that an innocent person may be convicted and unjustly imprisoned.

Finally, to a woman raped, although the act is an outrageous crime in which the body is violated, it is not the end of the world. Life goes on and as with most unpleasant actions to which a person's body may be subjected in which physical or psychic harm results, there is a tendency for the mind to forget the trauma experienced thus enabling the victim to adjust to whatever life situations she may encounter.

## SEXUAL VARIATIONS INVOLVING SOCIAL STRUCTURE

### Nudism and Nudity

Nudism is a form of behavior that seeks to promote general health and well-being by doing away with the wearing of clothes. Its advocates believe that nudism acts as a democratizing influence by eliminating distinctions between the social classes often symbolized by clothes, kings and peasants looking very much alike in the absence of clothes. They also maintain that nudism dignifies the body and acts to deemphasize sexualism by integrating the sex organs with other systems of the body and eliminating the shame and embarassment so frequently experienced on viewing a naked body. Nudism is encouraged as a natural and pleasurable way of life that promotes both physical and psychological well-being. Physical fitness is an important adjunct of their program, the importance of sunbathing, fresh air, and pure foods being emphasized. Early nudists were often vegetarians and abstained from alcohol and tobacco.

Nudism as a cult or social movement developed during the early decades of this century as a part of the revolt against the excessive restraints of Victorianism. It flourished especially in Germany after World War I and from there spread throughout Europe. It became popular in Holland, in Scandinavia, and in other countries where nude swimming in pools and on beaches became common. Nudist camps flourished and attracted a large clientele.

Nudism was introduced into the United States in 1929 but its growth has been limited by active opposition and sometimes persecution by groups opposed to its practice. Religious opposition has been pronounced based on the traditional concealment of sex organs and the mistaken concept that exposure of the undraped human body was obscene and pornographic and promoted sexual license. Nudists responded by pointing out that concealment of the sex organs draws attention to them and emphasizes sex and that nudist camps, rather than promoting promiscuity, actually contribute to the opposite.

Even though nudism has become more generally accepted and laws pertaining to its practice have been liberalized, the growth of the movement has been slow in the United States. In order to maintain a high degree of respectability and to prevent members from harassment and sometimes persecution, strict rules pertaining to admission and behavior in a nudist group are generally followed. Membership is usually limited to married couples or families. Singles, especially males, are usually not admitted. The reputation of nudist camps is highly regarded and behavior is in general circumspect. General rules for most camps include the following: no staring, no sex talk, no profanity, no body contact, no use of alcoholic beverages, and no attempts at concealment or covering of the body. Because of these rules, the behavior at nudist camps, instead of favoring promiscuity, tends to develop an atmosphere of asexuality. Sexual misbehavior is not condoned, in fact, deemphasis of genital sex is one of the aims of nudism.

Because our society does not condone the public exposure of the sex organs or the naked body, those engaging in nudism have been regarded as exhibitionists or voyeurs. Such is hardly the case, however, for exhibitionists secure their gratification from the reaction (fright, fear, disgust) of the victim and the person viewed by a voyeur is usually unaware that they are being observed and would be offended if they were.

The number of nudists in the United States is not known but it is estimated that around 25,000 regularly participate in nudism. Most male nudists cite magazines as their initial source of interest in nudist camps while females usually become interested through the urgings of their husbands. The motivation for the first visit to a nudist camp is usually from curiosity but invitations are generally limited to those who show a sincere interest in nudism and are recommended by practicing nudists. Most members of nudist groups are of the middle class and include many in executive and professional classes. Members of the lower social groups show little interest in nudism.

Among the reasons cited for membership in and participation in nudist activities are the friendliness and sociability of the group and the pleasure of relaxation and enjoyment of nature unencumbered by clothes. The morality of nudist camps is no problem as there are strict rules governing behavior. Important principles underlying nudism are that there is nothing shameful about exposing all the parts of the human body, that nudity and sexuality are unrelated, that the lack of clothing leads to a feeling of freedom and natural pleasure, and finally, that exposure of the body to the sun improves and enhances physical, mental, and social well-being.

There have been marked changes in attitudes of society toward nudity during the past decade. The relaxation of laws pertaining to obscenity and pornography has resulted in a flood of magazines, books, and art works in which the human body is portrayed in every conceivable form. The popular girl centerfold of male-oriented magazines has been adopted by women's magazines, which display the unadorned male to their readers. Their popularity is probably indicative that women are just as interested in (or curious about) the naked male body as males are about the female body. Nudity in the movies, in the theatre, in restaurants, in night clubs, and on the beaches has become increasingly common and is accepted often with little comment.

A recent development in nudism has been the occurence of a fad called *streaking,* participated in principally by male college students. This consists usually of a small group of men darting across a campus, through a public place, or through a building as a dormitory entirely naked or possibly wearing only sneakers. Whether streaking is a true form of social nudity or a form of exhibitionism has not been definitely determined. By some, it is regarded as the latest form of a spring ritual engaged in by college students, forerunners of which were swallowing goldfish, crowding into phone booths, engaging in panty raids, or exposing the buttocks from a car window (mooning) common during the 1950s and 1960s. A recent study of streaking identified it as a minority activity involving very few persons. The streakers were about equally distributed between upper and lower classmen and most were males. The motivation for streaking seemed to be an attempt to destroy conventional behavior by rejecting authority, to shock the conservative element and to attract attention to themselves.

To some who worry about the psychological well-being of the streaker, it might be more appropriate to worry about the psychological well-being of the hundreds or thousands who sometimes gather to witness an event of streaking.

*Homosexuality*
Homosexuality is a type of sexual behavior in which there is a sexual attraction between individuals of the same sex. It is one of the most common of sexual-variant behaviors and one that presents the most problems for some members of society. According to the Kinsey concept of homosexuality and heterosexuality, about 75 percent of men and 85 percent of women are exclusively heterosexual. About 25 percent of men and 15 percent of women have had varying amounts of both heterosexual and homosexual experiences. Two percent of men and 1 percent of women are exclusively homosexual. In a population of 230 million Americans the latter two percentages would mean that over 2 million men and over 1 million women are homosexual. Some have suggested that if those who spend part of their life as homosexuals are included, then maybe as many as 10 percent of the population could be considered homosexual. The status of homosexuality has been cleared up to a certain degree by the action of the American Psychiatric Association, which removed it from its "sick list," thus taking it out of the category of a mental illness. At the same time, recommendations were made that steps be taken at local, state, and federal levels to remove

discriminatory legislation against homosexuals and that homosexual activity between two consenting adults no longer be considered a crime.

Historically, during the early part of this century, sexual orientation was viewed as an either-or situation, either heterosexual or homosexual (Fig. 34). It wasn't until the end of the 1940's that Alfred Kinsey and his colleagues conceptualized sexual orientation on a seven-point continuum (Fig. 34). The scale ranged from 0 (exclusive heterosexual orientation) to 6 (exclusive homosexual orientation). While category 3 represents what would be considered a bisexual orientation, Kinsey and his colleagues did not uti-

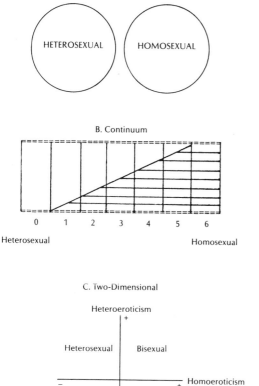

Figure 34   Historical Conceptualizations of Homosexuality.

lize this term. Kinsey's theory was the dominant theory of homosexuality until 1980. In 1980, Storms conceptualized homosexuality/heterosexuality in a two-dimensional model (Fig. 34). According to this model, one can have varying amounts of homoerotic or heteroerotic orientations. This conceptualization permits a better understanding of those who are bisexual or even asexual in their orientation. Homosexual activity, as a form of sexual play, is commonly engaged in by most preadolescent and adolescent boys and girls. On the attainment of maturity, sexual activities in most individuals are directed toward those of the opposite sex, and permanent heterosexual or "straight" relationships are established. Among homosexuals, however, the activity is directed toward one of the same sex. When the individuals are females, the terms lesbian and lesbianism are used.

Professors Bell and Weinberg of the Kinsey Institute completed a study in 1978 of overt homosexuals. They were able to classify 71 percent of the homosexuals in five general patterns of interpersonal relationships. The remaining 29 percent were too diverse to be grouped into any of the five patterns. About 28 percent of the lesbians and 10 percent of the males lived as "closed couples," which is similar to the pattern of married heterosexuals. Better than one-third of these couples had lived in this type of relationship for four or more years. A second category was termed "open couples," which included 17 percent of the lesbians and 18 percent of the males. They were less deeply committed to each other and were more likely to be involved in other sexual encounters. These couples were less happy and more lonely than the closed-couple homosexuals. A group of homosexuals that lived a "swinging singles" lifestyle were termed "functionals." This group consisted of 15 percent of the males and 10 percent of the lesbians. These individuals were much more interested in having multiple partners, were sexually well adjusted, and had few regrets that they were homosexual. A fourth group, termed dysfunctionals, were not living as couples and had both psychological and sexual problems. About 12 percent of the male homosexuals and 5 percent of the lesbians were in this group. These individuals were most likely to regret being homosexuals and came closest to fitting the "maladjusted homosexual" stereotype. Finally, some 16 percent of the males and 11 percent of the lesbians lived a lifestyle characterized by a lack of intimate involvement, and were termed "asexuals." They tended to be older and to be more likely to describe themselves as lonely. Bell and Weinberg's study gave us new insights into the diversity of homosexual relationships. Yet much of their data seemed to confirm some stereotypes that society has of homosexuals. They found that 45 percent of the lesbians were currently living in a couple relationship, compared to 28 percent of the males. Fifty-seven percent of the lesbians had fewer than 10 different sexual partners in their lives, whereas 57 percent of the males had more than 250 different sexual partners. Finally, 74 percent of the men as compared to 6 percent of the lesbians claimed that more than half of their sexual partners were strangers.

Activities engaged in by homosexual couples are the same as those engaged in by heterosexual couples, with the exception of vaginal intercourse. Mutual masturbation, oral-genital contact, and, among males, anal intercourse are the principal activities leading to orgasm. Kissing and petting are common preliminaries. Homosexual pairing usually involves the assignment of roles, one assuming the active, dominant, or "male" role; the other, the passive, receptive, or "female" role. The roles, in general, are not permanent, most homosexuals altering their roles according to their own or their partner's preferences.

Because homosexuals are subjected to legal punishment, socially condemned, and more or less isolated from our society, they tend to group together and form a loosely organized homosexual community, the so-called gay world. In most large cities there are gay bars and gay clubs where homosexuals congregate. These bars serve not only as a source of liquor and entertainment but also as a meeting place for those seeking sexual encounters. There are also other gathering places that attract homosexuals and soon become known as gay areas. These include beaches, parks, restaurants, public rest rooms, gyms, and other places.

Homosexuals fall into two general types, the covert and overt. Covert homosexuals keep their activities hidden to avoid penalties often associated with their discovery—ridicule, possible imprisonment, blackmail, job loss, and social ostracism. Most pass as heterosexuals and they may be married and have children. They avoid associating publicly with known homosexuals and, in general, lead the life of a typical heterosexual, their homosexual activities being carried on secretly. In contrast, the overt homosexual has no hesitancy in admitting his homosexuality and in practicing it openly. He is usually a part of the homosexual community and often expresses his defiance of the heterosexual world. He tends to be socially isolated and to have little involvement in heterosexual activities.

Homosexuality may result from forced lack of heterosexual contacts. This occurs in male-only or female-only associations, such as schools, military forces, and especially prisons, both male and female. This deprivation type of homosexuality may result when individuals who are normally heterosexual are deprived of their normal sexual outlets. Homosexuality in prisons constitutes a complex and difficult problem as new inmates are often forced to submit to homosexual acts by the more aggressive and often vicious inmates, some of whom are imprisoned for homosexual offenses.

The causes of homosexuality are obscure. Studies indicate that there are no significant differences in the physical or physiological makeup of most homosexuals as compared with heterosexuals. No genetic causal factors have been identified although some investigators support the Freudian view that there is a constitutional or genetic disposition toward homosexuality with subsequent development serving to favor or inhibit its manifestation. It has been believed that variations of the levels of male and female hormones might be a significant factor and some recent studies indicate that hormone levels in homosexuals do differ significantly from those of heterosexuals; however, hormone therapy is generally ineffective in changing sexual orientation. One of the most currently accepted theories on the cause of homosexuality is that it develops as a result of a person's positive and negative life experiences. However, there is little scientific evidence to show support for the learning theory of development of homosexuality.

There has been a concerted effort in recent years to change the attitude of society toward homosexuality and this has resulted in the development of a more permissive view of homosexual behavior and a relaxation or elimination of legal restraints between consenting adults. This has largely come from the gay liberation movement sponsored by organizations, both local and national, seeking to promote a better understanding of homosexuals and their behavior and a greater tolerance toward those who participate in this mode of life.

It is now recognized that some homosexuals lead satisfactory sex lives and are productive members of their community. However, it is generally considered that, in our society, the life of a homosexual is not as satisfactory or rewarding as that of a heterosexual. As a consequence, many homosexuals seek to have their sexual life modified as they miss the home and family life of a heterosexual. Since homosexuality is considered by some to be primarily an emotional disorder, treatment is sometimes suggested for those unhappy with their orientation. In psychiatry, an attempt is made to discover the causes and then to recondition the homosexual's behavior. Behavior techniques utilize aversion therapy, desensitization and assertive treatment, and sometimes group therapy. However, success in changing the pattern of life of a homosexual, whether by psychoanalysis or by behavioristic techniques, has been extremely limited.

### Bisexuality (Ambisexuality)

Bisexuality, sometimes called ambisexuality, is defined as an erotic attraction toward both men and women. It is difficult to estimate the incidence of bisexuality in our society. Masters and Johnson have estimated that bisexuals may constitute almost five percent of the adult population. Bisexuals usually explain their orientation in terms of a need for variety in their relationships. Some bisexuals rationalize their behavior as an indication of sexual openness, an indication that they are not prejudiced in their sexual activities. The exact "causes" of bisexuality are unknown; it is thought by some to be an outgrowth of a very open and experimental type of personal philosophy.

### Prostitution

A prostitute is a person who indiscriminately provides sexual services for a monetary reward. Prostitution, like homosexuality, while sometimes carried on as an individual activity, usually involves a socialized structure consisting of prostitutes and others associated with them in their activities. This includes panders, pimps, steerers, and customers and, since prostitution is generally an illegal activity, police are usually involved in its protection. Sometimes it is referred to as "organized vice." The subculture of prostitution involves a feeling of togetherness among those involved, a special language or trade lingo, and an apartness from the straight, heterosexual world.

Prostitution may involve individuals of both sexes although females are the principal participants. Prostitution exists because there is a demand for sexual services by men who are willing to pay for it. A male may seek a prostitute for various reasons, as when usual heterosexual outlets are not available or he may wish to engage in sexual activity without assuming the obligations imposed by marriage or other male-female associations. Furthermore, males may wish to engage in certain deviant practices that a prostitute will perform but which are considered undesirable in "normal" sexual associations. A prostitute or "hooker" seeks to please her customer or "trick." Another reason that prostitution exists is that it is an expedient and sometimes the only feasible way for a female to make a living and, in many cases, it is the kind of life that a prostitute likes. Houses of prostitution variously called brothels, bordellos, houses of ill fame, cathouses, or whorehouses exist. In addition, brothels may exist and operate as a legitimate business under various guises as sauna baths, Turkish baths, massage parlors, and the like.

Prostitution, sometimes called the world's oldest profession, has been regarded throughout history in various ways. In some ancient societies, prostitution was a part of religious activities and the utilization of the services of a prostitute was a form of worship. Women who served as sacred or temple prostitutes were regarded as having rendered a significant religious service. At other times, prostitution has been accepted as a part of a community's social organization and life and prostitutes were regarded as respectable persons, their services being esteemed.

At the present, prostitution is regarded in various ways throughout the world. In some countries it is accepted freely and no efforts are made to restrict it. In other countries it is legalized and protected, houses of prostitution being licensed and inspected. In the United States it is generally condemned as being morally degrading and sinful, a source of crime and corruption, and an important factor in the spread of venereal disease. Because it involves the commercial exploitation of sex, it is considered undesirable and most states have laws against its practice.

Female prostitutes, also called hookers, hustlers, harlots, or whores are of three general types: *bar girls* or *streetwalkers,* who move about soliciting customers in public places; *call girls,* who initiate contacts with customers or are contacted by telephone; and *brothel prostitutes,* those whose activities are restricted to a brothel under the administration of a "madam."

Other types of prostitutes include *dance hall* "hostesses," *fleabags* (those who frequent skid rows), and *camp followers,* prostitutes in the vicinity of military, naval, and air bases.

Prostitutes usually will perform any type of sexual activity their "trick" or "john" desires. In intercourse they usually do not experience an orgasm with their client, but as men usually expect them to, they may play the role of a passionate woman experiencing sexual ecstasy. Because orgasms usually indicate emotional involvement, the prostitute reserves this for her procurer or pimp. The pimp, who lives off of her activities, may serve as companion, confident, manager, and protector.

Prostitution is not limited to heterosexuals but occurs among homosexuals and lesbians as well, it being estimated that the number of homosexual prostitutes· exceeds the number of female prostitutes. Male hustlers exist who seek customers in various locations, such as street corners, hotel lobbies, theaters, bars (especially gay bars), or in any place where prospective customers might be encountered. Most are young and engage in prostitution as an easy way of making money. Customers are men of all ages but most are over 40.

Most contacts are one-incident affairs or one-night arrangements but most hustlers seek to develop more permanent relationships in which they are provided living expenses. They serve as "kept boys" often accompanying their patrons on long trips. Public baths, massage parlors, and health clubs are places that provide ready service for homosexuals. Female impersonators, often transsexuals who work night clubs, may supplement their income by engaging in various homosexual practices. These include fellatio, anal intercourse, masturbation, and sometimes sadomasochistic activities. There are some male houses of prostitution but they are less common than female brothels. Call boys are available in some of the larger cities.

Lesbian brothels are only infrequently encountered. They are frequented principally by active lesbians for whom submissive girls are provided. Demonstrations of lesbian love are often presented. Sometimes female heterosexual prostitutes may provide lesbian services.

# THIRTEEN
# VENEREAL DISEASES AND VARIOUS PATHOLOGICAL CONDITIONS INVOLVING THE SEX ORGANS

## VENEREAL DISEASES—SEXUALLY TRANSMITTED DISEASES (STD)*

Associated with sexual activity are venereal diseases (VD) so-called because they are spread primarily through sexual intercourse or other intimate sexual contacts, either homosexual or heterosexual. The principal sexually transmitted diseases, listed in order of their frequency of occurrence, are chlamydias, trichomoniasis, gonorrhea, venereal warts, genital herpes, syphilis, and acquired immune deficiency syndrome (AIDS). Venereal diseases constitute one of the major health problems in the United States today because of their alarming rate of increase and their effects on individuals of all ages and both sexes. Their effects are especially serious among women because they are a common cause of sterility, stillbirths, and the birth of diseased babies.

Following the introduction and use of penicillin as an effective treatment after World War II, it was predicted the near total control of gonorrhea and syphilis, the two principal diseases, would be achieved, since the incidence of both diseases declined to a low point in 1957. However, since this time, there has been a steady increase in the number of cases reported annually to the extent that the incidence has reached epidemic proportions. The United States Centers for Disease Control (CDC) estimated in 1985 that the number of cases of sexually transmitted diseases (STD's) in 1986 would be approximately as follows: AIDS, 15,000; syphilis, 90,000; genital herpes, 500,000; venereal warts, 1,000,000; gonorrhea, 1,800,000; trichomoniasis, 3,000,000; chlamydias, 3,000,000; other STD's, 2,450,000. The predicted total exceeded the total of all other communicable diseases reported. Pelvic inflammatory disease, a serious complication of STD's, occurs frequently.

*See also the following sections in the Preface to the Dover Edition: "Acquired Immune Deficiency Syndrome (AIDS)"; "Chlamydial Infections"; "Viral Hepatitis"; and "Molluscum Contagiosum" (pp. ix–xi).

There are several causes for the marked increase in venereal disease. Of importance are changes in sex habits, especially an increase in adolescent sexual relations, homosexuality, and oral sex; increased use of the Pill and IUD's for contraception with corresponding decrease in use of the condom and chemical contraceptives; increased resistance of the gonococcal organism to antibiotics and a greater incidence of asymptomatic cases; and decreased public health venereal disease control programs and inadequate sex education programs.

The control of venereal diseases depends on the treatment of all persons having the disease and all persons who have had contact with an infected person. Treatment of the latter prevents the disease from developing and prevents their serving as possible transmitters of the disease to others. In this way the chain of transmission from one person to another is broken. The existence of highly resistant strains of the disease and the high number of asymptomatic cases especially in males have greatly aggravated the problem of treatment. Because of social and moral implications of the diseases, there is a reluctance in seeking treatment, especially reporting contacts that can be traced and brought in for treatment. Attempted self-medication with delay in diagnosis and treatment are common.

Widespread educational campaigns are now being conducted by educational leaders, public health authorities, physicians, and others in schools, through the news media, and through public forums and lectures to inform the public and especially school children on the subject. However moralistic attitudes lead to difficulties in getting information to those who need it most.

Significant changes are taking place in manifestations of the diseases and the nature of their transmission. Of importance is the increase in extragenital infections especially those involving the mouth, throat, and rectum. Furthermore, homosexuals have been found to be an important source of VD infections, the transmission of gonorrhea and syphilis by homosexuals having increased significantly in recent years. This is related to the extreme promiscuity of some homosexuals and the fact that many are bisexual, thus contributing to the spread of the diseases. The increased indulgence in oral-genital contacts is also a contributing factor.

A few general facts about gonorrhea and syphilis should be noted. They are both serious diseases but they can be cured if treated early. If they are untreated, serious effects including sterility, invalidism, and possibly death may result. They may be acquired singly, or together, and may be acquired repeatedly since the body does not develop an immunity to either of them. Self-treatment should never be attempted. VD treatment should be supervised by a physician and all treatment is kept confidential. Birth control methods do not protect against VD with the exception of the use of a condom. A person may have VD and be able to transmit it without showing obvious symptoms. This is especially true of females in whom the signs are internal and not readily seen. However it is also true of males as a considerable percentage of "healthy" males have been found to be capable of transmitting virulent gonorrhea organisms.

## Gonorrhea

Gonorrhea, commonly called *clap,* is an infectious disease involving, primarily, the mucous membranes of the genitourinary tract (the urethra and adjacent structures) and sometimes parts of the gastrointestinal tract. It is caused by a pus-forming bacterium, *Neisseria gonorrhoeae,* which induces inflammation in these structures. Gonorrhea is transmitted principally through sexual intercourse although oral-genital contacts are increasing in importance as a method of transmission. Nonvenereal infection is possible in female infants and children below the age of puberty.

Symptoms usually appear two to eight days after infection although the period may be longer. In the male, these consist of a burning sensation on urination and the appearance of a thick, yellowish discharge. In the female, infection of the urethra, uterus, and vestibular glands may occur. The vagina is usually not infected, except in children. In many cases, symptoms or signs of acute infection are absent in both males and females. This asymptomatic condition is referred to as "silent" or hidden gonorrhea.

Immediate treatment is of the utmost importance because, if the infection proceeds unchecked, it may spread to other organs with serious consequences. Inflammation of the bladder and uterine tubes may occur. In females, the body cavity may become infected via the uterine tubes and peritonitis result. The infection may spread throughout the body and involve joint cavities causing arthritis or the eyes causing blindness. Sterility is common, resulting from blockage of the uterine tubes in the female or blockage of reproductive ducts in the male.

An infected woman who gives birth to a baby may transmit the organisms to a newborn baby. Eye infections are commonly the result, ending in blindness. This can be prevented by treating the baby's eyes immediately after birth with silver nitrate solution (1 percent) or an antibiotic.

In males, if treatment is delayed or inadequate, the infection may spread from the urethra to the prostate gland and seminal vesicles. The vas deferens and epididymis may also become involved with resulting sterility. Systemic involvement resulting in arthritis is common.

Gonorrhea can be readily cured by the use of antibiotics; however, treatment should be under medical supervision. If a person has been exposed to infection or if he or she suspects that a venereal disease has been acquired, he or she should go immediately to a doctor, health clinic, or VD clinic. This is important so that early treatment can be initiated, the source of the infection ascertained, and the spread of the disease limited.

Prompt treatment also prevents the development of complications that can lead to disability and possible sterility.


*Syphilis*
Syphilis is caused by a spirochete, *Treponema pallidum,* which is acquired by contact, usually sexual, with a person in an infectious stage. It occurs in three stages, the *first or primary stage* occurring as a small, dry, hard sore or ulcer (a *chancre*). This primary lesion usually develops within three weeks after infection but the period may vary from a few days to three months. This sore or lesion is usually painless and occurs where the germ enters the body, usually on or near the genital organs but it may occur elsewhere as on the lip, tongue, or breast. In the female it may occur within the vagina and not be readily noticeable. This sore is highly infectious, since it contains the causative organisms that can be identified by darkfield microscopic examination. The sore or chancre (pronounced shang' ker) disappears after a time leading one to believe that the infection is over but such is not the case.

Sooner or later (six weeks to six months or longer) after the primary infection, the *secondary stage,* in which the organisms have spread throughout the body, manifests itself. The spirochetes may infect any organ but the skin and mucous membranes are the tissues most frequently involved. Skin disorders (rash), pharyngitis (sore throat), white patches within the mouth, eye and liver involvement, and loss of hair with partial balding may result. Small moist lesions may appear almost anywhere.

All of these conditions tend to be self-limited and disappear. However, during this stage the organisms are widely distributed throughout body and are readily transmissible. This stage is considered more infectious than the primary stage because the lesions are more numerous and widespread. Because the skin symptoms resemble those of many other diseases, syphilis has been called the "great imitator."

If syphilis in its first two stages has not been treated and cured, the organisms may lie dormant for many years, producing no obvious symptoms. This comprises *latent syphilis,* which is diagnosed only by serologic tests. However the organisms continue to multiply and spread and are capable of infecting any organ of the body.

In the *third* and *final stage,* the lesions are chronic and destructive. The nervous system, heart and blood vessels, or any organ of the body may become infected. Gummas (gelatinous, rubberlike masses) may develop in the skin, bone, or visceral organs. The central nervous system may be involved with resulting locomotor ataxia or general paresis (paralysis of the insane). Syphilis in a pregnant woman is readily transmitted to a developing fetus and is a common cause of abortion, stillbirth, and death in early infancy. A newborn baby is readily infected (congenital syphilis).

Syphilis can be readily diagnosed by serologic tests. This is one of the reasons for premarital blood tests. It is also diagnosed by detection of the causative organism in syphilitic lesions. Syphilis responds readily to treatment by antibiotics, penicillin and tetracyclines being the drugs of choice. Treatment should always be under medical supervision; self-treatment and the use of quack remedies are to be avoided.

### Chancroid

Chancroid or soft chancre is worldwide in distribution since man is the only natural host for its causative agent, Hemophilus ducreyi, a bacterium. This infection is most prevalent in the West Indies, Africa, and Southeast Asia, and in overcrowded cities and seaports. In the United States, most of the cases—about 1500 annually—are seen in the Southeast.

The disease begins with a chancre, at the point of infection, that resembles the chancre of syphilis. After an incubation period of two to seven days, the chancre appears as a small, pus-filled, necrotic mass. In about a week's time, it develops into a soft, swollen, and painful ulcer (unlike the chancre of syphilis) from which the synonym "soft chancre" is derived. In untreated cases, the infection spreads to local lymph nodes especially those of the groin, causing severe abscesses at the site of the nodes.

Mixed infections are rather common with either syphilis, lymphogranuloma venereum, or granuloma inguinale. Immunity to the disease does not occur and the period of infectivity is not precisely known. Several drugs are effective against Hemophilus ducreyi but the agents of choice are sulfonamides.

### Granuloma Inguinale

This seems to be the least common of the venereal diseases only about 300 to 600 cases per year being reported in the United States. It is more common in the tropics and subtropics than in temperate climates. In the United States, it occurs about seven times more frequently in blacks than in whites, a condition thought to be associated with low socioeconomic groups. The causative agent is a microorganism, Donovania granulomatis, which causes a chronic, progressive, ulcerative disease. The moist ulcers tend to spread covering large areas of skin. They heal slowly and have a tendency, on healing, to scar formation, unlike LGV. If untreated, the disease may spread covering large areas of the body and may eventually become systemic with complications as arthritis and bone infections.

Treatment consists of the use of antibiotics as the tetracyclines, streptomycin, and erythromycin. No immunity is developed to granuloma inguinale.

*Lymphogranuloma Venereum (LGV)*
This disease is usually acquired, although not exlusively so, through sexual contact with an infected person. The disease is global in distribution, occurring mostly in tropical and subtropical areas. In the United States, most cases, which total 500 to 1000 a year, occur in the Southeast. The causative organism belongs to the bedsonia group, intermediate between viruses and bacteria.

The disease progresses through three stages. The primary lesion, which comprises the first stage, develops three days to two weeks after contact with an infected person. It consists of a small blister which bursts leaving a shallow, grayish ulcer surrounded by reddened skin. Although it appears painful, it is not.

The second stage of the disease, in which the organisms are spread through the lymphatics, occurs about two weeks after the appearance of the primary lesion. Characteristic lesions are enlarged inguinal lymph nodes (buboes) that break through the skin discharging a foul-smelling pus. Fever, chills, sweating, nausea, headache, and muscular and joint aches and pains are characteristic. In some individuals, even though untreated, LGV will regress. In others the remission is temporary, symptoms returning in the form of chronic, ulcerative lesions about the genitals and rectum. This comprises the third stage. Complications may develop leading to possible death from sepsis. In some cases, elephantiasis (grotesque swellings of the vulva, penis, or scrotum) may occur.

Immunity to the disease is not developed. Preferred treatment is the use of sulfonamides; in stubborn cases the tetracyclines are usually employed. Positive diagnosis of the disease is accomplished by use of the Frei skin test. Prolonged treatment is often necessary.

*Genital Herpes Simplex Virus Infection*
The causative agent of this disease is a virus of the *herpes group,* a group of DNA viruses that are the etiological agents of a number of animal and human diseases. A characteristic of these viruses is their tendency to establish latent infections, that is, after the original infection the disorder tends to be recurrent occurring repeatedly in the ensuing years. Among the human viruses of this group are herpes simplex (*Herpesvirus hominis*), including labial and genital types, *Herpesvirus varicella* (chicken pox), herpes zoster (shingles), Epstein-Barr virus (EBV), and cytomegalovirus (CMV). The Epstein-Barr virus has been implicated in infectious mononucleosis and a type of cancer (Burkitt's lymphoma) found in central Africa. Two of these viruses, herpes simplex and cytomegalovirus, have been identified as venereally transmitted diseases.

*Genital Herpes, or Herpes Simplex Virus, Type 2 (HSV-2)*

This disease is related to herpes simplex virus, Type 1, which is characterized by the formation of fluid-filled vesicles, commonly called *cold sores* or *fever blisters,* on the skin, lips, or mucous membranes and almost always occurring above the waist. The HSV-2 condition is characterized by similar symptoms but the lesions usually occur below the waist and affect the sex organs. Although herpes virus, Type 1, usually infects facial sites and herpes virus, Type 2, generally infects the genital organs, both viruses can attack either area and both types can move from one affected part of the body to another.

The disease is characterized in the male by the development of painful, fluid-filled blisters on the penis, scrotum, and skin of the thighs; in females, the vesicles are usually internal occurring on the lining of the vagina and on the surface of the cervix of the uterus. They may also occur on the labia, in the pubic region, and on inner surfaces of the thighs. The symptoms are generally of a local nature but sometimes systemic conditions involving the entire body as fever, enlargement of lymph nodes, and general bodily discomfort may result.

The causative agent is a virus usually transmitted by sexual intercourse. However it can be transmitted by oral and anal sex relations and by nonvenereal means. Although the condition is not new, it has, in recent years, increased in frequency and has now reached epidemic proportions, the number of cases now being estimated at over 500,000 per year. This places HSV-2 second in rank among venereal diseases, the number of cases outnumbering reported cases of syphilis and approaching first-place gonorrhea.

HSV-2 or genital herpes is usually a self-limited disease that, with or without treatment, disappears within a short time (one or a few weeks). The vesicles dry up and disappear and the infection appears to be cured. However the virus remains within the body and at any time it can begin to multiply and bring about a recurrence of the symptoms. Several attacks a year are common. Stress of any nature, physical or emotional, is usually a precipitating factor.

The importance of HSV-2 lies in the following aspects of the disease. There is no cure and the body fails, in most cases, to develop an immunity to it. As a consequence, the disease tends to recur and a person once infected remains a constant source of infection. Also a baby born of a woman who has an active infection is almost certain to acquire the infection usually with serious consequences. Death or the development of serious birth defects are commonly the result. The risk is so great that Cesarean section is recommended for women with active HSV-2 in late pregnancy. Of greater significance is the fact that among women who have the disease in which the cervix of the uterus is infected, the incidence of cancer of the cervix developing is possibly eight times higher than in noninfected women. It is also postulated that the infection in males may be related to cancer of the prostate.

New and improved diagnostic methods that include a skin test and a blood test to reveal the presence of HSV-2 antibodies have been recently developed. Also diagnosis can be made by examination of the material obtained and utilized in the Pap test. This is of value in alerting women to the possibility of increased susceptibility to cancer so that precautionary steps can be taken through more frequent examinations for the early detection of any abnormal growth.

At the present, as with most virus-caused diseases, there is no specific cure. Steps are being taken to develop a protective vaccine or therapeutic agent but to date, no effective drug has been developed in this country. Prophylactic or preventative measures necessitate the use of a condom by the male during sexual intercourse. Persons with an active infection should abstain from intercourse.

### Cytomegalovirus (CMV)

Cytomegalovirus, a salivary gland virus, requires close contact for transmission. It is presumed that CMV is transmitted orally through saliva or venerally through sexual contact. In the United States, a minimum of one per cent of all newborn babies have congenital CMV infection. The infection may be acquired early in pregnancy through transplacental transmission or it may be acquired directly during delivery. Typically, a baby manifesting congenital CMV infection is of low birth weight and has abnormal liver function and brain damage. The majority of infected infants are asymptomatic but about 10 percent will exhibit central nervous system involvement. CMV is the most common known viral cause of mental retardation and motor impairment. By adulthood, the majority of individuals will have acquired a CMV infection but most postnatal infections are asymptomatic.

### Genital Mycoplasmas

The mycoplasmas are microorganisms intermediate between bacteria and viruses combining the characteristics of both. They are the size of a large virus. Three strains of mycoplasmas have been isolated from human genital tracts. The T-strain mycoplasmas are found more frequently in the sexually active and are thought to be sexually transmitted. They have been linked to nongonococcal urethritis in men and reproductive failure in women. Controversy is widespread among medical men as to the role of T mycoplasmas in infertility, spontaneous abortion, premature births, and low birth weight in babies.

## OTHER SEXUALLY COMMUNICABLE DISEASES INVOLVING THE SEX ORGANS

Sometimes abnormal symptoms such as itching, a burning sensation during urination, the presence of an excessive discharge, or an excessively dry vagina may occur. These symptoms do not necessarily mean that a person has gonorrhea or syphilis, since there are a number of other conditions that may be responsible for these or other symptoms. The following are some conditions that might be encountered.

### Trichomoniasis

This is a condition resulting from infection by *Trichomonas vaginalis* (TV), a one-celled animal organism that lives in the vagina. It causes itching and irritation in the genital region and is associated with a characteristic cream-like discharge (*leukorrhea*). Occasionally the organism infects the urethra of the male. As a result, the male plays an important role as a reservoir of infection.

### Candidiasis (Moniliasis)

This is a condition resulting from infection by *Candida albicans,* a yeast organism. These organisms are normally held in check by the normal bacterial flora of the vagina but when this is altered as by an antibiotic treatment for another condition, or use of the Pill, excessive growth may occur causing itching and irritation accompanied by excessive discharge.

Both of the above conditions respond to chemotherapy. Diagnosis and treatment should be handled by a physician. Self-diagnosis and self-treatment should be avoided.

### Leukorrhea

This female disorder, commonly called "whites," is characterized by an abnormal, whitish discharge from the genital tract. It can be caused by a number of conditions. Infections by bacteria, protozoa, or fungi, including the two previously listed, are common direct causes. Other direct or predisposing causes include infection by parasitic worms, endocrine disturbances, lacerations such as those following childbirth, cervical inflammation, postmenopausal atrophy, hypersecretion, uncleanliness as that from soiled undergarments, irritation by chemicals or foreign objects, and other causes. Genital deodorant sprays may cause severe irritation of the vulva and vagina. They are designed for *external use only* and under no conditions should they be sprayed into the vagina.

*Crabs or Crab Lice*

These organisms, *Phthirus pubis,* inhabit the pubic region but may be found in other parts of the body where coarse hairs grow as in the beard, eyebrows, eyelashes, and armpits. They are sucking insects that feed on blood, their bites causing severe itching. They reproduce, each female producing about 25 eggs that grow to adults in about three weeks. They are acquired by contact with an infested person, thus they may be acquired during sexual intercourse. They may also be picked up from contaminated clothing, bedclothes, and possibly toilet seats. Gymnasiums are reservoirs of infection.

For treatment, consult a physician or health service. Avoid infection by scrupulous cleanliness. Bathe frequently; wear clean clothing (your own) and do not borrow from others.


*Scabies*

Scabies or the "itch" is a transmissible, parasitic infection caused by a minute arachnid, *Sarcoptes scabiei,* which infests man. Adult mites penetrate the skin forming sinuous burrows in the epidermis. Eggs deposited in the burrows hatch in three or four days and the larvae excavate new burrows in which they develop into adults, the life cycle being completed in 8 to 14 days. The infection, once acquired, usually persists indefinitely unless the condition is treated and all infectious organisms destroyed.

Following infection, the skin becomes sensitized and the activities of mites and their secretions cause an intense itching or pruritus. Common regions of infestation are the genital organs of males, the groin and buttocks, between the digits, on the flexor surfaces of the wrists, elbows, and knees, in the axilla, and on or beneath the breasts of females. Scratching may result in bleeding and the formation of scabs frequently followed by secondary bacterial infection.

The mites are commonly acquired through direct contact with an infected person or with their clothing or bedding. The occurrence of scabies transmitted venerally is becoming increasingly common consequently any persistent skin lesions accompanied by itching should be checked by a physician. Positive diagnosis is made by finding the parasites or their eggs in skin scrapings. The spread of the infection from one part of the body to another may occur through scratching and manual transfer of the mites. Treatment consists of the application of a scabicide (benzyl benzoate or sulfur ointment) to the infected parts and thorough cleaning of all clothing and bedding. Contact with possible infected persons should be avoided.

*Venereal Warts (Condylomata Acuminata, Moist Warts, or Genital Warts)*
These are soft, pink growths occurring singly or in grapelike clusters in the moist areas about the genitalia, rectum, or oral cavity. In the female, they are seen on the vulva and are found on the vaginal wall and cervix; in the male, they occur about the anus and on the penis beneath the prepuce. They may occur as single, minute bodies or large cauliflower-shaped masses. The warts are caused by a virus commonly transmitted by sexual contact. They are definitely not associated with gonorrhea or syphilis. They respond to therapy (chemotherapy, electrodesiccation, surgical excision) and in many cases undergo spontaneous regression and disappear.

## VARIOUS PATHOLOGICAL CONDITIONS INVOLVING THE SEX ORGANS

A number of other pathological conditions exist that may involve the various organs of reproduction. Among these are cancer, endometriosis, ovarian cysts, Peyronie's disease, and retrograde ejaculation. These conditions are discussed in the following pages.

### Carcinoma (Cancer)
Cancer is a disease of the cell. The initial event that leads to the development of cancer is an abnormal change in one or a few living cells. From an unknown cause, a single cell loses its ability to limit growth and multiplication. As a result, cells begin to multiply and a mass of cells develops. Such a mass is called a *tumor* or *neoplasm*. If the cells of a tumor remain localized and do not spread, it is called a *benign tumor* and is generally harmless. If the cells break loose from the original site and spread through lymph and blood vessels to other parts of the body, a process called *metastasis,* new growths develop and sooner or later a vital organ will be affected and death usually results. A tumor of this type is a *malignant tumor* or *cancer.* If the tumor consists of epithelial cells, it is called a *carcinoma*; if it consists of muscle or connective tissue cells, it is a *sarcoma*; if of fat cells, a *lipoma.* If it involves blood-forming tissue, it is called *leukemia.*

Cancer is a primary cause of death in the United States, ranking second to diseases of the heart. It will affect one of each four persons reaching the age of 70 and will be the cause of death of one of each six. Some cancers grow slowly; others spread rapidly. If detected early, many cancers can be cured or prevented from spreading. Methods of treatment employed are: *radiotherapy,* or the use of X-rays, radium, or other ionizing radiation; *chemotherapy* or the use of drugs, various chemicals, or hormones; and *surgery.*

Common cancers in both sexes involve organs of, or related to, the

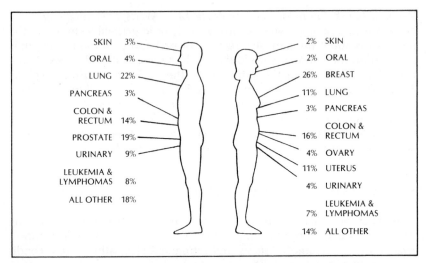

**Figure 35  Estimated cancer incidence by site and sex, 1986 (excluding nonmelanoma skin cancer and carcinoma in situ). (American Cancer Society.)**

reproductive system. In the female these are cancer of the breast and uterus and in the male, cancer of the prostate. The incidence of cancer in the various organs and tissues of the body is shown in Fig. 35.

*Cancer of the Breast*
Breast cancer is the second most common type of cancer in women, surpassed only by lung cancer. In 1985 approximately 119,000 new cases of breast cancer were diagnosed. About one out of 11 women will develop breast cancer at some time during her life. In 1985 about 39,000 women died of breast cancer. However when breast cancer is diagnosed early and treated promptly, survival rate is 85 percent or higher. Breast cancer can occur in a woman of any age but it is most common in women over 35 and especially common in women over 55. Breast cancer appears to run in families, where the daughters and sisters of breast cancer patients are two to three times more likely to develop cancer than women not related to a breast cancer patient. Breast cancer is also more likely to develop in women who bore their first child after the age of 25. The risk is also higher in women who begin to menstruate early or who have a late menopause. No relationship has been shown to exist between the taking of oral contraceptive pills and breast cancer, nor is there any relationship between the size of the breast (whether large or small) and the possibility of developing cancer.

There are a number of misconceptions about the causes of breast cancer. Although the actual cause or causes are unknown, it is fairly certain that the following factors or conditions *do not cause* cancer of the breast. Important ones are (1) sexual relations, (2) injury as from blows, (3) having babies or nursing babies, (4) infectious germs introduced through the nipple, (5) close association or contact with a cancer patient, (6) injections of or taking female hormones especially estrogen, and (7) taking oral contraceptives (the Pill).

Although cancer of the breast can be fatal, with early diagnosis and treatment prognosis is good. Every woman should have a physical examination *at least once a year* in which the breasts and the uterus are examined. Breast examination is by (1) *palpation* in which the breasts are felt by the examiner for any lumps (cancer starts as a painless lump), (2) by *mammography* or X-ray examination, and (3) by *thermography* or the use of heat-ray detectors in which a picture taken can reveal "hot spots" that may be indicative of cancer.

As most breast cancers are discovered by women themselves who note the presence of a lump in the breast, women are being encouraged to learn the procedure of *breast self-examination* (BSE) and to practice it regularly. The American Cancer Society has established a number of Cancer Detection Projects-Screening Centers throughout the country at which BSE is taught. They also publish and provide, without charge, instruction pamphlets. These are available from their units, which are established in nearly every community.

*Technique of Breast Self-Examination*
The breasts should be examined regularly *once a month* about a week after the menstrual period when the breasts are not enlarged and tender (Fig. 36). After the menopause, they should be checked on the first day of each month. The examination is performed in three steps or stages as follows.

*Step 1.*    Examine the breasts during a bath or shower. With fingers flat, rub the hand gently over each moistened breast. The arm on the side of the breast examined should be held with the hand behind the head. Check for any lumps, knots, or thickened tissues.

*Step 2.*    Examine breasts while standing or sitting before a mirror. With arms at your sides, inspect your breasts looking for any changes in shape, size, or contour. Check for dimpling of the skin, or any change in the nipple especially retraction. Gently squeeze each nipple and note if any discharge is forthcoming. Next raise your arms high overhead and look for the same things as when arms were pendant.

*Step 3.*    Examine breasts while lying down. Place a folded towel or a small pillow under your left shoulder and your left arm under your head. With your fingers of the right hand held together and flat, press gently

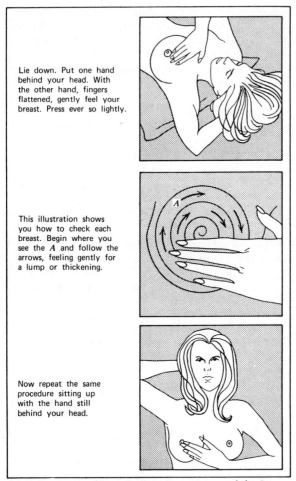

Lie down. Put one hand behind your head. With the other hand, fingers flattened, gently feel your breast. Press ever so lightly.

This illustration shows you how to check each breast. Begin where you see the *A* and follow the arrows, feeling gently for a lump or thickening.

Now repeat the same procedure sitting up with the hand still behind your head.

**Figure 36   Breast self-examination. (From** *Cancer of the Breast,* **1975, American Cancer Society.)**

against the left breast and apply a circular motion moving clockwise. Start at the outermost top portion of the breast and make several circling movements moving inwardly about an inch for each succeeding movement. Be sure every part of the breast is examined. Do not be alarmed if numerous masses of tissue are felt. Normal breast tissue is of a lumpy nature. If a suspected abnormal lump is found, check the opposite breast for a similar lump. If such is present it is probably normal tissue.

After the left breast is checked, examine the right breast with a pillow or folded towel under the right shoulder, using the same procedure as for the left breast.

If any abnormal structure in the nature of a lump or if any discharge is noted, see your physician, who with the assistance of a pathologist, can determine whether a growth is cancerous or not. Any discharge whether clear or bloody should be reported to a physician immediately.

Millions of women were made aware of the importance of routine breast examinations when it was found that both President Ford's and Vice-President Rockefeller's wives were victims of breast cancer. Their cancerous conditions were an unfortunate way of calling the public's attention to a condition most women are reluctant to talk about—mastectomy, or removal of the breast. Their involvement emphasizes the fact that cancer is no respecter of persons, that it affects the famous or the unknown, the rich or the poor, the old or the young. Many public figures have had cancer and are alive today cured of the disease. The factor of greatest importance is early diagnosis and prompt treatment.

### Cancer of the Uterus

Cancer of the uterus is the second most common type of cancer occurring in women. It may arise in two different locations. It may begin in the lower portion of the uterus called the *cervix* or *neck* or in the upper portion, the *corpus* or *body*.

*Cervical cancer* occurs principally in women who are sexually active at an early age, in women who have borne many children, and in women whose marital partners are uncircumcised males. Its incidence is also increasing in women who are infected with herpes simplex virus, Type 2, the genital type.

Cervical cancer begins by multiplication of cells in the epithelial lining at the junction of the cervix and the vagina or within the cervical canal. Growth at first is usually slow with no symptoms. After a considerable period of growth, the cells begin to penetrate the layers of the wall of the cervix and a flat, nodular, and often ulcerated lesion appears. Intermenstrual bleeding and often a vaginal discharge occurs. These are warning signals. Hysterectomy usually is an effective cure during this stage.

If the growth persists, its spread is fairly rapid. Cells penetrate into the muscle layers and soon involve the entire surface of the cervix and upper portion of the vagina. Bleeding, pain, and a vaginal discharge are characteristic. If allowed to continue, cells usually spread to the bladder and rectum and to adjoining tissues. They may enter the lymphatics and invade the pelvic lymph nodes and may pass by way of the blood stream to various organs of the body.

Fortunately there is a simple, painless, and inexpensive test by which cancer can be detected in its early, curable stages even before any symptoms have appeared. This is known as the *Pap test,* named after the late Dr. George N. Papanicolaou who originated the test in 1942. It consists of scraping lightly the surface of the cervix and the upper end of the vagina with a spatula and placing the material on a microscopic slide. The material is then prepared for microscopic examination by fixing and staining. Examination of the slide by a skilled pathologist will reveal the presence of any malignant cells. This test, followed by a biopsy that provides an accurate method of early detection of a malignancy, has greatly increased the cure rate for this type of cancer.

*Cancer of the body of the uterus* occurs more commonly in older women and the first symptom is irregular, vaginal bleeding. Women who have borne no children are commonly affected, as well as obese women and diabetics.

Treatment of uterine cancer is by surgery or radiation, or both. The treatment will depend on the size of the tumor, its location, the extent to which it has spread, and the general physical condition of the patient. Surgery involves the removal of involved organs. Radiation involves the use of X-rays or a cobalt therapy machine or the implanting of a capsule of radium through the vagina in the region of the cancerous tissue.

The incidence of cancer as a cause of death could be significantly reduced if every adult woman would have an annual checkup including a Pap test and if women, especially women approaching the menopause and postmenopausal women, would check with their physician on the appearance of any abnormal vaginal discharge or abnormal bleeding.

*Cancer of the Prostate Gland*

The prostate gland is an accessory male reproductive organ that lies just below the urinary bladder. Through it passes the first portion of the urethra, the tube from the bladder and the ejaculatory ducts that open into the urethra. Several short ducts that carry the prostatic secretion open into the urethra. The prostatic secretion is an important component of the semen.

Cancer of the prostate is the second most common form of cancer in men, exceeded only by lung cancer. Tumors of the prostate are very common but most are not cancers. The most common tumor is called *benign prostatic hypertrophy* or in everyday language, enlargement of the prostate, a condition that afflicts over half of the men in the United States over 50.

Cancer of the prostate consists of cells undergoing unrestricted growth. They grow into the prostatic tissue disrupting the normal structure and depriving normal cells of nourishment. They may spread into adjacent tissues or, through metastasis, spread throughout the body.

Enlargement of the prostate, whether caused by benign hypertrophy or prostatic cancer is usually accompanied by the following symptoms. First to be noted are disturbances in the flow of urine. Urgency, increased frequency, or smarting may occur. Urinary obstruction usually develops leading to inability to urinate or difficulty in beginning urination. Nocturia or desire to urinate frequently, especially at night, incomplete emptying of the bladder, and painful urination are characteristic. Blood may occur in the urine; bladder infections are common. Pain in the lower portion of back or in pelvic region may occur.

Every male over 40 should have a rectal prostate examination at least once a year. Direct digital examination will usually reveal an enlarged gland. If the condition is mild, palliative treatment to relieve congestion or infection may be effective. Massage may bring relief. If there is severe urinary obstruction, surgical excision is usually necessary. If an operation is performed, examination of the tissue (biopsy) will reveal if it is cancerous.

When cancer is present, immediate treatment should be instituted. If the cancer is limited to the prostate, complete removal of the prostate will prevent it from spreading to other tissues. If other tissues are involved, the cancer can be checked by suppressing the manufacture of male hormones (androgens). This can be accomplished by injection of female hormones especially estrogens or in extreme cases, castration or removal of the testes. Cortisone, which suppresses secretion of androgens by the adrenal cortex, is sometimes employed. In addition, the use of radium and X rays and the administration of radioactive gold or phosphorus is sometimes effective in destroying cancer cells and bringing relief from pain. Effective treatment depends on early diagnosis and prompt treatment. Yearly examinations are 'essential.

### Cancer of the Penis and Testicles

These two cancers are relatively rare in the United States. The most common neoplasm of the penis is a skin cancer occurring on the glans penis or the prepuce. Although its cause is unknown, retained smegma, which is known to be carcinogenic, repeated irritation, infection, or trauma are thought to be contributing factors. Cancer of the penis is almost unknown in circumcised males, consequently it is thought that circumcision is an important preventive of penile cancer. The retention of smegma beneath the prepuce for long periods is to be avoided. In uncircumcised males, strict habits of cleanliness involving retraction of the prepuce and frequent removal of smegma are essential.

### Cancer of the Testes

Although cancer of the testes is rare causing less than 1 percent of cancer deaths in men, its importance lies in the fact that its incidence is highest in men between the ages of 20 to 35. Neoplasms of the testes usually involve the sperm-forming tissue and are generally highly malignant. The cure rate for testicular cancer is high if it is discovered in its early stages.

Self-examination of the testes similar to breast examination for women is recommended as a routine practice. The initial tumor is usually a small, firm lump no larger than a pea usually located on the front or side of the testis. The testes should be carefully examined every two or three months. The best time for examination is immediately after a shower when the scrotal skin is relaxed. Feeling each testis gently between the fingers of both hands will generally reveal any abnormality.

*Endometriosis*
Endometriosis is the presence of endometrial tissue (tissue lining the uterus) in any abnormal location. Common sites where this tissue is found are on the surface of the ovaries, uterine ligaments, uterine tubes, uterus, and the rectovaginal septum. The tissue may be found on almost any surface within the pelvic cavity. This disorder is common in women between the ages of 20 and 40. It is of no significance after the menopause.

Two theories have been advanced as to the cause of endometriosis. One is the *retrograde menstruation theory* in which it is thought that during regular menstruation, contractions of the uterus force fragments of shed endometrium backwards through the uterine tubes into the abdominal cavity. The other theory postulates that *displaced embryonic cells* in early development come to occupy locations other than their normal location, the uterine lining. Endometriosis seems to occur more frequently in women of higher income groups, women who tend to marry late in life and have fewer children.

During each menstrual cycle, the endometrium of the uterus proliferates, becomes thicker, and, if pregnancy does not occur, is sloughed off and shed with some bleeding through the vagina at menstruation. This process in initiated and controlled by ovarian hormones. Misplaced endometrial tissue responds at the time of menstruation in the same way except that it is sloughed off into the pelvic cavity instead of being discharged to the outside. The accumulation of this material and often the development of adhesions may lead to severe abdominal pain or discomfort that occurs at the time of menstruation. Other clinical symptoms may include dysmenorrhea, dyspareunia, pain when defecating, and menstrual difficulties. Fibrous adhesions about the ovaries, uterine tubes, and uterus may distort these structures and are a common cause of infertility.

Treatment of endometriosis is by means of drugs or surgery. By the use of drugs, the menstrual cycle is suppressed for a period of six months or longer, thus reducing the number of painful episodes to one or two a year. For patients in which symptoms are severe, surgery may be necessary. This may involve not only the removal of endometrial tissue but removal of the ovaries and sometimes also the uterine tubes and uterus.

## Ovarian Cysts

A cyst is a closed, saclike structure that usually contains fluid or other material. Simple follicle or corpus luteum cysts are normal structures in the ovary. A follicle, in development, is a closed saclike structure that may become quite large. Normally it bursts and releases an ovum and the contained fluid. Within the follicular cavity, a corpus luteum, a mass of cells usually containing a small cavity, develops. The corpus luteum normally regresses and becomes a small white scar, the corpus albicans.

Sometimes, however, abnormal cystic structures develop that are actually neoplasms, most of which are benign. The line between normal follicles and corpora lutea and cystic follicles and corpora lutea is not clearly defined. Cystic structures are larger in general than normal structures.

Follicular cysts may originate from unruptured graafian follicles or from follicles that ovulate and immediately seal themselves retaining some of the follicular fluid. They tend to be multiple (polycystic) although they may exist as single cysts. They are generally located immediately beneath the ovarian capsule and may become quite large, two inches or more in diameter. Luteal cysts are less common, are often blood filled, and have a tendency to rupture. Neither type of cyst is serious, although both types are a common cause of abdominal pain. The cysts may fluctuate in size and many disappear spontaneously.

Bilateral polycystic ovaries are associated with anovulation, a characteristic of the *Stein-Leventhal syndrome*. Other symptoms associated with this syndrome are amenorrhea, obesity, hirsutism, and reduced fertility or sterility. This syndrome occurs commonly in girls in their late teens or early twenties.

The cause of the symptoms in the Stein-Leventhal syndrome is an enzymatic block in the conversion of androgens to estrogens in the ovaries. As a consequence, the pituitary produces increased amounts of FSH and LH in an effort to increase estrogen production. However a common effect is an increase in the level of androgens in the patient causing a considerable degree of masculinization, including hirsutism. An increase in the number and size of follicles in the ovaries also results.

Other types of cysts may develop in the ovaries. An unusual and not uncommon type is a *dermoid cyst,* a peculiar cystic structure that contains structures characteristic of the skin such as hair, nails, and oil glands. An explanation of this type of cyst is that it may represent an ovarian pregnancy or embryonic tissue may persist in the ovary, which is capable of responding to stimuli that induce development. Another theory is that a dermoid cyst is an internal twin in which development has been suppressed.

### Peyronie's Disease
This rare disease is characterized by the development of plaques of dense fibrous tissue in the fibrous covering of the corpora cavernosa, the two large masses of erectile tissue in the penis. The plaques may calcify causing the formation of bonelike material along the top side of or between the two corpora. Generally the condition is painless when the penis is flaccid, but as the disease progresses, the plaques may cause an arching or angular deformity of the penis. When this occurs, erections may be extremely painful and sexual intercourse difficult or impossible.

The cause of the disease is unknown. Its onset is gradual and in many cases spontaneous regression occurs. It occurs in men of all ages but is more common in men past 50. Various methods of treatment, most of questionable value, are employed including drugs, X rays, and surgery. Surgery usually results in interference to blood supply of the penis resulting in loss of erectile capacity.

### Retrograde Ejaculation
This is the condition in which the semen, on reaching the urethra within the prostate, is discharged backwardly into the bladder instead of passing out through the penile urethra. Externally only a drop or two may be present or no semen at all. Urination after ejaculation will yield a milky urine sample that will contain the semen of a retrograde ejaculation.

Retrograde ejaculation is the result of a malfunctioning of the internal sphincter muscle, a muscle surrounding the urethra at its junction with the bladder. Normally during ejaculation, this muscle contracts and prevents semen from entering the bladder. A disruption of the functioning of this muscle may result from injury as may occur during prostatic surgery, damage to the sympathetic nerves that control the sphincter, spinal cord injury involving centers of reflex control, or the use of certain drugs, as hypertensive agents, which interfere with sympathetic control.

It is now possible to retrieve sperm from the bladder in cases of retrograde ejaculation. The bladder is alkalinized, completely emptied of urine, and then ejaculation is accomplished manually. The bladder contents are then removed by voiding or the use of a catheter. The sperm are then concentrated by centrifugation and often motile sperm can be recovered. Healthy pregnancies resulting from artificial insemination using sperm recovered by this technique have been accomplished.

# FOURTEEN
# DRUGS AND SEX

One of the significant developments during the past two decades has been the spectacular increase in the use of drugs for social and recreational purposes. As a consequence the problem of drug use and abuse has become one of primary importance. In this section, the relationship of drugs to sex and sexual activity will be considered.

Specifically the term "drug" refers to any biologically active substance that is used for medicinal purposes or for its pleasurable effects. The effects of a drug depend on its nature, its purity, how much is consumed, its frequency of use, and the physiological nature of the person using it. *Drug abuse* refers to the chronic excessive use of a drug to the extent that health is impaired or social and vocational adjustments made difficult. Drugs that are primarily sought and used for pleasure fall in the category of mind-altering drugs. These include alcohol (the most widely used and abused), nicotine, marijuana, sedatives (barbiturates, tranquilizers), stimulants (caffeine, cocaine, amphetamines), psychedelics (LSD, DOM, mescaline), and narcotics (heroin, morphine, opium).

The effects of drugs on sex and sexual performance are extremely variable and the causal relationship is often difficult to determine. A drug that increases sexual desire and enhances sexual performance is called an *aphrodisiac*; one that reduces sexual activity, depresses the libido and delays or interfers with the orgasm is an *anaphrodisiac.* There are few drugs that specifically act in either way. Because a person's sexual activity depends primarily on his biological makeup, training, experiences, and beliefs, drugs generally play a secondary role in either increasing or decreasing it.

## Alcohol

This is the number one drug problem in the United States today as indicated by the number of alcoholics and problem drinkers estimated to number over nine million. Consequently it plays an important role in sexual activities, a role often misunderstood. It seems to act as an aphrodisiac stimulating sexual behavior all the way from petting to sexual intercourse. However, alcohol is not a stimulant. It is a depressant, exerting a sedative effect on the brain and slowing reflexes. When used moderately it does temporarily increase sexual responses but this is accomplished through the lowering or removal of inhibitions, reducing anxiety, diminishing guilt complexes, and overcoming fears concerning sexual performance.

When it is used to excess, it may have a marked inhibiting effect upon sexual performance. As Shakespeare in his play "Macbeth" said of alcohol, "it provokes and it unprovokes"; that is, it provokes the desire but it takes away the performance. In many males, too much alcohol is the cause of temporary impotence. This may cause an anxiety-ridden male to question his potency thus setting up the conditions that may give rise to long-lasting impotence. In females, alcohol is effective in releasing social inhibitions concerning sexual activities as evidenced by Ogden Nash's comment that "Candy is dandy but liquor is quicker." Indulging in alcohol is commonplace but under no circumstances should the use of alcohol become a necessary prerequisite for sexual intercourse.

*Nicotine*

One of the most commonly used drugs, *nicotine,* has been suspected of having adverse effects on sexual performance. Nicotine, along with coal tars, and other substances (e.g., carbon monoxide, present in tobacco smoke), have been related to the higher incidence of lung cancer and other respiratory disorders in smokers. Recent evidence points to it as a significant factor in impaired sexual performance.

In the male, heavy smoking is associated with low sperm count and low sperm motility, both of which are of significance in infertility. Also, the production of testosterone is reduced and this is possibly a factor in the decline in sexual capability commonly noted in heavy smokers. Although a cause-effect relationship between smoking and the above-mentioned conditions has not been scientifically established, the circumstantial evidence is so strong that in cases of reduced sex drive and infertility, one of the first recommendations usually made by the therapist is that smoking be discontinued.

In a pregnant woman, excessive smoking may have serious ill-effects on the development of a fetus. Mothers who smoke have a significantly higher number of miscarriages and stillborn babies and most babies born are below average in weight and are more susceptible to sickness and disease. The ill-effects of smoking usually are induced during the second half of pregnancy. Pregnant women are generally advised to refrain from smoking.

Sexual performance in both men and women nearly always improves following the termination of smoking. As nicotine acts to constrict blood vessels, blood flow to the sex organs is reduced thus limiting the response of erectile tissues to stimulation. Increased intake of carbon monoxide reduces the oxygen level in the blood that, with reduced lung capacity which is common in smokers, acts to impair general well-being and to reduce sexual response. Generally, improved health and improved sex lives almost invariably follow a reduction in smoking.

*Marijuana*
Reports pertaining to the effects of *marijuana* on sexual experiences are numerous and conflicting. By some it is claimed that marijuana reduces sexual interest and activity; others claim that it provokes sexual desire and leads to sexual excesses. It should be noted that a drug as marijuana affects different people in different ways. Although its use in the United States is illegal, it is widely used today. For some individuals, the excitement of using an illegal drug is enough to cause any activity in which they might engage whether it be eating, drinking, or sex to be a more exciting experience. Some of the effects reported are a marked feeling of relaxation; a heightened sensuousness in the perception of colors, music, and pictures; a more favorable sense of one's worth and increased sociability; increased sensory awareness; spatial and time distortion, far objects appearing near and a short time, as a minute, lasting indefinitely. Such would tend to heighten sexual pleasure causing the orgasm to appear prolonged.

There is no evidence that marijuana is a true sexual stimulant, an aphrodisiac, but its use does seem to make the sex act more pleasurable. Its effects can be attributed to the fact that it is a social drug, its use being learned in a social setting, and its utilization usually in a social setting. The reactions to the drug are the result of social expectations, beliefs, and experiences. In short, it is a self-fulfilling prophecy in which one becomes sexually aroused because one has been taught to expect this behavior. The effect is largely a placebo effect in which a drug produces in a patient the effects that the patient expects.

Dosage and the extent of use may alter the effects of marijuana on sexual relations. Moderate social users usually report positive effects; chronic users tend to have a diminished interest in sex. Recent studies have shown that in heavy users of marijuana, testosterone levels in men fell significantly and users reported a lower frequency of coitus and orgasm, reduced potency, and difficulties in control of ejaculation. In general, the effects of marijuana, like those of other mind-altering drugs, vary greatly depending on dosage, frequency of use, the setting, age of user, user expectations, and other factors.

*Barbiturates (Amytal, Luminal or Phenobarbital, Nembutal)*
These are hypnotics or sedatives, that depress the central nervous system. In moderate dosage, through reduction of inhibitions and relief from

anxiety and tension, they may apparently stimulate sexual feelings. In higher dosage, sedation or sleep is induced with the loss of sexual ability. Barbiturates, being the principal ingredients of sleeping pills, diminish sexual appetite as well as performance.

## Amphetamines
These include pep pills and diet pills, stimulants that have a pronounced effect on the central nervous system. They are used to overcome fatigue and listlessness and to promote alertness and wakefulness. A tolerance to their pleasurable effects tends to develop, necessitating increased dosage to achieve their effects. Prolonged use of amphetamines prevents orgasm in the male without impairing the ability to have an erection, thereby permitting prolonged intercourse without ejaculation. Chronic use usually leads to impotence that can be reversed following a period of abstinence from the drug. In the female, menstrual disorders and frigidity have been attributed to its chronic use but further research is needed to confirm the relationship.

## Methamphetamine (Methedrine or "Speed")
This is the most popular type of amphetamine associated with sexual use. The injection of "speed" into the veins induces an almost instantaneous onset of the drug's euphoric properties. The reaction, called a *flash* or a *rush* by "speed freaks" has been described as a "full body orgasm." The drug tends to delay and intensify orgasm in sexual intercourse. High dose, intravenous injection of methamphetamine tends to eliminate normal sexual inhibitions to the extent that group sex, bisexuality, and troilism become common behavior. Psychic dependence often develops leading to several days continuous use (a "run") during which the "speed freak" develops a sustained state of anxiety with extended fears and suspicions, which may result in a paranoidlike state with a total lack in sexual interest. A major problem in studying the effects of amphetamine abuse is the development of malnutrition and weight loss due to the appetite-depressant effect of the drug. These, accompanied by fatigue, markedly affect orgasmic response. Females using amphetamines generally report a reduction in vaginal secretions so that, in prolonged intercourse, dyspareunia develops.

## Cocaine
Another drug which has stimulating effects is *cocaine* or *"coke,"* which is either sniffed ("snorted") or injected intravenously ("mainlining"). An intravenous injection commonly results in spontaneous erections in males

sometimes resulting in priapism (long-term erection) which may last 24 hours or longer. Cocaine is sometimes applied directly to the glans of the penis where it acts as an anesthetic thereby prolonging intercourse without ejaculation. While this may lead to a pleasurable experience for both partners for a time, it frequently leaves the female with a raw and inflamed vagina which may necessitate medical attention.

### Opiates

The effects of opiates in the relief of pain and the induction of a state of euphoria have been known since ancient times. It had also been observed that opiates had a tendency to lower the addict's interest in sex. Two of the most powerful and widely used opiates are morphine and heroin, both addictive drugs.

In some aspects of sexual performance as premature ejaculation, heroin may have a favorable effect. The drug in certain cases enables a premature ejaculator to hold off ejaculation long enough for satisfactory intercourse with his partner. However, the heroin addict's sexual relationships most often dissolve into a nonsexual partnership based purely on the acquisition of "dope." Because of the high cost of maintaining the heroin habit, estimated at $50 to $100 a day, women often sell their services in prostitution while male addicts may become involved with homosexual prostitutes or, more commonly, pimping. Consequently, sex becomes a means to an end, that is, acquiring drugs, and not an end in itself.

### Methadone (Dolophine)

A synthetic, opiate analgesic, this drug is commonly used in maintenance therapy for opium addicts. The daily maintenance dosage of methadone has three effects: it prevents withdrawal symptoms; it reduces the desire for opiates; and it blocks the pleasurable effects of opiates. Physical dependence usually develops after continuous use with withdrawal symptoms being more prolonged but less pronounced than for various opiates. The effect of methadone on sexual performance in the male most often reported is occasional impotence. In the female, methadone, like heroin, alters menstruation causing amenorrhea and irregularity. Following withdrawal, regularity in menstrual cycles may be regained after a period of time.

### Psychedelic or Mind-Manifesting Drugs

These include LSD (lysergic acid diethylamide), mescaline, psilocybin, STP (DOM), and MDA. LSD has pronounced hallucinatory effects markedly increasing sensory awareness and distortions in the perception of time and space. As a consequence, an orgasmic experience may seem considerably longer than it actually is. However LSD is not an aphrodisiac and because of its extremely powerful mental effects, it tends to act more as an anaphrodisiac. *Mescaline*, obtained from the peyote cactus, and *psilocybin*, obtained from certain mushrooms, both produce effects similar to those produced by LSD but less pronounced. Alterations in consciousness and sensory distortions occur and sexual performance is generally diminished. *STP* often produces severe psychotic reactions and it has seldom been reported as producing satisfactory sexual experiences. *MDA*, an amphetamine-related psychedelic drug, has been reported as having aphrodisiac properties, that is, stimulating sex drive, but cases of impotence have also been reported by users of the drug.

### Amyl Nitrate and Other Drugs

This drug, an inhalant dilator, is used by physicians to bring about dilatation of the coronary blood vessels for relief in heart attacks. It is also used illicitly for its alleged sexual stimulation properties. Amyl nitrate is commonly used by men who break the inhaler or "popper" at the beginning of intercourse or just before orgasm. An increased awareness and intensification and prolongation of the orgasm are commonly reported as results. Women seem to be less interested in using amyl nitrate for orgasmic experience. The drug seems to have had its origin for sexual abuse in the homosexual culture but it is now frequently used by heterosexuals.

Other drugs which have been regarded as sexual stimulants include Ritalin, Preludin, L-dopa, and PCPA. *Ritalin* and *Preludin* are both stimulants of the central nervous system and are used to combat mild depression. Preludin is also an appetite suppressant. Neither has been shown to have a specific stimulating effect on sexual activity. *L-dopa* is used in the treatment of muscular tremors and depression in Parkinson's disease. It was noted that in elderly men, sexual activity seemed to be stimulated and a degree of sexual rejuvenation experienced. However, when tested on younger men, its effects were minimal and short lived. It was concluded that L-dopa did not have aphrodisiac properties but that its beneficial effects were principally the result of improvement in general well-being. *PCPA* (P-chlor-phenylalanine) acts as an aphrodisiac in several forms of lower animals but its effect in humans is still questionable.

### Anaphrodisiacs

Seldom do people desire an *anaphrodisiac,* a sexual depressant, however some tranquilizers, as Thorazine, have been reported as diminishing sexual desire. A compound with a greatly inflated reputation is potassium nitrate (saltpeter). It is a diuretic, greatly increasing urine output that may account for the misconceptions associated with its use in male dormitories and in the military. A recently discovered compound, *cyprosterone*, seems to be effective in allaying sexual desire. *Medroxyprogesterone acetate* has similar effects.

## Aphrodisiacs

Among the *aphrodisiacs* (sexual activators), cantharides or *Spanish fly* is the most notorious. This substance, prepared from the dried bodies of a beetle, *Cantharides vesicatoria*, when taken internally, is extremely toxic causing severe enteritis and irritation of the urinary tract. The sexual organs are stimulated and it may cause painful erection of the clitoris or penis. It is not however a true sexual stimulant.

Other substances that allegedly have sexually stimulating properties include *yohimbine*, an alkaloid from the yohimbe tree of Africa. It is used in the treatment of impotence because it stimulates reflex centers in lower portions of the spinal cord. There are other substances to which are ascribed sexual-stimulating qualities but most are too toxic to use with safety or they are ineffective. Some foods, as bananas and oysters, have been considered as being sexually stimulating but no foods are known that have such a specific effect. Odd preparations as powdered rhinocerous horn, because of its resemblance to the penis, or ginseng root, because its shape is like a little man, have been employed but are obviously ineffective. Vitamin E, while of importance in reproduction and development, is not a specific sexual stimulant. In the final analysis, sexual stimulation comes largely from the mind and if a person thinks and especially believes that a certain food or chemical substance will increase his or her sexual effectiveness, it possibly becomes a self-fulfilling prophesy.

With respect to drugs and sex in general, the more frequently marijuana is used and the greater the number of drugs experimented with, the greater the number of partners an individual is likely to have had intercourse with. This is to be expected when one realizes that drug use grows out of an adoption of a specific life-style in which alteration of behavior is a characteristic. A part of this liberal life-style is the liberal orientation toward sexual practices, a part of the life-style and not an outgrowth of drug use. Leo Hollister, M.D., of California, has made a pertinent observation that appropriately summarizes this section on drugs and sex: "That drugs of such divergent pharmacological properties should all enjoy a reputation as sexual superchargers suggests that their effects are mainly in the eye of the beholder. One can't help wondering if there is not some pervasive myth of the perfect orgasm."

# FIFTEEN
# SEX AND THE LAW;
# OBSCENITY AND
# PORNOGRAPHY

## SEX AND THE LAW*

From a legal standpoint, premarital, extramarital, and oral-genital sex, homosexual activities, and sexual relations with animals or prostitutes are illegal in most states. Studies relative to the incidence of the above types of sexual offense indicate that 90 percent or more of all adult males in the United States have committed sexual crimes one or more times in their lives. In spite of the number and frequency of sexual offenses, only a small number of the offenders are ever arrested and of these only a few are tried and convicted.

It would appear then that the laws dealing with sexual behavior are not intended to "preserve public order" but rather are legalized attempts to legislate sexual morality. The constitutionality of most sex laws has been questioned on the basis that they are efforts by the state to enforce moral or religious views of a particular religion. However, serious efforts to alter firmly established laws dealing with sexual behavior are difficult to bring about and progress in updating the sexual code has been limited.

Laws dealing with sexual behavior are objected to for a number of other reasons. It is claimed that they violate an individual's right to privacy, that some are vague and difficult to interpret, that they do not provide equal protection for all, and that some have penalties that could be considered cruel and unusual punishment. Furthermore, the violation of sex laws daily by millions of Americans without fear of detection or punishment leads to a general attitude of disrespect for all laws. The difficulty in detecting violations often leads to undesirable police practices, as homosexual entrapment, and the existence of criminal penalties often favors or makes possible extortion and sometimes police corruption. The proscription of certain activities, as prostitution and homosexuality, leads to the development of a deviant subculture of questionable value to any

*See also the section "Sexual Harassment" in the Preface to the Dover Edition, pp. xi–xii.

society. Moreover, it cannot be shown that any significant secular harm results from private adult sexual behavior (or misbehavior) and the public welfare is not promoted by the existence of criminal penalties.

For these reasons and others, the American Law Institute has developed a *Model Penal Code,* which recommends that all laws governing sexual activities performed in private between consenting adults be abolished. The Institute considers that it is undesirable for the state to attempt to establish moral and religious standards and enforce compliance in a population in which widely different and often conflicting views regarding the morality of various kinds of sexual behavior are held.

There are three general categories of sexual offenses to which criminal penalties are applied. The first includes actions between consenting adults performed in private. This includes activities as premarital sex (fornication), adultery, oral-genital sex, and cohabitation and certain sexual acts between husband and wife. The view now commonly held by authorities on criminal law and becoming generally accepted by the public is that these offenses should be decriminalized. A second category involves offenses against individuals such as rape, sexual molestation of children, incest, and offenses that constitute a public nuisance as exhibitionism and voyeurism. There is little question among legal authorities that these offenses constitute crimes and should be subject to legal sanctions.

The third category includes offenses in which commercial exploitation of sex is involved. This includes prostitution and commercial pornography. Opinions as to the advisability and the extent to which these activities should be subject to criminal penalties are divided.

The penalties for violations of sexual offenses vary widely in different states. Certain offenses may be misdemeanors in one state and felonies in another, or vice versa. A felony is a serious crime punishable by a fine and imprisonment, and in some cases by death; a misdemeanor is a less serious offense punishable by a fine or imprisonment in a place other than a state prison. There is also a wide disparity in the penalties imposed on violators of sex laws, depending on sex, race, and jurisdiction in which the offense is committed. A black man convicted of raping a white woman in a southern state is usually subjected to extreme penalties while the sentence for a white man raping a white woman is much less.

An understanding of most of the sex laws of our society depends on an understanding of the influence the Judeo-Christian tradition has had in the development of our legal code. Any sexual activity except intercourse for procreation was considered evil and sinful as well as that of having more than one wife, hence laws were enacted to punish offenders. Activities as oral sex (cunnilingus and fellatio), anal intercourse, and sexual relations with animals (sodomy) were regarded as unnatural or contrary to the laws of nature, hence were called "crimes against nature." Also included with these acts were homosexuality, pederasty, and lesbianism, sexual activities between two persons of the same sex being regarded as "unnatural." These acts were often classified as felonies and severe penalties imposed.

Additional activities prohibited were *prostitution,* the performance of sexual acts for compensation; *indecent exposure,* the exposure of the genital organs in a public place; *indecent assault,* the taking of sexual liberties, short of sexual intercourse, with another against their will; and *abduction* the unlawful taking away or detention of a child or woman for the purpose of prostitution or marriage. Furthermore the depicting of sexual acts on the stage or screen or specific descriptions in literature, except for artistic purposes, were in general prohibited under laws pertaining to obscenity and pornography.

Also, most states had laws that prevented, greatly limited, or prohibited the sale of contraceptive materials or devices or the dispensing of information concerning their use except when prescribed by physicians. Connecticut even prohibited the use of contraceptives by married couples. This law, however, was declared unconstitutional in 1965. Most states had laws that prohibited abortions except to save the life of or preserve the health of the mother. These also have been declared unconstitutional.

### Adultery

Adultery is voluntary sexual intercourse between a married person and someone who is not that person's spouse. Sometimes adultery is limited to intercourse with a married woman, intercourse by a married man with a single woman being generally considered as fornication. It is interesting to note that intercourse by a married man with a prostitute is rarely considered adultery.

Penalties for adultery have generally been severe. In Puritan times, adultery and other sexual offenses as incest were punishable by death, although this punishment was rarely inflicted. Lesser punishments including the branding of the victim with the letter "A" or sentencing the victim to prison were invoked. Although penalties in general have been reduced with the passage of time, adultery is still considered a criminal offense in most states and punishable by a fine or imprisonment or both. In most states in which an offense is required for the granting of a divorce, adultery is regarded as sufficient grounds for divorce action.

There is a trend now to exclude adultery from the list of criminal offenses. Such is recommended in the Model Penal Code.

### Fornication and Cohabitation

Fornication is sexual intercourse between unmarried, consenting adults; cohabitation is the living together of a couple as man and wife without legal or religious sanction. Sexual intercourse between unmarried adults is prohibited in 35 of the 50 states, yet studies indicate that the majority of men and women in our society engage in premarital coition. It is obvious that only rarely are individuals prosecuted for this offense. In most states fornication is considered a misdemeanor and subject to a small fine or jail term. In Arizona, however, fornication or cohabitation is a felony and a person may be imprisoned for up to three years.

An anomalous situation exists with respect to fornication laws and unmarried, pregnant girls or mothers. It is obvious that they have violated fornication laws yet they are seldom if ever prosecuted for such. Furthermore, most states in a way condone fornication through the establishment of welfare agencies to render aid to girls who become pregnant and little or no effort is made to apprehend or punish the male involved.

It is interesting to note that a recent survey of divorcees indicated that at least 90 percent had been sexually active during the year following their divorce with a median frequency of coitus of twice a week. On the basis of this study, one can only conclude that most divorcees are illegally engaging in fornication.

With respect to fornication, the Model Penal Code does not include it as a criminal offense. Similarly, seduction is not included. Seduction is the act of a male engaging in sexual intercourse with a female under the promise of marriage. Most states have laws that enable a male to avoid prosecution by marrying the girl seduced although if she is under 18, he would be guilty of statuatory rape.

### Marital Sexual Activities

The law is not only concerned with sexual activities outside of marriage but it also deals in various ways with relationships between husband and wife. Some of the legal problems that may arise include the following. It is generally held that a spouse is entitled to the exclusive right of sexual intercourse with his or her mate and sexual relations by either spouse with another person is generally considered grounds for divorce (see "Adultery"). Impotence in a male present at the time of marriage or refusal of a woman to engage in sexual intercourse with her husband after marriage are grounds for annulment. Sterility of either spouse discovered after marriage, unless caused by postmarital accident or disease, or failure to consummate a marriage are grounds for annulment or divorce.

In most states, activities such as oral-genital contacts, anal intercourse, mutual masturbation, and intercourse with animals are prohibited such being considered as "crimes against nature." Most of these offenses are classified as felonies and may constitute grounds for divorce. Even the frequency of intercourse is regulated to some extent and, in some states, only certain coital positions are legal. Rape laws specifically exclude wives from rape by their husbands even though force may be used on an unwilling wife. Although rape cannot be charged against a husband, the wife may charge the husband with assault and battery. If a married couple engages in sexual intercourse while divorce proceedings are in progress, the proceedings are automatically terminated.

*Homosexuality*

In most states homosexuality is not a crime but the sexual acts that homosexuals engage in such as oral-genital sex, anal intercourse, and mutual masturbation are criminal acts subject to punishment regardless of the sex of the participants. Most are felonies and carry severe penalties, although felony charges are usually reduced to misdemeanors. Since most homosexual activity is carried on in private, arrests can only be made when the act is committed in a public place. The most common act is fellatio. Many arrests are the result of entrapment, a questionable procedure in which a nonuniformed police officer entices a homosexual to perform a lewd or lascivious act, the act being observed by a second police officer who makes the arrest and serves as witness. However most arrests of homosexuals are simply for solicitation or loitering in a public place and not for any specific sexual act. For this reason, the number of arrests for homosexuality is relatively small in comparison to the number of sexual acts committed. It is estimated that at least 300,000 homosexual acts are committed for each conviction in the United States.

At present, there is increasing acceptance of homosexuality as a way of life and a trend is developing toward removing legal penalties against it. Laws are generally enacted to prevent and deter crime and to rehabilitate the offender. Considering the number of acts committed, it is obvious that the laws do little to prevent homosexuality and as a deterrent, it is also quite obvious to any homosexual that the chances of being caught are very small. With respect to rehabilitation, the environment of a prison encourages homosexuality, as homosexual relations are entered into much more freely within a prison than in the outside world. Men who are normally heterosexual out of prison often become aggressors in a prison environment forcing their activities on younger, passive men whom they treat as females. New inmates are often subjected to continuous harassment and either submit to homosexual acts or are placed in solitary confinement for their protection. Usually they accept the protection of one man to whom sexual favors are granted. Obviously imprisonment does little to rehabilitate a homosexual.

In addition to severe legal penalties, homosexuals and lesbians are subjected to various types of discrimination. In most cities they can be barred or evicted from private housing with no legal recourse. In the armed services, known homosexuals are discharged with a less than honorable discharge, although this procedure is being challenged. Homosexuals are generally barred from teaching although recent court decisions have declared that homosexuality is not sufficient cause for disqualification. Most homosexuals are denied entrance into professional schools, such as schools of medicine, dentistry, law and divinity.

Most business firms have, in the past, refused to hire known homosexuals but a number of large corporations have lifted the ban and now employ them. The U.S. Civil Service Commission has ruled that homosexuals may not be barred from federal employment although some agencies such as the CIA and the FBI still refuse to employ them. In both government and industry, security clearances will not be granted to homosexuals for fear of blackmail. An effort is now being made to pass a federal law barring any discrimination on the basis of sexual behavior but, although sponsored by a number of congressmen, its passage is unlikely in the immediate future. Most churches refuse to ordain avowed homosexuals although there is a trend toward greater understanding of the homosexual's problems and difficulties.

The homosexual or gay movement is now pushing hard for civil rights maintaining that no one should be prevented from carrying on the usual activities of life as holding a job, engaging in professional activities, or living in a certain locality, because he or she happens to engage in certain sexual activities disapproved of by the majority of the populace. They further maintain that any sexual activities carried on in private between consenting adults should be as free from police harassment as heterosexual activities.


## Prostitution

Female prostitution, in which women perform sexual acts with men for compensation, is the only sexual activity for which women in significant numbers are arrested. These women are arrested primarily for solicitation which usually includes any activity that a prostitute uses to entice customers, especially that occurring in a public place.

The question as to whether prostitution should be removed from the criminal code and whether it should be legalized is constantly debated. The problem of prostitution has always been difficult to deal with primarily because of moral principles involved. In general, there are two methods of dealing with prostitution. The first is that of *prohibiting prostitution* thus making engaging in prostitution a crime. The most important objection to this method is that the rights of consenting adults to engage in sexual activities of their choosing without interference by the state is denied. In addition, prohibition leads to the exploitation of prostitutes by pimps, politicians, dope dealers, and underworld characters and the necessity of operating underground further forces prostitutes into association with dangerous criminal elements of our society. The enforcement of laws against prostitution generally necessitates the existence of a vice or morals police squad, which often resorts to questionable trapping procedures in making arrests and which is especially susceptible to corruption. The criminal attitude directed toward the prostitute creates in her an antisocial personality making reform or rehabilitation difficult and the punishment of the victim and not the initiator of the "crime," usually the male involved, leads to gross inequities of justice.

In the United States, the prohibitionist approach is generally followed. Most states have laws in which prostitution is prohibited. Owners, landlords, or operators of brothels are subject to penalties, and procuring or pimping is prohibited. Agents who assist procuring by providing information for obtaining females may be prosecuted, and in general anyone associated with or involved in the business of prostitution may be subject to criminal penalties. A few states have laws that penalize the person who patronizes a prostitute but they are rarely enforced.

A study of prostitution throughout history brings out the fact that the elimination of prostitution is practically impossible, or if achieved, could only be accomplished by the gross denial of basic human rights.

Ruling out then the possibility of complete eradication of prostitution, a second approach is that which involves *regulation*. Under this system, the practice of prostitution is limited to brothels that are located in restricted areas. Steps are taken to control venereal disease through registration and required medical examination of prostitutes and to see that they are not exploited. The main object of this method, which legalizes prostitution, is to keep prostitutes out of the respectable sections of a community.

Reasons often cited for continuing the present status of prostitution, in which it is considered a criminal activity, are that it is a threat to the family as a social unit, that it is a major source of venereal disease, that it fosters abnormal sexual behavior, that it is a cause of juvenile delinquency, that it creates a forced or slave labor situation, that it leads to political and judicial corruption, and that it is morally wrong.

The advocates of decriminalizing prostitution maintain that prostitution can never be totally suppressed, that making prostitution a criminal act tends to drive it underground and force the prostitute to develop underworld associations in which exploitation by drug pushers, hoodlums, and other undesirable characters is promoted. With respect to prostitutes being a threat to the family, few sociologists regard prostitution as a significant factor in the breakup of marriages. As a major source of venereal disease, researchers estimate that less than 5 percent of new VD cases results from prostitution. There is no evidence that it is a primary causative factor of juvenile delinquency and, with respect to prostitution providing a "white slave market," rarely are cases encountered where prostitutes have been forced into the profession. Most become prostitutes of their own volition or from force of necessity as in the case of drug addiction in which prostitution is resorted to in order to pay for narcotics.

That prostitution leads to political and judicial corruption is a recognized fact. Vice squad members, lawyers, bondsmen, magistrates, and judges are sometimes involved in bribery and the fixing of cases. Framing or entrapping prostitutes is a common practice. Such practices may be followed by extortion and oftentimes nonprostitutes may become innocent victims.

Even though there is a trend toward recognizing female prostitution as a legitimate profession, one that exists because there is a demand, one that could provide a sizable tax revenue, and one in which the workers (the prostitutes) could be granted Social Security and unemployment benefits, the likelihood that significant changes in laws dealing with prostitution will be made in the immediate future is small. The Model Penal Code still regards prostitution as a criminal activity and no changes are recommended.


### Homosexual and Lesbian Prostitution

Male homosexual prostitution is much more common than is generally recognized. There seems to have been a pronounced upsurge in recent years in homosexual prostitution not only in the United States but throughout the world. This is probably the result not only of an increase in the number of homosexuals but in the increased openness of their activities. Although lesbian prostitution is also considered to be on the increase, it is encountered less frequently as lesbians tend to frequent female prostitutes who are willing to accept female customers.

Most male prostitutes or hustlers are young men in their twenties or are juveniles, often members of gangs who are forced by older members of the gang into prostitution. Hustlers frequent all-night movie houses where older men are enticed into accosting them. Sexual relations, fellatio, or masturbation may be performed sometimes in the theater, in the theater's restroom, or at the customer's hotel room or apartment. Other places where contacts are made by male prostitutes are along certain streets or areas which are frequented by homosexuals, in public parks, in rest rooms in bus and railway stations and on public highways, in gay bars, bath houses, massage parlors and other places which become known to homosexuals.

Male prostitutes generally do not consider themselves to be homosexuals as they tend to be motivated primarily by the desire to acquire money without working rather than for sexual satisfaction. They are often heterosexual and regard themselves as normal individuals. Their customers are usually older men, and fees that are usually less than those charged by female prostitutes depend on the age and attractiveness of the customer and his willingness to pay.

Male houses of prostitution that cater to homosexuals exist in most of the larger cities, although they are less common than female brothels. Clubs featuring female impersonators who frequently engage in prostitution exist in many cities. Lesbian brothels that cater to women are rarely encountered.

## OBSCENITY AND PORNOGRAPHY

*Obscenity* refers to the state or quality of being obscene, that is, offensive to accepted standards of morality and decency. *Pornography* is the use of written or visual materials with the deliberate intent of arousing erotic sensations.

During the nineteenth century and into the early part of the present century, Victorian concepts that regarded almost any matter dealing with sex as being disgusting, immoral, or sinful were commonly held in our society. Sexual feelings, impulses, and drives were in general denied and sex, or matters pertaining to sex, were not discussed in public. Even topics such as pregnancy and childbirth were not mentioned in mixed company. Children grew up with little knowledge of sex or sexuality and literature on the subject was limited. Textbooks of anatomy and physiology used in public schools usually omitted the male and female reproductive systems and diagrams or figures of the reproductive organs were conspicuously absent. Literature that dealt in detail with the sexual act and related subjects as contraception was practically unavailable to the general public. Men and women dealt with each other more or less as though neither possessed sex organs nor had any feelings involving sex.

A federal law existed that declared any obscene, lewd, or lascivious book or any printed matter of an indecent nature unmailable. Because of this and other laws, the importation or the publication of books that were declared to be lustful was prohibited. A condition of censorship existed preventing freedom of expression of thought or dissemination of knowledge in the field of sexual matters. The First Amendment of the Constitution, which protected the right of free expression of political ideas, did not apply to the field of arts and letters.

A pronounced change in sexual attitudes and behavior occurred rather suddenly in the 1920s following, and to some extent the result of, World War I. With a decline in moral standards that usually occurs in wartime, associated with the development of the automobile, which, removed dating and courtship relationships from the home, and the general breakdown in attitudes toward conventional behavior, which the Jazz Age and resistance to prohibition engendered, an age of liberalism with respect to sex came into existence. Instead of sex being repressed, it emerged into the open. Sex came to be discussed openly and freely between the sexes. New attitudes toward marriage were developed with the advocacy of trial marriages and greater freedom in sexual relationships.

As a result of these changes, Western society changed from one that tended to ignore sex, or pretended that it did not exist, to one which placed great emphasis on it. However, in the field of arts and letters, severe restraints in the form of obscenity laws, both federal and state, prevented the free flow of information or the distribution of any books or printed matter that could be categorized as stimulating lust. A form of censorship existed which even prevented the distribution of information on contraception.

During a period roughly from the mid-1950s to the mid-1960s, a radical change occurred. In a series of important cases (Roth, Albert, Ginsberg), the U.S. Supreme Court, through its decisions pertaining to libel and obscenity, changed the law with respect to freedom of expression. In the field of obscenity, the court established certain limitations on the word "obscene," which had not been clearly defined in any federal law. It established three criteria that must be met before the distribution of material could be prohibited on the basis that it was "obscene." These criteria that were to apply to lower federal courts and state courts are (1) the dominant theme of the material, taken as a whole, must appeal to a "prurient" interest in sex; (2) the material must be patently offensive with respect to "contemporary community standards"; and (3) it must lack "redeeming social value."

The interpretation of this law makes it questionable whether any written material that has any literary merit whatsoever could be judged to be obscene for adults. As a result, censorship with its restrictions on freedom of expression and the role of the government in promoting virtue in sexual behavior came virtually to an end.

These decisions opened the floodgates to the flow of literature pertaining to sex. In a society in which sex was largely considered a procreational activity and the relationships between the sexes more or less specifically delineated, sex has become an obsession. There is hardly a newspaper or magazine that does not, in nearly every issue, carry articles dealing with some aspect of sexual behavior. Subjects which were once principally limited to scientific publications are now presented to the general public. Topics such as impotence, frigidity, abortion, adultery, teenage pregnancy, contraception, sterilization procedures, orgasmic difficulties, homosexuality, and related subjects are published freely. Books dealing with every aspect of sexual behavior, are now readily available. In the visual arts, especially the cinema, practically every form of sexual behavior and misbehavior, are readily displayed. In television, there is some degree of restraint but limitations as to what can be shown have markedly changed toward liberalization. In the theater and in literature, there is little or no restraint as to what is presented to the public.

As a result of these changes, there has been a marked increase in traffic in obscene and pornographic materials. Whereas obscenity involves simply the transgression of accepted standards of morality, pornography consists of the display of visual or written material with the deliberate intent of arousing erotic sensations. It is more simply stated as "smut for smut's sake." Obscene and pornographic materials include sexually orientated motion picture films in which sexual activity is depicted, art films, exploitation films or "skin flicks," books, newspapers, and magazines written for the mass market, and printed materials for the "adult only market." Hardcore pornographic materials are readily available but usually sold "under the counter." The latter includes "stag films," photo sets, and picture magazines that depict nearly every kind of sexual activity.

The traffic in obscenity and pornography reached such an extent, variously estimated at from $500 million to $2.5 billion per year, that it became a matter of national concern. As a result, Congress, in 1967, established an advisory Commission on Obscenity and Pornography to investigate the situation and to make recommendations with respect to the regulation of traffic in obscene and pornographic materials. It was directed especially to study the effects of and to determine whether the materials were harmful to the public, especially minors, and to study its relationship to crime and other antisocial behavior.

In 1970, after two years of study and an expenditure of about two million dollars, the Commission released its report containing its findings, conclusions, and recommendations. One of its conclusions was "In summary, empirical research designed to clarify the question has found no evidence to date that exposure to explicit sexual materials plays a significant role in the causation of criminal behavior among youth or adults. The Commission cannot conclude that exposure to erotic materials is a factor in the causation of sex crime or sex delinquency."

Some of its other findings were that (1) pornography can have an educational value for the sexually ignorant if the information presented is accurate; (2) that there is no evidence to show that youngsters exposed to pornography have a negative change in moral character, attitudes, or sexual orientation; (3) that imprisoned sexual offenders have histories of less exposure to pornography than adults normally their age and more often have histories of sexual repression in their younger years growing up in very strict families; (4) that women are virtually as interested in erotic materials as are men; (5) that the majority of individuals exposed to erotic material did not change the frequency or form of their sexual activity; (6) that most Americans are introduced to explicit sexual material during adolescence not through the purchase of the material but through the sharing of it with friends in social situations; and (7) that those who frequent erotic cinemas and adult book stores are usually middle-aged, middle-class, white married males.

The Commission further recommended that (1) there should be no censorship of pornography for consenting adults since pornography has not been shown to be dangerous to society; (2) that pornographic materials should be denied to children under 16 not primarily because there is any evidence that it is harmful but because parents so desire it; (3) that sex education programs in the public schools be developed to provide quality sex education; and (4) that there should be ongoing research into the effects of pornography upon society.

The Commission studied the situation as it exists in foreign countries. Evidence exists that in Denmark where pornography is given free rein, there is doubt as to whether pornographic literature, films, and pictures have had any significant effect on the sexual behavior of children or adults. The same view is arrived at by official commissions in Sweden, West Germany, Great Britain, and Israel.

However not all members of the Commission agreed with the majority report. In a minority report prepared by certain members of the Commission, it was held that the report was slanted and biased in favor of protecting the business of obscenity and pornography, that the findings of the Commission were based on inadequate and manipulated evidence. It questioned the findings that pornography is harmless and of no significance in the causation of criminal behavior.

As a result of differences of opinion concerning the report of the Commission, the findings of the Commission were rejected by the President and by Congress before the report was published and made available to the public. The report does seem to indicate however that society worries unnecessarily about the harmful effects of pornography on minors and of its significance in sex crimes. The basic question seems to be: Which is potentially most dangerous to our society: unlimited license in the distribution of obscene and pornographic material or censorship that would be necessary to control it? Does society have a legitimate concern in maintaining certain moral standards?

The Supreme Court has affirmed the rights of states to regulate commerce in obscene materials. Pornography and obscenity are illegal in all states, however penalties for pornography vary widely and are enforced only to a limited degree. Freedom of expression with respect to sexual matters is no doubt here to stay. The question seems to be: Will pornography add to human happiness and the enjoyment of sex? The Victorian concept was love without sex; the modern concept is sex without love. There should be a middle ground.

# SIXTEEN
# SOME COMMON
# FALLACIES AND
# MISCONCEPTIONS
# CONCERNING SEX

*That menstruation is to get rid of the unfertilized egg.*  Fact—in menstruation, the lining of the uterus, which has developed to receive the developing embryo, is discharged. An unfertilized egg normally does not enter the uterus; it degenerates within the uterine tube.

*That if menstruation starts at an early age, the menopause will occur earlier.*  Fact—just the opposite. The earlier the menarche, the later the menopause.

*That there is a particular time during the menstrual cycle when a woman is more receptive in sexual relations.*  Fact—women are receptive sexually at any time during the menstrual cycle except during the menses when sexual relations are avoided for esthetic reasons. The particular time most desirable varies with individuals depending on physiological, psychological, and environmental factors.

*That a woman, like a man, experiences only one orgasm during intercourse.*  Fact—a woman may have one, a few, or many orgasms repeatedly. Or she may experience none. It is a physiological reaction dependent on many factors.

*That two types of orgasms are experienced by a female, one clitoral, the other vaginal.*  Fact—physiologically and psychologically, there is only one kind of orgasm. This orgasm is primarily induced by stimulation of the clitoris or surrounding structures, or both.

*That the menopause is the end of a woman's sexual life.*  Fact—not necessarily. A woman is still capable of responding sexually and sexual desire may be intensified as fear of pregnancy and the necessity of using contraceptives are eliminated.

*That both men and women after 50 should forget about sex and "act their age."* Fact—since the sex organs in men are still functional during the fifties, sixties, and even seventies and as women can still experience orgasms, it is possible and even desirable, for physical and mental well-being, that sexual activity be continued.

*That excessive sexual activity in the male will reduce his capacity to produce spermatozoa and eventually lead to sterility or impotence.* Fact— there is a continuous production of spermatozoa from puberty to old age. The capacity of the testes to produce sperm is unlimited. The testes produce as many sperm as necessary to meet the physiological demand— they do not wear out.

*That masturbation in boys results in physical and mental weakness, loss of manhood, and is the cause of pimples and other skin disorders.* Fact— there is no evidence that masturbation is the cause of any of these disorders. The seminal fluid plays no role in the makeup of body tissues so its loss would have no effect upon the makeup of muscles, brain, or skin. It is a "vital fluid" in only one respect: fertilization of the egg.

*That masturbation in adults is harmful.* Fact—there is no evidence that any ill effects either anatomically or physiologically result from mastur-bation. Psychologically sometimes undesirable reactions may result.

*That there is a relationship between the size of the penis and female sexual response.* Fact—there is little or no relationship between the size of the penis (its length or diameter) and the sexual response of a female. As the female orgasm is dependent more on stimulation of accessory struc-tures, the clitoris and labia, rather than the vagina, any penis capable of penetrating the vagina is usually adequate in inducing the desired response.

*That the size of a woman's vagina is of vital significance in sexual satis-faction.* Fact—its size has little effect on the response of either the male or female.

*That the presence of an intact hymen is an indicator of vir-ginity.* Fact—The hymen of most females is usually broken before inter-course by the insertion of objects as vaginal tampons, or through masturba-tion or vigorous physical activity.

*That the size of the breasts is related to the capacity to experience sexual response.* Fact—women with small breasts can experience as full and complete sexual response as women with large breasts.

*That pregnancy cannot occur unless a woman experiences an orgasm.* Fact—an orgasm is not an essential prerequisite to pregnancy. In artificial impregnation, no orgasm is experienced.

*That pregnancy cannot occur at first intercourse.* Fact—it most definitely can and it often does.

*That regular douching is essential for personal cleanliness.* Fact—it is not essential and the frequent injection of cleansing or antiseptic substances may have harmful effects on the vaginal mucosa or the normal bacterial flora.

*That venereal disease can be acquired without engaging in sexual relations.* Fact—sexual intercourse or contact with an infected mucous membrane as the lips are the usual means of transmission. Acquiring the causative organism from objects as toilet seats, doorknobs, or by shaking hands rarely if ever occurs. However, female infants and children before puberty may become infected through nonvenereal means.

*That when cured of a venereal disease, one is immune to further infection.* Fact—just the opposite. Immunity is not developed for either gonorrhea or syphilis and reinfection by either or both is possible.

*That oral sex is a perversion, abnormal, and dirty.* Fact—only if a person thinks it is. For inducing an erection in a man of questionable potency or an orgasm in an unresponsive woman, oral-genital stimulation may be proper and oftentimes desirable.

*That if semen is swallowed during oral sexual activities, illness or pregnancy may result.* Fact—swallowing semen is not harmful if it does not contain agents which cause sexually transmitted diseases; it is a nontoxic fluid. A woman cannot become pregnant from swallowing it since it is digested and absorbed like other organic matter taken into the body.

*That premarital coitus has a detrimental effect on future marital happiness.* Fact—there is little evidence as to whether premarital sex is either beneficial or detrimental to marital happiness. It depends on the individuals involved and their attitudes toward it.

*That simultaneous climax or orgasm is the most advantageous form of sexual response.* Fact—although most sex manuals insist that simultaneous orgasm is essential for mutual satisfaction, such is not always attainable or even desirable. When simultaneous orgasm becomes the ultimate goal of intercourse, sex relations may become too mechanical and not provide the maximum in pleasure or satisfaction.

*That a woman does not need to get a Pap test until middle age or later.* Fact—any woman, regardless of age, who is sexually active and especially one who uses the Pill for contraception or a woman infected with genital herpes should get a Pap test annually. This is a simple, painless test for the detection of carcinoma (cancer) of the cervix.

*That a woman should scream and run away from an exhibitionist to avoid possible rape.* Fact—exhibitionists are seldom dangerous individuals. Their satisfaction is principally in observing the reaction of the woman—usually fright and disgust. The best reaction is to calmly ignore the action and turn and walk away.

*That male homosexuals are deficient in male hormones (androgens) and should respond to androgen therapy.* Fact—although there is questionable evidence that male homosexuals do have a reduced level of androgens, androgen therapy has not been effective in treatment. While androgens are primarily responsible for sex drive, they do not give direction to the drive. The determination of a person's sexual object choice is largely determined by experiences in early development.

That promiscuity in men and women is primarily the result of a high sex drive (libido). Fact—promiscuity is often a symptom of emotional immaturity. Promiscous individuals need constant reassurrance of their masculinity or femininity, and knowledge that they are needed and are desirable. Sexual promiscuity is often the outward manifestation of escape and defense mechanisms resulting from loneliness, depression, isolation, and alienation.

That sex drive in children is principally due to inheritance. Fact— evidence indicates that the development of sex drive in children depends more on socialization than on genetic factors.

That athletes make the best sexual partners. Fact—there is no correlation between excellence in sexual performance and excellence in physical activity. Also there is no correlation between size of the penis and body size. Since sexual satisfaction depends on many emotional factors unrelated to physical size or accomplishment, athletes do not necessarily make the best partners.

That married couples or consenting adults may legally engage in any type of sexual relations they desire. Fact—in most states they cannot. Most states still have statutes that make fornication, adultery, oral-genital contact, anal intercourse, and homosexual activities criminal offenses. Efforts are being made to correct this situation. In 1964, the International Penal Law Association adopted resolutions favoring the elimination of criminal penalties for such activities. Efforts to put corrective measures into effect have been slow.

That men paralyzed from the waist down are sexually impotent. Fact—many male paraplegics are capable of engaging in sexual relations and fathering children. Most male erectile and ejaculatory functions are reflex activities not under control of the motor centers of the brain.

That women paralyzed from the waist down are incapable of obtaining or giving sexual satisfaction. Fact—depending on the specific cause of the paralysis, it is possible for some paralyzed women to participate in and respond normally to various forms of sexual activity. Pregnancy and normal delivery are even possible.

That if a woman has a "vaginal" odor, she should use a feminine deodorant spray. Fact—there is a remarkable constancy in vaginal odors even when women have a common vaginal infection as trichomoniasis. If an odor is present, it is probably not from the vagina but more likely from the entire vulval area. Sweat glands in the pubic area may contribute to the development of an odor if good hygiene of the vulval area is not maintained. Soap and water as cleansing agents are preferable to feminine hygiene sprays.

That men's fantasies are mainly of raping women. Fact—information from questionnaires reveals that men do not frequently fantasize raping women. They more frequently fantasize seducing women or being seduced by an attractive female.

*That women do not have sexual fantasies.* Fact—most women have sexual fantasies. It is a normal part of one's sexuality and fantasizing should not lead to feelings of guilt. Women most often fantasize about being with a fictitious lover or being taken by force by a male aggressor.

*That if a mother breast feeds her infant, she will have larger breasts than a mother that does not nurse her infant.* Fact—during pregnancy the secretory alveoli of the breast do not increase in number, only in complexity. When lactation ceases, either from not nursing or following weaning of an infant, the breasts decrease in size to near prepregnancy size.

*That sexual intercourse uses up a lot of calories and therefore is a good way to lose weight.* Fact—while sexual intercourse can be considered as a positive form of exercise, it usually does not entail a large expenditure of energy. It has been estimated that the average sexual encounter "burns up" about 100 Calories for each partner. One pound of human fat contains the equivalent of 3500 Calories when burned, consequently in order to lose one pound of body weight, it would be necessary to engage in coitus 35 times.

*That the performance of an athlete is diminished by engaging in sexual intercourse the night before.* Fact—there is no scientific support to the theory that ejaculation is physically weakening and interferes with an athlete's performance on the following day. In a recent interview, Dr. William Masters stated that after a sexual experience, an athlete should be able to perform maximally if he is allowed a sufficiently long recuperative period, a period of from *one to five minutes.*

*That nocturnal emissions do not occur in mature men.* Fact— nocturnal emissions usually begin about a year or so after puberty and reach their greatest frequency in the late teens or early twenties. However, some men continue to have nocturnal emissions throughout their entire adult life.

*That most sex offenders are oversexed men with uncontrollable sex urges.* Fact—the "typical" sex offender is usually sexually immature. He lacks normal masculine aggressiveness and usually feels inferior to other men. He is generally timid, passive, and full of anxiety concerning relationships with women. His frequency of sexual intercourse and number of sexual partners indicate that he is undersexed rather than oversexed.

*That most prostitutes are lesbians.* Fact—while some lesbians become prostitutes and some prostitutes become lesbians, it would be an overgeneralization to assume that most prostitutes are lesbians.

*That maternal impressions or prenatal influences as fright or an unpleasant sight during pregnancy will result in an unborn child being "marked."* Fact—this belief is without foundation. There are no nerve connections between a mother and the fetus so the developing fetus cannot be directly affected by any thoughts, ideas, or fears experienced by the mother.

*That children should be permitted to view parents during coitus thereby developing in the child a wholesome respect for sexual intimacies.* Fact—children are usually extremely disturbed on viewing their parents engaged in coitus. The degree of disturbance depends on a variety of factors including the age of the child, the scene perceived by the child's mind, and the reaction of the parents to the child's intrusion. Repeated exposure to parental sexual activities should definitely be avoided.

*That physicians are generally well-trained and emotionally equipped to deal with the sexual problems of their patients.* Fact—except for their knowledge of reproductive anatomy and physiology, most physicians have a large void in their knowledge of human sexuality. Also many physicians, especially older ones, have rather prudish biases toward sex that they let show to their patients. However, several books have recently been published on human sexuality directed toward physicians and medical students. These in addition to courses and workshops in human sexuality for physicians have brought about a change making most physicians effective counselors on sexual matters.

# APPENDIX A
# SOURCES
# OF INFORMATION

American Association of Marriage and Family Counselors, 225 Yale Ave., Claremont, Cal. 91711

American Association of Sex Educators and Counselors, 815 Fifteenth St., N.W., Washington, D.C. 20005

American Fertility Society, 1810 Ninth Ave. So., Birmingham, Ala. 35205

American Institute of Family Relations, 5287 Sunset Blvd., Los Angeles, Cal. 90027

American Social Health Association, 1740 Broadway, New York, N.Y. 10019

Association for Family Living, 32 W. Randolph St., Chicago, Ill. 60601

Association for Study of Abortion, Inc., 120 West 57th St., New York, N.Y. 10019

Association for Voluntary Sterilization, Inc., 708 Third Ave., New York, N.Y. 10017

E. C. Brown Center for Family Studies, 1802 Moss St., Eugene, Ore. 97463

Bureau of Community Health Services, Health Services Administration, DHEW, Washington, D.C. 20201

Community Sex Information and Education Service, Inc., P.O. Box 4246, New Orleans, La. 70118

Education Foundation for Human Sexuality, Montclair State College, Upper Montclair, N.J. 07043

Erickson Education Foundation, 1627 Oreland Ave., Baton Rouge, La. 70808

Institute for Family Research and Education, 760 Ostrom Ave., Syracuse, N.Y. 13210

Institute for Sex Education, 18 South Michigan Ave., Chicago, Ill. 60603

Institute for Sex Research, Inc., Indiana University, Bloomington, Ind. 47401

Maternity Center Association, 48 East 92nd St., New York, N.Y. 10028

Mattachine Society, 243 West End Ave., New York, N.Y. 10023

National Council on Family Relations, 1219 University Ave., S.E., Minneapolis, Minn. 55414

National Council on Illegitimacy, 44 East 23rd St., New York, N.Y. 10010

Planned Parenthood Federation of America, Inc., 810 Seventh Ave., New York, N.Y. 10019

SIECUS (Sex Information and Education Council of the United States), 1858 Broadway, New York, N.Y. 10023

Society for the Scientific Study of Sex, Inc., 12 East 41st St., New York, N.Y. 10017

# APPENDIX B
# JOURNALS
# AND NEWSLETTERS

## I. JOURNALS

*Alternative Lifestyles,* Human Sciences Press, New York.

*American Journal of Obstetrics and Gynecology,* American Gynecological.Society, C. V. Mosby Co., St. Louis.

*Andrologia,* Grosse Verlag GmbH, Berlin, West Germany.

*Archives of Sexual Behavior,* Plenum Publishing Corp., New York.

*Bibliography of Reproduction,* Reproduction Research Information Services, Ltd., Cambridge, England.

*Biology of Reproduction,* Academic Press, New York.

*British Journal of Venereal Disease,* British Medical Association, London, England.

*Child and Family,* National Commission on Human Life, Reproduction, and Rhythm, Oak Park, Ill.

*Contraception,* Geron-x Inc., Los Altos, Cal.

*Family Coordinator,* National Council on Family Relations, Minneapolis, Minn.

*Family Planning Perspectives,* Planned Parenthood Federation of America, New York.

*Fertility and Sterility,* Harper & Row, Inc., New York.

*International Journal of Andrology,* Scriptor Publisher, Copenhagen, Denmark.

*International Journal of Fertility,* Waverly Press, Baltimore.

*Journal of Divorce,* Haworth Press, Inc., New York.

*Journal of Homosexuality,* Haworth Press, Inc., New York.

*Journal of Marriage and the Family,* National Council on Family Relations, Minneapolis, Minn.

*Journal of Reproduction and Fertility,* Blackwell Scientific Publications, Oxford, England.

*Journal of Reproductive Medicine,* Medical Medice Association, Mt. Pleasant, Ill.

*Journal of Sex and Marital Therapy,* Behavioral Publications, New York.

*Journal of Sex Education and Therapy,* American Association of Sex Educators, Counselors, and Therapists, Washington, D.C.

*Journal of Sex Research,* Society for the Scientific Study of Sex, New York.

*Medical Aspects of Human Sexuality,* Clinical Communications, Inc., New York.

*Obstetrical and Gynecological Survey,* Williams and Wilkins, Baltimore.

*Obstetrics and Gynecology,* Harper & Row, Inc., New York.

*Sex Roles,* Plenum Publishing Co., New York.

*Sexuality and Disability,* Human Sciences Press, New York.

*Sexuality Today,* ATCOM, Inc., New York.

*Sexually Transmitted Diseases,* J. B. Lippincott, Philadelphia.

## II. NEWSLETTERS

*Androgyny Newsletter,* National Androgyny Center, San Diego, Cal.
*Emko Newsletter,* Emko, St. Louis.
*Family Planning Digest,* Center for Family Planning Program Development, New York.
*Reports of Population and Family Planning,* Population Council, New York.
*Sex News,* P. K. Houdek, Kansas City, Missouri.
*SIECUS Report,* SIECUS, New York.

# APPENDIX C
# SPECIAL REFERENCES
# (INEXPENSIVE PAPERBACKS)

Barbach, L. For Each Other—Sharing Sexual Intimacy. Signet, New York, 1982.

Beauvoir, S. de. The Second Sex. Bantam Books, Inc., New York, 1949.

Brecker, R., and E. Brecker, eds. An Analysis of Human Sexual Response. New American Library, New York, 1966.

Bullough, V. L. Homosexuality—A History. Signet, New York, 1979.

Bullough, V. L., and B. Bullough. Sin, Sickness, and Sanity—A History of Sexual Attitudes. Signet, New York, 1977.

Cisda, J. B., and J. Cisda. Rape: How to Avoid It. Books for Better Living, Chatsworth, Cal., 1974.

The Diagram Group. Sex: A User's Manual. Berkley Publishing Group, New York, 1981.

Dodson, F. How to Father. Signet, New York, 1974.

Farrell, W. The Liberated Man. Bantam, New York, 1975.

Fasteau, M. F. The Male Machine. Delta, New York, 1975.

Ford, C. S., and F. A. Beach. Patterns of Sexual Behavior. Harper & Row, New York, 1951.

Gochros, H. L., and J. Fischer. Treat Yourself to a Better Sex Life. Spectrum Books, Englewood Cliffs, N.J., 1980.

Goldberg, H. The New Male. Signet, New York, 1979.

Hartman, W., and M. Fithian. Any Man Can. St. Martin's Press, New York, 1984.

Hatcher, R. A., G. K. Stewart, F. Stewart, F. Guest, D. W. Schwartz, and J. A. Jones. Contraceptive Technology—1985-86. Irvington, New York, 1985.

Hendrick, C., and S. Hendrick. Liking, Loving, and Relating. Brooks/Cole, Monterey, Cal. 1983.

Hite, S. The Hite Report. Dell Publishing Co., Inc., New York, 1976.

Levine, L., and L. Barbach. The Intimate Male. Anchor Press/Doubleday, New York, 1983.

Lips, H. M., and N. L. Colwill, The Psychology of Sex Differences. Spectrum, Englewood Cliffs, N.J., 1978.

McCarthy, B., and E. McCarthy. Sexual Awareness—Enhancing Sexual Pleasures. Carroll and Graf Publishers, Inc., New York, 1984.

McCary, J. L. Sexual Myths and Fallacies. Schocken Books, New York, 1974.

Menning, B. E. Infertility—A Guide for the Childless Couple. Prentice-Hall, Inc., New Jersey, 1977.

Money, J., and A. A. Ehrhardt. Man and Woman, Boy and Girl. Mentor, New York, 1972.

Newill, R. Infertile Marriage. Penguin Books, Baltimore, 1974.

Noonen, J. T. Contraception. New American Library, New York, 1967.

Pietropinto, A., and J. Simenauer. *Beyond the Male Myth.* Signet, New York, 1978.

Planned Parenthood. *How to Talk with Your Child About Sexuality.* Doubleday and Company, Inc., New York, 1986.

Robbins, J., and J. A. Robbins. *An Analysis of Human Sexual Inadequacy.* New American Library, New York, 1970.

Seaman, B. *Free and Female.* Fawcett, Greenwich, Conn., 1972.

Segal, J. *The Sex Lives of College Students.* Dell, New York, 1984.

Shapiro, H. I. *The Birth Control Book.* Avon, New York, 1978.

Stockard, J., and M. M. Johnson. *Sex Roles—Sex Inequality and Sex Role Development.* Prentice-Hall, Englewood Cliffs, N.J., 1980.

Tripp, C. A. *The Homosexual Matrix.* Signet, New York, 1975.

Weitz, S. *Sex Roles—Biological, Psychological, and Social Foundations.* Oxford University Press, New York, 1977.

Westheimer, R. *Dr. Ruth's Guide To Good Sex.* Warner Books, New York, 1983.

Wheeler, M. *No-Fault Divorce.* Beacon Press, Boston, 1974.

Zilbergeld, B. *Male Sexuality.* Bantam, New York, 1978.

# APPENDIX D
# GENERAL REFERENCES

Abse, D. W., et al., eds. *Marital and Sexual Counseling in Medical Practice.* Harper & Row, New York, 1974.

Allgeier, E. R., and A. R. Allgeier. *Sexual Interactions.* D. L. Heath, Lexington, Mass., 1984.

Allgeier, E. R., and N. B. McCormick, eds. *Changing Boundaries: Gender Rôles and Sexual Behavior.* Mayfield, Palo Alto, Cal., 1982.

Ard, B. N. *Treating Psychosexual Dysfunction.* Jason Aronson, New York, 1974.

Avers, C. J. *Biology of Sex.* Wiley, New York, 1974.

Barnard, M. U., B. J. Claney, and K. E. Krantz. *Human Sexuality for Health Professionals.* W. B. Saunders, Philadelphia, 1978.

Bell, A. P., and M. S. Weinberg. *Homosexualities—A Study of Diversity Among Men and Women.* Touchstone, New York, 1978.

Bell, A. P., M. S. Weinberg, and S. K. Hammersmith. *Sexual Preference—Its Development in Men and Women.* Indiana University Press, Bloomington, 1981.

Brecher, E. M. *The Sex Researchers.* Little, Brown, Boston, 1969.

Brecher, E. M., and Editors of Consumer Reports Books. *Love, Sex, and Aging.* Little, Brown, Boston, 1984.

Brown, F., and R. T. Kempton. *Sex Questions and Answers.* McGraw-Hill Book Co., New York, 1970.

Brownmiller, S. *Femininity.* Fawcett Columbine, New York, 1984.

Bruess, C. E., and J. S. Greenberg. *Sex Education: Theory and Practice.* Wadsworth, Belmont, Cal., 1981.

Burgess, A. W. *Child Pornography and Sex Rings.* Lexington Books, Lexington, Mass., 1984.

Burt, J. J., and L. A. Brower. *Education for Sexuality.* Saunders Co., Philadelphia, 1975.

Butler, R. N., and M. I. Lewis. *Love and Sex After 40.* Harper & Row, New York, 1986.

Byrne, D., and W. A. Fisher, eds. *Adolescents, Sex, and Contraception.* Lawrence Erlbaum, Hillsdale, N.J., 1983.

Cameron-Bandler, L. *Solutions—Practical and Effective Antidotes for Sexual and Relationship Problems.* Future Pace, San Rafael, Cal., 1985.

Chafetz, J. S. *Masculine/Feminine or Humor?* F. E. Peacock Publishers, Itasca, Ill., 1974.

Coleman, J. C. *Intimate Relationships, Marriage, and Family.* Bobbs-Merrill, Indianapolis, Ind., 1984.

Cook, E. P. *Psychological Androgyny.* Pergamon, New York, 1985.

Cox, D. J., and R. J. Daitzman, eds. *Exhibitionism: Description, Assessment, and Treatment.* Garland, New York, 1980.

Crooks, R., and K. Baur. *Our Sexuality*. 2nd edition. Benjamin/Cummings, Menlo Park, Cal., 1983.

Day, B. *Sexual Life Between Blacks and Whites*. World Publishing, New York, 1972.

De Lamater, J., and P. MacCorquodale. *Premarital Sexuality: Attitudes, Relationships, Behavior*. University of Wisconsin Press, Madison, 1979.

Eysenck, H. J., and D. K. B. Nias. *Sex, Violence, and the Media*. Harper & Row, New York, 1978.

Farber, M., ed. *Human Sexuality—Psychosocial Effects of Disease*. Macmillan, New York, 1985.

Feldman, P., and M. MacCulloch. *Human Sexual Behavior*. John Wiley & Sons, New York, 1980.

Finkelhor, D. *Sexually Victimized Children*. Free Press, New York, 1979.

Fisher, S. *The Female Orgasm*. Basic Books Inc., New York, 1973.

Francoeur, R. T. *Becoming a Sexual Person (Brief Edition)*. John Wiley & Sons, New York, 1984.

Francoeur, R. T., and A. K. Francoeur. *The Future of Sexual Relations*. Prentice-Hall, Englewood Cliffs, N.J., 1974.

Frazier, N., and M. Sadker. *Sexism in School and Society*. Harper & Row, New York, 1973.

Gagnon, J. H. *Human Sexualities*. Scott, Foresman, Glenview, Tex., 1977.

Gagnon, J. H., and C. J. Greenblat. *Life Designs—Individuals, Marriages, and Families*. Scott, Foresman, Glenview, Ill., 1978.

Geer, J., J. Heiman, and H. Leitenberg. *Human Sexuality*. Prentice-Hall, Englewood Cliffs, N.J., 1984.

Giles-Sims, J. *Wife Battering—A Systems Theory Approach*. Guilford Press, New York, 1983.

Godow, A. G. *Human Sexuality*. C. V. Mosby, St. Louis, 1982.

Golanty, E. *Human Reproduction*. Holt, Rinehart & Winston, New York, 1975.

Goldsmith, S. *Human Sexuality: The Family Source Book*. C. V. Mosby, St. Louis, 1986.

Gordon, S., and C. W. Snyder. *Personal Issues in Human Sexuality*. Allyn & Bacon, Boston, 1986.

Gosselin, C., and G. Wilson. *Sexual Variations: Fetishism, Sadomasochism, Transvestism*. Simon & Schuster, New York, 1980.

Gotwald, W. H., and G. H. Golden. *Sexuality—The Human Experience*. Macmillan, New York, 1981.

Green, R. D., ed. *Human Sexuality: A Health Practitioner's Text*. 2nd edition. Williams and Wilkins, Baltimore, 1979 (first edition, 1975).

Green, R. D., and J. Money, eds. *Transsexualism and Sex Reassignment*. Johns Hopkins Press, Baltimore, 1969.

Greenberg, J. S., C. E. Bruess, and D. W. Sands. *Sexuality—Insights and Issues*. Wm. C. Brown, Dubuque, 1986.

Griffin, S. *Pornography and Silence*. Harper & Row, New York, 1981.

Griffitt, W., and E. Hatfield. *Human Sexual Behavior*. Scott, Foresman, Glenview, Ill., 1985.

Groth, A. N., and H. J. Birnbaum. *Men Who Rape—The Psychology of the Offender*. Plenum Press, New York, 1979.

Hafex, E. S. E., and T. N. Evans. *Human Reproduction: Conception and Contraception*. 2nd edition. Harper & Row, New York, 1980.

Harmatz, M. G., and M. A. Novak. *Human Sexuality*. Harper & Row, New York, 1983.

Hawkins, D. F., and M. G. Elder. *Human Fertility Control—Theory and Practice*. Butterworths, New York, 1979.

Hettlinger, R. F. *Human Sexuality: A Psychosocial Perspective*. Wadsworth, Belmont, Cal., 1975.

Himes, N. E. *Medical History of Contraception.* Schocken Books, New York, 1970.

Hite, S. *The Hite Report on Male Sexuality.* Alfred Knopf, New York, 1981.

Hogan, R. *Human Sexuality—A Nursing Perspective.* Appleton-Century-Crofts, New York, 1980.

Hoult, T. F., L. F. Henze, and J. W. Hudson. *Courtship and Marriage in America.* Little, Brown, Boston, 1978.

Hunt, M. *Sexual Behavior in the 1970's.* Playboy Press, Chicago, 1974.

Hyde, J. S. *Understanding Human Sexuality.* 3rd edition. McGraw-Hill, New York, 1986.

Janda, L. H., and K. E. Klenke-Hamel. *Human Sexuality.* D. Van Nostrand, New York, 1980.

Jones, K. L., L. W. Shainberg, and C. D. Byer. *Sex and People.* Harper & Row, New York, 1977.

Jones, R. E. *Human Reproduction and Sexual Behavior.* Prentice-Hall, Englewood Cliffs, N.J., 1984.

Kaplan, H. S. *The New Sex Therapy.* Brunner/Mazel, New York, 1974.

Karlen, A. *Sexuality and Homosexuality: A New View.* W. W. Norton, New York, 1971.

Katchadourian, H., and D. T. Lunde. *Fundamentals of Human Sexuality.* Holt, Rinehart & Winston, New York, 1975.

Kempe, R. S., and C. H. Kempe. *The Common Secret—Sexual Abuse of Children and Adolescents.* W. H. Freeman, New York, 1984.

Kilmann, P. R., and K. H. Mills. *All About Sex Therapy.* Plenum, New York, 1983.

Kinsey, A. C., et al. *Sexual Behavior in the Human Male.* Saunders Co., Philadelphia, 1948.

Kinsey, A. C., et al. *Sexual Behavior in the Human Female.* Saunders Co., Philadelphia, 1953.

Knox, D. *Human Sexuality—The Search for Understanding.* West Publishing, St. Paul, Minn., 1984.

Knox, D. *Marriage—Who? When? and Why?* Prentice-Hall, Englewood Cliffs, N.J., 1974.

Kolodny, R. C., W. H. Masters, and V. E. Johnson. *Textbook of Sexual Medicine.* Little, Brown, Boston, 1979.

Lasswell, M., and T. E. Lasswell. *Marriage and the Family.* D. C. Heath, Lexington, Mass., 1982.

Leiblum, S. R., and L. A. Pervin, eds. *Principles and Practice of Sex Therapy.* Guilford Press, New York, 1980.

Lockwood, D. *Prison Sexual Violence.* Elsevier, New York, 1980.

Luker, K. *Abortion and the Politics of Motherhood.* University of California Press, Berkeley, Cal., 1984.

Luria, Z., and M. D. Rose. *Psychology of Human Sexuality.* John Wiley & Sons, New York, 1979.

MacDonald, J. M. *Indecent Exposure.* C. C. Thomas, Springfield, 1973.

MacKinnon, C. A. *Sexual Harassment of Working Women.* Yale University Press, New Haven, Conn., 1979.

Maier, R. A. *Human Sexuality in Perspective.* Nelson-Hall, Chicago, 1984.

Malamuth, N., and E. Donnerstein, eds. *Pornography and Sexual Aggression.* Academic Press, New York, 1984.

Marmor, J., ed. *Homosexual Behavior.* Basic Books, New York, 1980.

Masters, W. H., and V. E. Johnson. *Homosexuality in Perspective.* Little, Brown, Boston, 1979.

Masters, W. H., and V. E. Johnson. *Human Sexual Inadequacy.* Little, Brown, Boston, 1970.

Masters, W. H., and V. E. Johnson. *Human Sexual Response.* Little, Brown, Boston, 1966.

Masters, W. H., and V. E. Johnson. *Pleasure Bond.* Little, Brown, Boston, 1975.

Meeks, L. B., and P. Heit. *Human Sexuality.* Saunders, Philadelphia, 1982.

Meiselman, K. C. *Incest—A Psychological Study of Causes and Effects with Treatment Recommendations.* Jossey-Bass, San Francisco, 1986.

Milbauer, B. *The Law Giveth: Legal Aspects of the Abortion Controversy.* Atheneum, New York, 1983.

Mims, F. H., and M. Swenson. *Sexuality: A Nursing Perspective.* McGraw-Hill, New York, 1980.

Money, J., and H. Musaph, eds. *Handbook of Sexology II—Genetics, Hormones and Behavior.* Elsevier, New York, 1977.

Mueller, G. O. W. *Sexual Conduct and the Law.* 2nd edition. Oceana Publications, Dobbs Ferry, N.Y., 1980.

Murstein, B. I. *Love, Sex and Marriage: Through the Ages.* Springer Publishing Co., New York, 1974.

Nass, G. D., R. W. Libby, and M. P. Fisher. *Sexual Choices—An Introduction to Human Sexuality.* 2nd edition. Wadsworth, Monterey, Cal., 1984.

Nass, G. D., and G. W. McDonald. *Marriage and the Family.* 2nd edition. Addison-Wesley, Reading, Mass., 1982.

Offir, C. W. *Human Sexuality.* Harcourt Brace Jovanovich, New York, 1982.

Oliven, J. F. *Clinical Sexuality.* Lippincott, Philadelphia, 1974.

Parrinder, G. *Sex in the World's Religions.* Oxford University Press, New York, 1980.

Parsons, J. E., ed. *The Psychobiology of Sex Roles.* McGraw-Hill, New York, 1980.

Peel, J., and M. Potts. *Textbook of Contraceptive Practice.* Cambridge University Press, Cambridge, 1970.

Petras, J. W. *Sex: Male / Gender: Masculine.* Alfred Publishing Co., Washington, N.Y., 1975.

Petras, J. W. *Sexuality in Society.* Allyn & Bacon, Boston, 1973.

Pope, K. S., and Associates. *On Love & Loving.* Jossey-Bass, San Francisco, 1980.

Qualls, C. B., J. P. Wincze, and D. H. Barlow. *The Prevention of Sexual Disorders.* Plenum, New York, 1978.

Rathus, S. A. *Human Sexuality.* Holt, Rinehart & Winston, New York, 1983.

Read, D. A. *Healthy Sexuality.* Macmillan, New York, 1979.

Reiss, I. L. *Journey Into Sexuality—An Exploratory Voyage.* Prentice-Hall, Englewood Cliffs, N.J., 1986.

Renshaw, D. C. *Incest—Understanding and Treatment.* Little, Brown, Boston, 1982.

Richmond-Abbott, M. *Masculine & Feminine—Sex Roles Over the Life Cycle.* Addison-Wesley, Reading, Mass., 1983.

Roberts, E. J., ed. *Childhood Sexual Learning: The Unwritten Curriculum.* Ballinger, Cambridge, Mass., 1980.

Rosen, R., and E. Hall. *Sexuality.* Random House, New York, 1984.

Rubin, H. H., and B. W. Newman. *Active Sex After Sixty.* Arco Publishing Co., New York, 1969.

Russell, D. E. H. *Rape in Marriage.* Macmillan, New York, 1982.

Russell, D. E. H. *The Secret Trauma—Incest in the Lives of Girls and Women.* Basic Books, New York, 1986.

Russell, D. E. H. *Sexual Exploitation—Rape, Child Sexual Abuse, and Workplace Harassment.* Sage, Beverly Hills, Cal., 1984.

Sandler, J., M. Myerson, and B. N. Kinder. *Human Sexuality: Current Perspectives.* Mariner Publishing, Tampa, Fla., 1980.

Schaffer, K. F. *Sex Roles and Human Behavior.* Winthrop, Cambridge, Mass., 1981.

Schulz, D. A. *Human Sexuality.* Prentice Hall, Englewood Cliffs, N.J., 1979.

Shope, D. *Interpersonal Sexuality.* Saunders Co., Philadelphia, 1975.

Sorenson, R. C. *Adolescent Sexuality.* World Publishing Co., New York, 1973.

Spanier, G. B. *Human Sexuality in a Changing Society.* Burgess, Minneapolis, 1979.

Starr, B. D., and M. B. Weiner. *The Starr-Weiner Report on Sex and Sexuality in the Mature Years.* Stein & Day, New York, 1981.

Strong, B., and R. Reynolds. *Understanding Our Sexuality.* West Publishing, St. Paul, Minn., 1982.

Strong, B., S. Wilson, M. Robbins, and T. Johns. *Human Sexuality—Essentials.* 2nd edition. West Publishing, St. Paul, Minn., 1981.

Stuart, I. R., and L. E. Abt. *Interracial Marriage.* Grossman Publishers, New York, 1973.

Szasz, T. S. *Sex by Prescription.* Anchor Press, Garden City, N.Y., 1980.

Unger, R. K. *Female and Male—Psychological Perspectives.* Harper & Row, New York, 1979.

Weinberg, M. S., and C. J. Williams. *Male Homosexuals.* Oxford University Press, New York, 1974.

Wheeler, M. *No-Fault Divorce.* Beacon Press, Boston, 1974.

Witters, W. L., and P. Jones-Witters. *Human Sexuality—A Biological Perspective.* D. Van Nostrand, New York, 1980.

Woods, N. F. *Human Sexuality in Health and Disease.* 3rd edition. C. V. Mosby Co., St. Louis, 1984.

Zubin, J., and J. Money, eds. *Contemporary Sexual Behavior.* Johns Hopkins Press, Baltimore, 1973.

# APPENDIX E
# ATLASES AND DICTIONARIES

Beigel, H. G. *Sex From A to Z.* Stephen Daye Press, New York, 1961.

Dickinson, R. L. *Human Sex Anatomy: A Topographical Hand Atlas.* Williams and Wilkins, Co., Baltimore, 1949.

Ellis, A., and A. Abarbanel, eds. *The Encyclopedia of Sexual Behavior.* Jason Aronson Inc., New York, 1973.

Gillette, P. J. *The Complete Sex Dictionary.* Award Books, New York, 1969.

Goldstein, M., et al. *The Sex Book, A Modern Pictorial Encyclopedia.* Herder and Herder, and Bantam Books, New York, 1973.

Haeberle, E. J. *The Sex Atlas.* Seabury Press, New York, 1978.

Kramarae, C., and P. A. Treichler. *A Feminist Dictionary.* Pandora Press, Boston, 1985.

Lingeman, R. R. *Drugs From A to Z.* McGraw-Hill, New York, 1974.

Netter, F. H. *Reproductive System, Vol. 2, Ciba Collection of Medical Illustrations.* Ciba, Summit, N.J., 1965.

Reissner, A., and C. Wade. *Dictionary of Sexual Terms.* Associated Booksellers, Bridgeport, Conn., 1967.

Steen, E. B. *Dictionary of Biology.* Barnes and Noble, a Division of Harper & Row, Inc., New York, 1971.

Valensin, G. *Sex from A to Z.* Berkeley Publishing Corp., New York, 1969.

Wentworth, H., and S. B. Flexner. *Dictionary of American Slang.* Thomas Y. Crowell Co., New York, 1960.

# APPENDIX F
# DICTIONARY
# OF SEXUAL TERMS

## PREFACE

This "Dictionary of Sexual Terms" defines not only words used in everyday language, including colloquial and slang terms, but also the more specialized terms. Basic terms in human reproduction and development, as well as abbreviations and acronyms for sexual terms, are also included. This is, so far as we know, the only comprehensive dictionary covering this field now available. The greatly increased interest in the broad field of sex and sexuality and the greater openness with which topics pertaining to sex are discussed and appear in print make a dictionary of this type indispensable.

Cross-references are used to expand the meaning wherever necessary. Specific pronunciations are not given, but each word is divided into syllables with the syllable bearing the primary accent designated, so that, in the majority of cases, correct pronunciation is attained with little difficulty.

## ABBREVIATIONS USED IN THE DEFINITIONS

| | | | |
|---|---|---|---|
| *Abbr.* | abbreviation | *n.* | noun |
| *adj.* | adjective | *pl.* | plural |
| C. | Celsius, Centigrade | q.v. | which see |
| e.g. | for example | *sing.* | singular |
| F. | Fahrenheit | *v.* | verb |
| i.e. | that is | | |

**ab'do·men** A large cavity of the trunk of the body extending from the diaphragm to the brim of the pelvis.

**ab·dom'i·nal** Of or pertaining to the abdomen.

**ab·dom'i·nal sec'tion** A cesarean section, q.v.

**ab·er'rant** Exceptional, abnormal; deviating from normal in development, form, structure, or behavior.

**ab·er·ra'tion** A departure from the usual or normal course. *See* deviation, perversion.

**ab'i·gail** 1. A lady's maid. 2. *Slang.* Nickname for a stuffy, middle-aged homosexual.

**ab·la'tion** The surgical removal or excision of a part of the body.

**ab·nor'mal** Extraordinary, unusual; not typical or average.

**a·bort'** To terminate a pregnancy prematurely; to miscarry.

**a·bor'ti·cide** 1. The killing of a fetus within the uterus. 2. An agent that kills a fetus and brings about an abortion.

**a·bor·ti·fa'cient** A drug or agent used to induce an abortion.

**a·bor'tion** The removal or expulsion of the products of conception (embryo or fetus and its embryonic membranes) before the age of viability, usually 28 weeks, now arbitrarily defined as a fetus weighing less than 500 gm. *See* miscarriage, premature birth.

   **a., crim'i·nal** An illegally induced abortion; one performed contrary to the laws of the state.

   **a., ha·bit'u·al** An abortion that occurs repeatedly at about the same time in successive pregnancies.

   **a., in·com·plete'** An abortion in which there is retention of a portion of the products of conception.

   **a., in·duced'** An abortion brought about by chemical or mechanical means; one that does not occur naturally; an artificial abortion.

   **a., in'ter·im** An abortion occurring between the thirteenth and sixteenth weeks.

   **a., in'tra-am·ni·ot'ic pros·ta·glan'din** An abortion induced by the injection of a prostaglandin into the amniotic sac.

   **a., lunch hour** Menstrual extraction, q.v.

   **a., meth'ods of in·duc'ing an** *See* abortion, intra-amniotic prostaglandin; dilatation and curettage; dilatation and evacuation; menstrual regulation; saline injection method; vacuum curettage.

   **a., missed** An abortion in which the fetus dies but the products of conception remain within the uterus for two weeks or longer before being extruded.

   **a., spon·ta'ne·ous** An abortion that occurs naturally; one not induced.

   **a., suc'tion** An abortion in which the conceptus is removed by aspiration.

   **a., ther·a·peu'tic** An abortion that is performed (a) when pregnancy or birth of the fetus threatens the life or health of the mother or (b) when the likelihood is great that the fetus conceived will be grossly abnormal.

   **a., threat'ened** An abortion that is likely to occur, usually indicated by vaginal bleeding or spotting accompanied by mild cramps.

**a·bor'tion·ist** One who performs abortions.

**a·bor'tion pill** The morning-after pill, q.v.

**a·bor'tus** An aborted fetus; the products of an abortion.

**ab·rup'ti·o pla·cen'tae** The premature separation of the placenta from the uterus.

**ab'scess** A collection of pus contained within a cavity formed from the disintegration of tissues.

**ab'sti·nence** Voluntary self-denial or abstention from indulgence in food, alcoholic drink, or sexual relations; continence. *See* celibacy.

**ac·ces'so·ry** Auxiliary, assisting, secondary.

**ac·ces'so·ry sex glands** In a male, the seminal vesicles, prostate gland, and bulbo-urethral glands that produce the seminal plasma.

**ac·couche·ment'** The act of childbirth; parturition.

**ac·cou·cheur'** One who assists in childbirth; a midwife or an obstetrician.

**AC-DC** *Slang.* Bisexual; ambisexual.

**ac'i·nus** A small, saclike secreting unit of a gland, especially one with a narrow lumen. *Compare* alveolus.

**ac'ne** An inflammatory disease of the skin, involving areas where the sebaceous glands are numerous and active and characterized by the development of comedones, pustules, and oftentimes inflamed nodules and cysts. It is common at puberty, affecting 80 percent or more of teenagers. It is attributed primarily to androgenic hormones.

**ac·quired' im·mune' de·fi'cien·cy syn'drome** *Abbr.* AIDS. A new disease of unknown cause and high virulence, resulting from a defective immune system, which leaves patients unable to resist infections, especially a certain type of pneumonia, and the development of cancer, particularly a rare form, *Kaposi's sarcoma.* About 75 percent of the patients are homosexual or bisexual men. The mortality rate is high.

**ac′ro·sin** A trypsin-like enzyme, produced in the acrosome of a spermatozoon, that destroys or inactivates the decapacitation factor present in the male reproductive tract.

**ac′ro·some** A small, dense, granular body in a spermatid that forms a caplike structure (acrosomal cap) covering the head of a spermatozoon. It contains hyaluronidase and a protein-digesting enzyme that aid the sperm in their ability to penetrate the corona radiata and zona pellucida of an ovum.

**ac′tive part′ner** In heterosexual or homosexual intercourse, the one who inserts the penis. *Compare* passive partner.

**A′dam's PJ's** *Slang.* The person is nude.

**ad·e·no′sis** Any disease of glands or the abnormal development and functioning of glandular tissue, especially the presence of red, mucus-secreting, glandular tissue on the surface of the cervix and the vagina.

**ad·nex′a** Accessory parts or appendages.

**ad·o·les′cence** The period of development extending from puberty to maturity; the period of youth.

**ad·o·les′cent** 1. *adj.* Undergoing adolescence. 2. *n.* An adolescent person.

**ad·re′nal gland** One of two endocrine glands located directly above each kidney, each consisting of a *medulla* and a *cortex.* The medulla secretes catecholamines, *epinephrine* (adrenalin) and *norepinephrine* (noradrenalin); also *dopamine.* The cortex secretes a number of steroid hormones including *glucocorticoids* involved in carbohydrate metabolism, *mineralocorticoids* involved in sodium, potassium, and water metabolism, and *sex hormones* (androgens, estrogens, and progesterone). The medulla is regulated by nerve impulses transmitted through the autonomic nervous system; the cortex is controlled by trophic hormones, especially the adrenocorticotrophic hormone (ACTH) secreted by the anterior pituitary. Also called *suprarenal gland.*

**ad·ren·arch′e** The time in the development of a child when pubic and axillary hair appear brought about by the increased output of adrenocortical sex hormones. Usually occurs about the age of eight or nine.

**ad·re·no·gen′i·tal syn′drome** A pathological condition characterized by excessive secretion of sex hormones, especially androgens, by the adrenal cortex. In both sexes, before puberty, it results in precocious development; in adult females, virilism.

**a·dult′** A mature individual; one who has attained full growth and development; of legal age.

**a·dul′ter·er** A person who commits adultery, especially a man.

**a·dul′ter·ess** A woman who commits adultery.

**a·dul′ter·y** Voluntary sexual intercourse between a married person and a person other than his or her lawful spouse. It may be *conventional,* which is characterized by deception and is unknown to the spouse, or *consensual,* which is known to and consented to by the spouse.

**aes·thet′ic, es·thet′ic** Of, pertaining to, or having a sense of appreciation for the beautiful.

**af·fec′tion·ate** Expressing fond feelings, love, or affection.

**af·fi′ance** To pledge in marriage; to betroth.

**af·fin′i·ty** 1. A natural affection for or attraction to a person or thing. 2. A relationship by marriage.

**af′ter·birth** The material expelled from the uterus following the birth of a baby, consisting of the placenta and embryonic membranes. Also called *secundines.*

**af'ter·pains** Cramplike pains, sometimes severe, that occur after the expulsion of the afterbirth. They result from uterine contractions following delivery.

**af'ter·play** Activities, such as kissing and caressing following coitus, that provide psychological and emotional satisfaction greatly enhancing coital pleasure, especially in the female. *Compare* foreplay.

**a·ga·lac'ti·a** Absence of, or failure of, the secretion of milk.

**a·gam'ic** Asexual; reproducing by means other than the union of sex cells.

**a·ga'pe** Pure Christian love; unselfish brotherly love of one person for another without sexual implications; nonsexual love.

**a·gap'tism** Continent marriage; marriage without sex.

**age of con·sent'** The age at which a minor is legally considered capable of consenting to sexual intercourse. It varies from 14 to 19; in most states it is 18.

**age of vi·a·bil'i·ty** The age at which a fetus is capable of extrauterine existence, generally considered to be 28 weeks, now arbitrarily defined as a fetal weight of more than 500 gm.

**ag'nate** Related through male descent or through the father's side only.

**a·gon'ad·ism** The absence of ovaries or testes, or the condition resulting from their failure to develop and function.

**AI** Artificial insemination, q.v.

**AID** Artificial insemination with sperm furnished by donor.

**AIDS** Acquired immune deficiency syndrome, q.v.

**AIH** Artificial insemination with sperm furnished by husband.

**al·bu·mi·nu'ri·a** The presence of serum albumin and other proteins in urine.

**al·go·lag'ni·a** A condition in which sexual feelings are aroused and sexual gratification is enhanced by experiencing pain or inflicting pain on another person. *See* masochism, Sacher-Masoch, Sade, sadism, sadistic, sadomasochism.

**al·go·phil'i·a** Masochism.

**al'i·mo·ny** Money paid to a woman from her husband's estate or income by court order to support her during divorce proceedings or following divorce or legal separation. In some states a husband may receive alimony.

**al·lan'to·is** An extraembryonic membrane that in reptiles, birds, and many mammals functions in respiration and excretion. In humans, it is a vestigial structure consisting of a tiny endodermal tube present in the yolk stalk. Its blood vessels, however, become important umbilical arteries and veins.

**al'ter** To sterilize or desex an animal by castration or spaying; applied especially to cats.

**al·ve'o·lus** A small saclike dilatation, hollow, or cavity, as (a) the terminal secreting portion of an alveolar gland (*compare* acinus); (b) an air cavity of the lung; (c) the socket of a tooth.

**am'a·tive** Passionate, amorous, disposed to love.

**am'a·tive·ness** Propensity to love; sexual desire.

**am'a·to·ry** Of, pertaining to, or associated with love, especially sexual love; expressive of love.

**am'a·zon** A tall, vigorous, aggressive woman with masculine traits. *Compare* termagant, virago.

**am·bi·sex'u·al** Of, pertaining to, or associated with both males and females; common to both sexes. Also *ambosexual.*

**am·bi·sex·u·al'i·ty** The practice of both heterosexual and homosexual behavior.

**am·biv'a·lence** The coexistence of contradictory feelings and attitudes such as love-hate or tenderness-cruelty toward a person, thing, or idea.

**am·bo·sex'u·al** *See* ambisexual.

**AMCA** Tranexamic acid, q.v.

**a·me'li·a** The congenital absence of a limb or limbs.

**a·men·or·rhe'a** The absence of or suppression of menstrual flow.

**A·mer'i·can cap** A condom that covers only the glans of the penis.

**am·i·no·glu·te·thi'mide** A luteolytic, once-a-month oral contraceptive.

**am·ni·o·cen·te'sis** Puncture of the amniotic sac, usually for removal of some of its contents by means of an injection needle or catheter. *See* karyotyping.

**am'ni·on** An extraembryonic membrane that forms a fluid-filled sac enclosing the embryo or fetus. It, together with the chorion with which it fuses, constitutes the "bag of waters" that bursts at parturition.

**am·ni·ot'ic** Of or pertaining to the amnion.

**am·ni·ot'ic flu'id** The fluid that surrounds the embryo or fetus during intrauterine development. Examination of this fluid is utilized in detecting fetal and placental abnormalities. Prenatal sex of the fetus can be determined by examination of the cells within the fluid.

**am·ni·ot'ic sac** The amnion, q.v.

**am'o·rous** Showing or expressing love; of or associated with love, especially sexual love.

**a·mour'** A love affair, especially an illicit one.

**a·mour' pro'pre** Self-love; self-esteem; self-respect.

**am·plex'us** Sexual embrace without true intercourse, as occurs in frogs, fertilization of the eggs taking place externally.

**am·pul'la** A dilated portion of a tube or duct, as (a) that in the vas deferens just before its junction with the ejaculatory duct; (b) the middle portion of the uterine tube located between the infundibulum and isthmus.

**am'u·let** An object, such as an ornament, gem, or relic, usually worn about the neck and believed to possess magical powers protecting a person against evil or harm or aiding one in love. Amulets were sometimes worn for contraceptive purposes.

**am'yl ni'trate** A drug inhaled before or during the sex act for its presumed stimulant properties.

**an·a·cli'sis** The development of a libidinal attachment to a person on the basis of resemblance of that person to an early-childhood protective figure, such as a parent or teacher.

**a'nal** Of, pertaining to, or associated with the anus.

**a'nal ca·nal'** The terminal portion of the large intestine that extends from the rectum proper to the anus.

**an·al·ge'si·a** Relief from pain and suffering, especially without the loss of consciousness; insensibility to pain.

**a'nal in'ter·course** Insertion of the penis into the rectum through the anus; buggery; sodomy.

**a·nal'o·gous** Resembling or corresponding to another structure in function but not in fundamental structure or embryonic development. *Compare* homologous.

**a'nal stage** According to Freud, the second stage in the psychosexual development of a child, during which the anus is the erogenous zone invested with libido. It follows the oral stage and precedes the phallic stage.

**a'nal vir'gin** A young man who has never been the recipient of the penis in anal intercourse.

**an·aph·ro·dis'i·ac** A substance that allegedly reduces or allays sexual desire or feelings, as cyprosterone, which is effective in males.

**a·nat'o·my** The study of the structure of an organism or a part of it.

**an·dro·blas·to'ma** A rare and usually benign tumor of the testis. Also called *Sertoli cell tumor.*

**an·dro·cen'tric** Male-centered; dominated by males.

**an′dro·gen** 1. A substance that promotes the development and functioning of the male genital organs and the development of male secondary sexual characteristics. 2. A male hormone, as testosterone.

**an·dro·gen′e·sis** The activation of an egg by a spermatozoon followed by development without active participation of the egg nucleus.

**an·dro·gen′ic** Producing masculine characteristics or effects.

**an·drog′en·ous** Producing males only.

**an·drog′en·ized** Masculinized.

**an′dro·gyne** A person who possesses both male and female characteristics; a hermaphrodite or pseudohermaphrodite.

**an·drog′y·nous** Bisexual; possessing both male and female characteristics; hermaphroditic.

**an·drog′y·ny** 1. A condition in which an individual possesses both male and female characteristics, especially the possession of female characters by a male. *Compare* gynandry. 2. A condition in a society that shows little differentiation between sex roles.

**an′droid** Resembling a male.

**an·drol′o·gist** A physician who specializes in diseases of men, especially those involving the urinary and reproductive systems; a specialist in male fertility. *Compare* gynecologist.

**an·dro·pho′bi·a** Excessive fear of or dislike of males. *Compare* gynephobia.

**an′dro·sperm** A Y-bearing, male-producing sperm. *See* Y chromosome.

**an·dro·stene′di·one** An androgen produced by the testes which, with testosterone, is an active, circulating, androgenic hormone.

**an·dros′ter·one** An androgenic steroid found in human urine, considered to be a metabolite of testosterone.

**an·es·the′sia** The loss of feeling or perception, especially insensibility to touch or pain.

    **a., general** Loss of sensation accompanied by loss of consciousness.

    **a., local** Anesthesia limited to a local area.

    **a., regional** Anesthesia of a region brought about by blocking the passage of sensory impulses over nerves from that area. Also called *blocking* or *conduction anesthesia.*

    **a., spinal** Anesthesia resulting from a lesion of the spinal cord, or produced by injection of an anesthetic into the subarachnoid space surrounding the spinal cord.

**an·es·thet′ic** An agent that induces anesthesia.

**an·es′trum** The interval between estrus periods (periods of heat) in mammals.

**an·he·do′ni·a** Inability to experience pleasure in acts that normally give pleasure.

    **a., ejaculatory** An ejaculation that is not accompanied by an orgasm.

**a·ni·lin′gus** The application of the tongue to the anus.

**an′i·mal** 1. *n.* Any living organism that is not a plant or a protist. 2. *n.* An inhuman person or being, one who is bestial or brutish; a sexually aggressive individual. 3. *adj.* Pertaining to sensual or physical qualities as opposed to spiritual values.

**an·nul′** To declare null and void, as a marriage; to invalidate.

**an·nul′ment** The legal invalidation of a marriage; a declaration that a marriage was never valid.

**a·nom′a·ly** Any deviation from the usual or normal; any organ or structure that is abnormal in structure, form, or location.

**an·or′chi·a, an′or·chism** The absence of testes.

**an·or·gas′mic** Incapable of experiencing an orgasm.

**an·or·gas′my** The inability to experience an orgasm or to reach a climax during coitus.

**an·ov'u·lar** Not accompanied by the discharge of an ovum from the ovary.

**an·ov·u·la'tion** The failure of an ovary to release an ovum.

**an·ov'u·la·to·ry cy'cle** A menstrual cycle in which an egg is not discharged from the ovary.

**an·te·na'tal** Present at or occurring before birth; prenatal.

**an·te·par'tum** Before delivery.

**an·te'ri·or** Toward the front end of a quadruped; in humans, toward the front side of the body. Opposite of *posterior.*

**an·te·ver'sion** The forward tipping or tilting of an organ, as the uterus.

**an·ti·an'dro·gen** A substance that counteracts the effects of an androgen, as cyprosterone, a synthetic steroid.

**an'ti·bo·dy** A substance, usually a gamma globulin, that acts as an immunizing agent. It occurs naturally or is formed as a result of the presence of an antigen. Depending on their action, antibodies are classified as agglutinins, precipitins, lysins, opsonins, etc. See antigen-antibody reaction.

**an·ti·fer·til'i·ty** Limiting excessive production of offspring; contraceptive.

**an'ti·gen** A substance, usually a protein, that induces the formation of an antibody.

**an'ti·gen-an'ti·bo·dy re·ac'tion** That which occurs when an antigen reacts with a specific antibody. It may result in agglutination, precipitation, toxin neutralization, destruction of cells by lysins, or complement fixation.

**an·ti·ov'u·la·to·ry** Inhibiting or suppressing ovulation.

**an·tique' deal'er** *Slang.* A person interested in having sex with an elderly person, especially in a homosexual relationship.

**an·ti·sper·ma·to·gen'ic** Preventing the development of or the functioning of spermatozoa.

**a'nus** The outlet of the alimentary canal; the external opening of the anal canal.

**a·pan'dri·a** An aversion to, or an extreme dislike for males.

**a·par·eu'ni·a** Inability to engage in sexual intercourse; abstinence.

**a·phal'late** Lacking a penis.

**a·phan'i·sis** Lacking all aspects of sexuality.

**aph·ro·dis'i·a** 1. Sexual desire, especially when excessive. 2. Coitus or sexual union.

**aph·ro·dis'i·ac** A substance that stimulates sexual desire and, in the male, induces an erection. See cantharides.

**Aph·ro·di'te** The ancient Greek goddess of love.

**ar·rhe·no·blas·to'ma** A malignant ovarian tumor secreting androgens that have a masculinizing effect on adult females and on a developing female fetus.

**a·re'o·la** The pigmented region surrounding the nipple of a mammary gland. It enlarges and becomes darker during pregnancy.

**ar·ti·fi'cial** Accomplished by or made by humans; not natural.

**ar·ti·fi'cial in·sem·i·na'tion** *Abbr.* AI. The injection of semen into the vagina or uterus as a means of bringing about a pregnancy. The sperm may be provided by the husband (AIH) or by another person called a donor (AID). The semen used may be fresh or frozen.

**ASA** Aspermatogenic antigen, q.v.

**as·cet'ic** One who practices asceticism; a person who lives a simple, austere life.

**as·cet'i·cism** A mode of living in which a person dedicates himself to high moral principles and values and abstains from the normal pleasures of life, denying himself that which provides material comforts and satisfaction. Such usually involves a life of solitude and the practice of celibacy, fasting, and self-mortification.

**a·sex'u·al** Without sex or sex organs; sexless.

**a·sper·ma·to·gen'ic an'ti·gen** *Abbr.* ASA. A substance that when injected into laboratory animals induces infertility by immobilizing sperm.

**a·sper'mi·a, a·sper'ma·tism** The absence of spermatozoa in the semen; failure of the testes to produce spermatozoa or failure of the ducts from the testes to transport sperm.

**as·pi·ra'tion** The withdrawal of gasses or fluids from a cavity by suction or negative pressure.

**ass** 1. A vain, stupid, arrogant person. 2. *Vulgar.* a. The buttocks. b. The anus or rectum. c. Sexual intercourse, referred to as a *piece of ass.*

**as·sault'** An unlawful, intentional threat or an attempt by force to inflict physical harm on another person. *See* sexual assault.

**As·wi'na-Mud'ra ex'er·cise** *See* Kegel's exercises.

**at'o·ny, a·to'ni·a** Absence of or an abnormally low degree of muscular tone.

**at'ro·phy** The reduction in size of an organ or structure resulting from lack of nourishment or reduced functional activity. This condition may be physiological or pathological. Some degree of degeneration may accompany atrophy.

**aunt, aunt'ie** *Slang.* 1. An aging homosexual. 2. An old prostitute. 3. A madam.

**au·to·er'o·tism, au·to·e·rot'i·cism** Obtaining sexual satisfaction through solitary sexual activities, such as masturbation, nocturnal emissions, and sexual fantasies; autosexuality.

**au·to·im·mu'ni·ty** A condition in which a humoral or cell-mediated response occurs against the body's own cells, which act as autoantigens, as when decreased semen quality and infertility occur due to the presence of sperm antibodies.

**au·to·ma·nip'u·la'tion** Self-manipulation of the sex organs; masturbation.

**au·to·sex·u·al'i·ty** Autoeroticism.

**au'to·some** Any chromosome that is not a sex chromosome.

**a·ver'sion ther'a·py** Therapy directed toward correcting an undesirable behavior by associating the behavior with an unpleasant stimulation or by treating the patient so that an unfavorable response or reaction results from the undesirable behavior.

**ax'il·lar·y** Of or pertaining to the axilla or armpit.

**a·zo·o·sper'mi·a, a·zo·o·sper'ma·tism** The absence of spermatozoa in the semen, which may result from testicular failure or from an obstruction in the transporting ducts.

**Ba'al** A male fertility and nature god of ancient Semitic peoples.

**ba'by,** *pl.* **ba'bies** 1. A very young child, an infant. 2. An immature or childish person. 3. *Slang.* An attractive girl or young woman; a sweetheart.

**ba'by pro** *Slang.* A child prostitute.

**bac'u·lum** A penis bone, q.v.

**bag** *Slang.* 1. A woman's douche bag. 2. A promiscuous woman or prostitute. 3. An unattractive girl or woman. 4. An old, gossipy, shrewish woman. 5. A woman's rubber diaphragm or pessary. 6. The scrotum. 7. A condom.

**bag of wa'ters** The amniotic sac and its contained fluid.

**ba'gnio** A brothel.

**bairn** A child (Scottish).

**bal·a·ni'tis** Inflammation of the glans penis or glans clitoridis.

**bal'a·nus** The glans penis or glans clitoridis.

**ball** *Slang.* 1. *n.* A wild, unrestrained, uninhibited party. 2. *v.* To have sex with.

**bal·lotte'ment** A diagnostic sign of pregnancy in which, when the finger is inserted into the vagina and brought into contact with the uterus, the movement of the fetus within the amniotic sac can be detected.

**balls** *Slang.* The testes.

**band-aid sur'gery** Laparoscopic sterilization.

**B and D** Bondage and discipline.

**bang** *Slang.* To have sex with.

**bare** Naked, nude, without appropriate covering.

**Barr bod'y** The sex chromatin body. *See* nuclear sexing.

**bar'rel** *See* vagina barrel.

**bar'ren** Incapable of producing offspring; infertile, sterile; childless.

**Bar'tho·lin, glands of** The vestibular glands, q.v.

**bas'al bod'y tem'per·a·ture** *Abbr.* BBT. The temperature of the body upon waking in the morning or after several hours of complete rest. Normal BBT in health is about 98.6° F. (37° C.) taken orally. Axillary temperature is lower, rectal temperatures about 1° higher.

**bas'ket** *Slang.* The scrotum.

**bas'tard** 1. An illegitimate child. 2. A child born of unwed parents. 3. *Slang.* A despicable, untrustworthy person; a generally disliked person.

**bat** *Slang.* An unattractive girl or woman.

**bat'tle-ax** *Slang.* A sharp-tongued, mean, belligerent woman.

**bawd** 1. A prostitute. 2. A woman who manages a brothel; a madam.

**baw'dry** Obscene or ribald language dealing with sex.

**bawd'y** Indecent, obscene, lewd.

**bawd'y·house** A house of prostitution.

**BBT** Basal body temperature, q.v.

**beat off, beat the meat, beat the dum'my** *Slang.* To masturbate.

**beau** The sweetheart or lover of a girl or young woman.

**be·get'** To procreate; to bring forth offspring, especially with reference to a male.

**be·hav'ior mod·i·fi·ca'tion** The application of the principles of behavioral psychology, especially operant psychology, to redirect actions in a desired manner or direction.

**be·hind'** *Informal.* The buttocks.

**belle** An attractive or beautiful girl or young woman.

**belle girl** *Slang.* A call girl who has an arrangement with bellboys of hotels and motels for the referral of potential customers.

**bel'ly** The abdomen or abdominal cavity.

**bel'ly but'ton** *Informal.* The navel or umbilicus.

**bel'ly but'ton sur'gery** Band-aid surgery, q.v.

**be·loved'** 1. *adj.* Held in high esteem and affection. 2. *n.* One that is beloved.

**bent-nail syn'drome** Peyronie's disease, q.v.

**ber·dache'** An American Indian transvestite who assumes the role of a woman.

**bes·ti·al'i·ty** Sexual relations between a human being and an animal other than a human.

**bes·ti·o·sex·u·al'i·ty** Sexual relationships with lower animals. *See* bestiality.

**be·troth'** To promise to marry.

**be·troth'al** An engagement or a promise to marry.

**bev'y** A group, especially of girls or women.

**B-girl** A bar girl, especially a nonprofessional prostitute who waits in bars to attract customers.

**bi·det'** A bowl-shaped, porcelain fixture usually found in the bathroom and provided with running water used for cleansing the region of the alimentary and urogenital openings. It is used for hygienic, esthetic, therapeutic, medicinal, and other purposes.

**big'a·mist** A person who commits bigamy; one who is married to two persons at the same time.

**big'a·my** The act of marrying another person while still being legally married.

**Big-O** *Slang.* An orgasm.

**Bill'ings meth'od** A method of determining the time of ovulation, based on changes in the cervical mucus.

**bi·o·feed'back** The enabling of a person to voluntarily control the activities of organs innervated by the autonomic nervous system, such as heartbeat, blood pressure, skin temperature, and the like, this being accomplished through the utilization of monitoring devices.

**bi'op·sy** The examination microscopically of tissue excised from a living body, this usually being performed for diagnostic purposes.

**bird song** *Slang.* Cries and sounds uttered during intercourse, especially at the time of orgasm.

**birth** The expulsion of a fetus from the uterus; the act of being born or that which is born. Also called *delivery, labor, parturition.*

**birth ca·nal'** The canal consisting of the cavities of the uterus and vagina through which the fetus passes at birth. Also called *parturient canal.*

**birth con·trol'** The voluntary control of the number of offspring, accomplished principally through the utilization of contraceptive techniques. *See* contraception.

**birth con·trol', meth'ods of** The principal methods involved in birth control include the following: (a) abstinence from intercourse; (b) withdrawal or coitus interruptus; (c) use of a condom to prevent spermatozoa from being deposited within the vagina; (d) use of spermicidal substances (foams, jellies, creams, pastes) to kill or inactivate sperm; (e) use of a sponge, diaphragm, or cervical cap to prevent sperm from entering the uterus; (f) use of oral contraceptives (the Pill) or contraceptive implants to prevent ovulation; (g) use of the rhythm method to restrict time of intercourse to infertile or "safe" periods; (h) use of an intrauterine device (IUD) to prevent implantation of the developing blastocyst. Other methods include postcoital douching, use of morning-after pill or minipill, and utilization of abortion. Permanent contraceptive techniques include castration (surgical removal of testes or ovaries); inactivation of testes or ovaries by radiation, heat, or chemicals; hysterectomy in the female; ligation of transport ducts (tubal ligation in the female, vasectomy in the male).

**birthmark** A nevus, q.v. *See* maternal impression.

**birth trau'ma** An injury sustained by an infant at birth. It may be of a physical or psychic nature.

**bi·sex'u·al** 1. Of or pertaining to both sexes. 2. Possessing both male and female sexual organs; hermaphroditic. 3. Engaging in sexual relations with individuals of both sexes; heterosexual and homosexual.

**bi·sex·u·al'i·ty** Sexual attraction to and sexual behavior directed toward persons of both sexes.

**bitch** 1. A female dog or the female of other canines (wolf, coyote, fox) or various other carnivores (ferret, otter). 2. *Slang.* (a) A malicious, unpleasant woman; (b) a spiteful, selfish, and unscrupulous woman; (c) a lewd, immoral woman; (d) a promiscuous woman, a prostitute. *See* harlot, slut.

**blad'der** A membranous sac or vesicle that serves as a receptacle for a fluid. *See* urinary system.

**blas'to·cyst** A spherical structure forming an early stage in the development of a mammal. It consists of a layer of cells, the *trophoblast,* to which is attached internally an *inner cell mass;* its cavity is the *blastocoel.* It follows the morula stage and corresponds to the blastula of lower forms. In humans, the blastocyst exists free within the uterus for two or three days and then implants within the uterine endometrium.

**blas'to·mere** One of the cells resulting from the cleavage of a fertilized egg; a segmentation cell. *See* macromere, micromere.

**blas'tu·la** The stage in embryonic development that follows the morula; it consists of cells arranged in a single layer forming a hollow sphere. Its cavity is the *blastocoel.*

**blood-tes'tis bar'ri·er** The Sertoli cells of the testes, which are bound together by tight junctions. These cells prevent the passage of albumin and other proteins from the blood into the seminal fluid.

**blood'y show** A slight bloody discharge from the vagina that occurs shortly before birth of a baby.

**blow** *Slang.* 1. To have oral sex; to perform fellatio or cunnilingus on someone. 2. To take a drug by inhalation, especially cocaine or heroin.

**blow job** *Slang.* Fellatio or cunnilingus.

**blue** 1. Puritanical, as *blue* Sunday. 2. Risqué or off-color; characterized by indecency and obscenity. 3. Containing cursing and swearing, as *blue* language; profane. 4. Pornographic, as *blue* movies.

**blue balls** *Slang.* A venereal disease, especially gonorrhea.

**blue films** Pornographic, stag films depicting sexual acts, usually shown to private groups.

**boar** An uncastrated male hog.

**bod'y stalk** An embryonic structure that connects the embryo to the chorion. In humans, it contains the allantois and blood vessels that become incorporated into the umbilical cord.

**bond'age** The tying up or binding of a sex partner in order to increase sexual response; ligotage.

**bond'ing** The union of a male and female into a more or less permanent sexual relationship. *See* pair-bond.

**boo'boos** *Slang.* The testes.

**boobs** *Slang.* The breasts, especially the well-developed breasts of a young woman.

**bor·del'lo** A brothel or house of prostitution.

**born** Brought into existence or being; brought forth by birth.

**bos'om** The breasts of a human, especially those of a woman.

**bou·doir'** A woman's private sitting or lounging room, dressing room, or bedroom.

**bounce test** A test for determining the time of ovulation. Six days before the expected time of ovulation, the subject bounces on a hard surface by sitting down forcibly three or four times every evening and morning. Occasionally the pain of ovulation (*mittelschmerz*) will be experienced.

**bo'var·ism** The conception of oneself as other than one is, such as in the amorous reveries of an unhappily married woman in which she pictures herself married to a romantic male.

**bowd'ler·ize** To expurgate a literary work by deleting or modifying passages considered to be indelicate, obscene, or otherwise objectionable; to alter a work of art by covering up or changing features considered to be objectionable.

**box** *Slang.* The female genitalia, especially the vagina.

**boy** 1. A male child or youth. 2. *Slang.* (a) An effeminate man; (b) a catamite; (c) a menial worker.

**boyfriend** The sweetheart of a girl or woman.

**breach of prom'ise** Failure to honor a commitment or promise, especially the promise to marry.

**bread** *Slang.* The vagina.

**break'through bleed'ing** Bleeding from the vagina that occurs between menstrual periods, a cause of the "spotting" that sometimes accompanies the use of oral contraceptives.

**breast** 1. The mammary gland in humans; the mamma. 2. The anterior portion of the body from the neck to abdomen; the front of the chest.

**breast works** *Slang.* The female mammary glands, considered as sexual objects.

**breech** The rear end of the body; the buttocks.

**breech birth** A breech presentation. *See* presentation.

**breed** 1. *v.* To produce offspring; to give birth or hatch young; to develop new and improved strains of animals or plants; to procreate. 2. *n.* A race, lineage, or strain.

**breed'ing** 1. Producing new forms of animals or plants; causing animals or plants to reproduce. 2. Training in good manners and proper forms of personal and social conduct.

**bri'dal** 1. *adj.* Of or pertaining to a bride or a wedding. 2. *n.* A wedding or marriage ceremony.

**bride** A newly married woman or one about to be married.

**bride'groom** A newly married man or one about to be married.

**broad** *Slang.* 1. A promiscuous woman or prostitute. 2. A coarse, vulgar woman.

**broad lig'a·ment** A fold of peritoneum that extends laterally from each side of the uterus, connecting it to the body wall.

**brood** 1. A large number of young, especially those hatched at one time and cared for by the mother. 2. The children of a single family, especially when numerous.

**broth'el** A house of prostitution, a bordello.

**broth'er** A male offspring who has the same father and mother as another; a male sibling.

**broth'er-in-law** The brother of one's spouse; the husband of one's sister; the husband of one's spouse's sister.

**broth'ers** *Slang.* Butch lesbians.

**brown** *Slang.* Anal intercourse.

**brown leath'er** A newcomer to the leather crowd in a gay bar.

**bu'bo,** *pl.* **bu'boes** A swollen lymph node, especially one in the region of the groin or axilla.

**buck** 1. The male of the deer, antelope, rabbit, and various other mammals. 2. *Slang.* (a) A high-spirited and impetuous young man; (b) a young black male.

**buff** 1. The bare skin, used principally in the phrase *in the buff,* meaning naked. 2. *Slang.* A girl or young woman.

**bug'ger** *Vulgar.* 1. One who practices buggery, a sodomite. 2. A contemptible or disreputable person.

**bug'ger·y** 1. Anal intercourse. 2. Intercourse with animals; bestiality.

**bul·bo·u·re'thral gland** One of two small glands, located anterior to the prostate gland, whose ducts open into the male urethra. They secrete a viscid fluid that serves for lubrication. Also called *Cowper's gland.*

**bull** 1. An uncastrated male bovine animal, especially of domestic cattle. 2. The adult male of various large mammals, as the elephant, moose, whale, seal, walrus. 3. *Slang.* (a) A large, strong, solidly built, aggressive male; (b) empty, insincere, exaggerated talk.

**bull dyke, bull dike** *Slang.* A lesbian with pronounced masculine appearance and mannerisms; one who usually assumes the aggressive male role.

**bum** *British slang.* The buttocks.

**bund'ling** The practice of a man and woman lying in the same bed with their clothes on, a custom common in early America among courting couples. It was resorted to in order to save firewood and candles. Couples were often separated by a *bundling board.*

**buns** *Slang.* The buttocks.

**bur·lesque'** 1. A literary or dramatic work that treats a serious subject in a light and frivolous manner, subjecting it to ridicule and laughter. 2. A vaudeville entertainment featuring slapstick humor, bawdy songs, ribald stories and jokes, and striptease acts. Also *burlesk.*

**buss** A loud, smacking kiss.

**bust** The breast or chest, especially a woman's bosom.

**butch** *Slang.* 1. A bull dyke. 2. An active lesbian who assumes the role of a male.

**butt** *Slang.* The buttocks or rear end; the rump.

**but'tered bun** *Slang.* A woman who has had intercourse recently with a man other than her present partner.

**but'tocks** The two fleshy parts of the body that lie posterior to the hip joints. They constitute the *gluteal prominences.*

**but'ton** *Slang.* The clitoris.

**ca·det'** *Slang.* A pimp.

**cake** *Slang.* A sexually attractive young girl or woman.

**cal'en·dar tech·nique'** A method for determining the safe period for sexual intercourse based on these assumptions, namely, that ovulation will occur 14 (+2 or –2) days prior to the onset of menstruation; that an unfertilized ovum survives no longer than 24 hours; and that spermatozoa remain functional no longer than 48 hours.

**call girl** A prostitute who contacts customers, or is contacted, by telephone.

**call house** A brothel that employs call girls.

**cal·li·py'gous, cal·li·pyg'i·an** Possessing beautiful buttocks; with shapely and well-formed buttocks.

**camp'ing** *Slang.* Effeminate behavior.

**cam'py** *Slang.* Displaying conspicuously mannerisms associated with homosexual behavior.

**can** *Slang.* 1. The human rump or buttocks. 2. A toilet or restroom.

**can'cer** A malignant tumor characterized by the tendency of its cells to spread and invade new sites, a process called *metastasis;* a neoplasm. *See* carcinoma, sarcoma.

**can·did·i'a·sis** An infection of the genital tract or genital organs by *Candida albicans,* a yeastlike fungus. The skin, mouth, rectum, and other parts of the body may be involved. Also called *moniliasis. See* thrush.

**can·thar'i·des** Spanish fly, the dried, pulverized bodies of beetles, *Cantharis (Lytta) vesicatoria,* reputed to have aphrodisiac properties. It is a gastrointestinal and urinary-tract irritant and highly toxic when taken internally.

**ca·pac·i·ta'tion** The process by which spermatozoa, in their passage through the female genital tract, acquire the ability to penetrate the zona pellucida of a recently ovulated ovum.

**ca'pon** 1. A castrated rooster. 2. *Slang.* An effeminate male, usually a homosexual.

**ca·pon·ette'** A capon produced by injecting or feeding a synthetic sex hormone. Also *capette.*

**cap'tive pe'nis** A rare situation in which the penis is held tightly within the vagina, making withdrawal difficult.

**car·cin'o·gen** A substance that induces the development of a cancer.

**car·cin·o·gen'ic** Inducing or causing the development of a cancer.

**car·ci·no'ma** A malignant, epithelial tumor or epithelioma; a cancer.

**ca·ress'** 1. *n.* The act of expressing affection or love as by embracing or kissing. 2. *v.* To touch, pat, or stroke lightly.

**ca·rez'za** Karezza, q.v.

**car'nal** Sensual, worldly; not spiritual.

**car'nal a·buse'** Contact of the genital organs between a male and a female below the age of consent, with or without penetration; rape of a female child.

**car·nal'i·ty** Sensuality; excessive interest in worldly pleasures.

**car'nal·ize** To sexualize.

**car'nal knowl'edge** Legal term for sexual intercourse, especially that involving a girl below the age of consent.

**Cas·a·no'va** A man who engages in unscrupulous amorous adventures; a libertine. From Giovanni Casanova (1725–1798), an Italian adventurer who was known for his scandalous behavior and became famous as the author of *Mémoires,* a cynical record of his amours and seductions.

**cas·so·lette'** *French* for perfume box.

**cas'trate** To render an individual incapable of reproducing by depriving of gonads; in the male, to emasculate; in the female, to spay. *See* castration.

**cas·tra'tion** 1. Removal of the testes (orchiectomy) or ovaries (ovariectomy). *See* gelding, spay. 2. Destruction or inactivation of the testes or ovaries, such as that resulting from irradiation, the effects of drugs, parasitic organisms, or various pathological conditions.

**cas·tra'tion com'plex** A child's fear or delusion of losing his genital organs, especially a boy's fear of being deprived of his penis for exhibiting sexual interest in his mother, an element of the Oedipus complex.

**cas·tra'to,** *pl.* **cas·tra'ti** A male singer, especially in the eighteenth century, castrated before puberty in order to preserve a voice of female range and timbre.

**cat** *Slang.* 1. A prostitute. 2. A mean, spiteful woman. 3. A man who dresses flamboyantly and seeks women for sexual pleasure. *See* tomcat.

**cat·a·me'ni·a** Menstruation; the menses.

**cat'a·mite** A boy kept for sexual purposes, especially in pederasty.

**caul** The embryonic membranes that cover the head of a fetus at birth.

**caus'tic** Burning, corrosive, destructive to human tissues.

**cau'ter·ize** To destroy tissues by the application of a cautery or a caustic agent.

**cau'ter·y** 1. A device or agent that coagulates tissues, as a white-hot or red-hot wire or iron, an electric current, or a caustic substance such as silver nitrate. 2. The process of cauterization. *See* cauterize.

**cel'i·ba·cy** 1. The state of remaining unmarried, especially for religious reasons. 2. Abstention from sexual intercourse.

**cel'i·bate** One who practices celibacy.

**ce·li·ot'o·my** Opening the abdomen; abdominal section; laparotomy.

**cer'e·bral** Of or pertaining to the cerebrum or the brain.

**cer·e·bro·vas'cu·lar** Pertaining to the cerebrum or brain and its associated blood vessels.

**cer·e·bro·vas'cu·lar ac'ci·dent** *Abbr.* CVA. The stroke syndrome, a condition resulting from an acute vascular lesion involving the blood vessels of the brain, such as a hemorrhage, embolism, thrombosis, or ruptured aneurysm. Usually characterized by temporary or permanent unconsciousness, often accompanied by paralysis on one side of the body (hemiplegia).

**cer'vi·cal ca·nal'** The passageway that connects the cavities of the uterus and the vagina. Its internal opening into the uterus is the *internal os;* its external opening to the vagina is the *external os.*

**cer'vi·cal cap** A thimble-shaped rubber or plastic cap shaped to fit snugly over the cervix of the uterus. It is used for contraception. Also called *check pessary.*

**cer'vi·cal in·com'pe·tence** *See* incompetent cervix.

**cer'vi·cal mu'cus** Mucus, produced by the cervical glands of the uterus, whose ducts open into the cervical canal. The amount and consistency of the mucus are dependent upon female hormones. Estrogen results in a thin, watery, and elastic mucus that favors the passage of sperm through the cervix; progesterone results in a thick, nonelastic mucus that acts to impede the entry of sperm into the uterus. *See* ferning, spinnbarkeit.

**cer'vi·cal sec'tion** A cesarean section in which the cut is made in the lower uterine segment.

**cer·vi·ci'tis** Inflammation of the cervix of the uterus.

**cer'vix** The cylindrical, lower portion of the uterus that projects into the upper end of the vagina. Its canal serves as a pathway for menstrual fluid and spermatozoa and serves as a part of the birth canal at parturition.

**ce·sar'e·an sec'tion** Delivery of a fetus through a surgical incision made through the abdominal wall into the body of the uterus. Also spelled *caesarean section.*

**chan'cre** (Pronounced shang'ker.) 1. An ulcer-like lesion that forms at the point of entrance of certain infective organisms. A *hard* or *true chancre* is the initial lesion of syphilis. 2. A *soft chancre,* chancroid, q.v.

**chan'croid** A venereal disease caused by a bacterium, *Hemophilus ducreyi,* the lesions usually occurring on the genitalia.

**change of life** 1. In females, the menopause or female climacteric. 2. The male climacteric.

**char'i·ty girl** A sexually promiscuous young woman, especially one who dispenses her services without charge.

**chaste** Virtuous; abstaining from unlawful sexual intercourse.

**chas'ti·ty** The state of being chaste; celibacy; virginity.

**chas'ti·ty belt** A beltlike device designed to prevent sexual intercourse, worn by women during the Middle Ages.

**cheat** *Slang.* To be sexually unfaithful to one's spouse or sexual partner; to commit adultery.

**check pes'sary** A cervical cap, q.v.

**cher'ry** *Slang.* The hymen, especially as a symbol of virginity.

**chi-chi** *Slang.* 1. The female breasts considered as sexual objects. 2. A sexually attractive woman.

**chick** *Slang.* An attractive girl or young woman.

**chick'en** *Slang.* 1. A young person, especially a girl. 2. A young boy or girl who is the client of a pimp. 3. Any boy under the age of consent who is heterosexual and not familiar with homosexual ways.

**chick'en hawk** *Slang.* A homosexual who preys on young boys.

**chick'en pluck'er** *Slang.* A homosexual who likes to deflower young boys.

**chick'en queen** *Slang.* A homosexual who prefers sex with underage boys. Also called *chicken hawk.*

**child,** *pl.* **child'ren** 1. An unborn or recently born human; a fetus, baby, or infant. 2. A boy or girl from birth to puberty.

**child'bear·ing** The act of bearing and bringing forth children; parturition.

**child'bed** The situation of a woman giving birth to a baby; parturition.

**child'bed fe'ver** Puerperal fever, q.v.

**child'birth** The act or process of bringing forth a child; parturition. *See* natural childbirth.

**child'hood** The time, state, or period of being a child.

**child mo·les·ta'tion** Sexual contact of an adult with a prepubescent child. *See* pederasty, pedophilia.

**chip'pie, chip'py** *Slang.* A sexually promiscuous girl or young woman, especially a delinquent girl.

**chip'py joint** *Slang.* A brothel.

**chlo·as'ma** The development of pigmented areas on the skin, occurring in patches, especially on the face and neck. Commonly occurs during pregnancy and menstruation, and in various uterine and ovarian disorders. Sometimes called the *mask of pregnancy*. *See* melasma.

**chor'dee** The downward or ventral curvature of the penis, resulting usually from hypospadias or from urethral infection, usually gonorrhea. Also called *gryposis penis.*

**cho'ri·on** The outermost of the membranes that surround a developing embryo or fetus. From it develops the fetal portion of the placenta.

**cho·ri·on'ic vil'li** Minute, fingerlike processes that grow out from the chorion into the endometrium of a gravid uterus, forming treelike structures containing fetal blood vessels. They constitute the most important part of the placenta, as the interchange of all substances between the mother and fetus occurs on the surface of chorionic villi.

**chro'mo·some** A self-duplicating body present in the cells of higher animals and plants, especially noticeable during stages of cell division when each appears as a rod-shaped structure that stains intensely. In somatic cells they appear as homologous pairs, the number varying with species but remaining constant within a given species. The number in humans is 46 (23 pairs). Chromosomes serve as a repository of genetic information and they are of importance in the determination of sex. *See* autosome, sex chromosome, sex determination, X chromosome, Y chromosome.

**cir·cum·ci'sion 1.** In a male, the surgical removal of the prepuce or foreskin from the penis. Among Jews and Muslims, it is practiced as a religious rite. In the United States, it is commonly performed for hygienic reasons, although, in cases of phimosis, it is obligatory. It is usually performed a few days after birth. **2.** In a female, removal of a fold of skin covering the glans clitoridis.

**clan** A group of families tracing common descent.

**clap** *Slang.* Gonorrhea.

**clea'vage 1.** The series of cell divisions of a zygote that occurs immediately after fertilization resulting in the formation of a morula. Also called *segmentation.* **2.** The cleft between a woman's breasts.

**cli·mac'ter·ic 1.** In a *female,* the menopause, usually accompanied by a "change of life," a period covering several years beginning between the ages of 45 and 50, characterized by marked physical, physiological, and psychological changes resulting from the cessation of ovarian function. Ovulation ceases, estrogen and progesterone production are reduced, and infertility results. Common symptoms include hot flashes, headaches, fatigue, increased irritability, and sometimes severe depression. Atrophic changes in the uterus and lining of the vagina occur. Sexual feelings and desire are generally not altered. *See* menopause. **2.** In a *male,* a period comparable to that occurring in females, seen occasionally in men, characterized by loss of libido and sometimes marked by emotional changes. It usually occurs between the ages of 50 and 60. Testicular function is generally not lost.

**cli'max** The period of greatest emotional intensity during sexual intercourse; an orgasm.

**clit** *Slang.* The clitoris.

**clit·o·ri·dec'to·my** Removal of the clitoris.

**clit'o·ris** A small erectile structure embedded in the tissues of the vulva anterior to the junction of the minor labia. It is highly sensitive and most women respond erotically to its stimulation. It is homologous to the penis of the male.

**clom'i·phene cit'rate** Clomid, a nonsteroidal, antiestrogenic compound (a fertility drug) that induces ovulation in certain females and spermatogenesis in oligo-spermic males. It acts on the hypothalamus bringing about the release of releasing factors for FSH and LH.

**clone** All the progeny that have descended asexually from a single ancestor.

**clon'ing** Reproduction without sex, accomplished in vertebrates, as the frog, by removing the haploid nucleus from a fertilized egg and replacing it with a diploid nucleus obtained from a cell of a frog embryo.

**closed mar'riage** A form of marriage in which the principal characteristics are possession or ownership of mate, rigid role behavior, enforced togetherness with absolute fidelity, and usually the sacrifice of self-identity for one or the other. *Compare* open marriage.

**clos'et queen** *Slang.* 1. A male homosexual who suppresses or does not admit his homosexual desires or feelings; one who hides his activities. 2. An inactive homosexual.

**clutch** 1. *v.* To seize and hold tightly. 2. *n.* A number of eggs produced and incubated at one time; a hatch of eggs; a brood of chickens.

**cock** 1. A rooster. 2. The male of any bird, especially a gallinaceous bird. 3. *Slang.* The penis, especially when erect.

**cock ring** A rubber or plastic ring bearing small projections that is worn about the penis during intercourse. It presumably intensifies female feelings.

**cocks'comb** 1. The comb or caruncle of a cock. 2. A small, triangular projection at the upper pole of the cervix seen in females exposed *in utero* to DES. Also *cock's comb, coxcomb.*

**cock's man** A man who is constantly seeking sexual intercourse, performing with as many women as possible.

**cock'suck·er** *Vulgar.* A person, especially a homosexual, who takes the penis into the mouth; one who performs fellatio. Often a general term of disparagement.

**cod'piece** A pouch or cover located at the crotch of a man's tight-fitting breeches, worn by men during the fifteenth and sixteenth centuries.

**co-ed** 1. A female student attending a college or university. 2. *Abbr.* for *coeducational,* meaning appropriate for or utilized by both sexes.

**cog'nate** Related by birth; having the same ancestry.

**co·hab·i·ta'tion** The state of living together in a sexual relationship without being married.

**co'i·tal** Of or pertaining to coitus or sexual intercourse.

**co'i·tal po·si'tions** Any number of positions that a couple may assume when engaging in sexual intercourse. The most common position is *face-to-face,* with the woman on her back and the man above, or vice versa, or with the couple lying on their sides. *See also* coitus, croupade, cuissade, flanquette, missionary position, rear entry, spoon position.

**co·i'tion** Coitus.

**co'i·tus** Sexual intercourse between a male and a female in which the penis is inserted into the vagina.

**co'i·tus à la vache** Rear-entry coital position with the female in a knee-chest position; coitus from behind.

**co'i·tus in·ter·rup'tus** Withdrawal of the penis from the vagina prior to ejaculation.

**co'i·tus à la flor'en·tine** Sexual intercourse in which the female grasps the root of the penis and forcibly pulls the skin of the penis, along with the prepuce if the man is uncircumcised, back along the shaft.

**co'i·tus ob·struc'tus** Delaying an ejaculation by firm pressure on a spot between the scrotum and the anus.

**co'i·tus res·er·va'tus** Coitus without ejaculation; especially prolonged intercourse and intentional suppression of ejaculation. Also called *karezza*.

**cold** Lacking in sexual feeling or passion; unresponsive, frigid.

**co·los'trum** The first milk secreted by the mammary gland after delivery. It is a thin, yellow fluid with a high mineral and protein content and rich in immunoglobulins.

**col·pi'tis** Inflammation of the vagina.

**col·po·plas'ty** Plastic surgery involving repair of the vagina.

**col'po·scope** An instrument used for visual examination of the vagina and the cervix of the uterus.

**co-mar'i·tal sex** The sharing of mates or partners or the incorporation of extramarital sex in marriage; swinging.

**come, cum** *Slang.* 1. *n.* Semen. 2. *n.* The viscid fluid produced in the female vagina at sexual climax. 3. *v.* In the male, to ejaculate and experience an orgasm. 4. *v.* In the female, to reach a sexual climax and experience an orgasm.

**come a·cross'** *Slang.* To grant sexual favors or to give in sexually, with reference to a female.

**come a·round'** *Slang.* To menstruate, especially when menstruation is delayed and pregnancy expected.

**com'ing out (of the clos'et)** *Slang.* 1. Entry of a homosexual into the gay community. 2. In the gay world, the first homosexual experience. 3. The time when a person changes his or her identity to that of a homosexual.

**com'mu·nal mar'riage** *See* group marriage.

**com'mune** A group consisting of several families and sometimes single adults who live together and share common interests. Members generally share common tasks, as the upkeep of a house or apartment, food production and preparation, child care and rearing. Indiscriminate sexual relationships involving group sex may occur in some communes. *Compare* group marriage.

**com·pan'ion** A homosexual prostitute, especially to an older customer with whom he establishes a full-time relationship.

**com·pat'i·ble** Capable of living harmoniously and agreeably with another person.

**Com'stock, An'tho·ny** (1844–1915) An American author and reformer active in the promulgation of laws pertaining to vice and immorality. He was excessively prudish in matters pertaining to morality in art.

**com·stock'er·y** Prudery; overzealous moral censoring of works in the field of fine arts and literature.

**Com'stock Law** A law passed by Congress in 1877 which prohibited the mailing across state borders of obscene material, which included birth-control information and contraceptive devices.

**con·ceive'** To become pregnant.

**con·cep'tion** Fertilization of an ovum; the act of becoming pregnant; impregnation.

**con·cep'tus** The products of conception, consisting of the embryo or fetus and associated structures such as the extraembryonic membranes and placenta.

**con·cu'bi·nage** Cohabitation of persons without legal sanction.

**con'cu·bine** A woman who cohabits with a man without being legally married to him. *See* mistress.

**con·cu'pis·cence** An extremely strong desire, especially excessive sexual desire or lust.

**con·cu'pis·cent** Sensual, lustful, desirous of sex.

**con'dom** A very thin sheath of latex or animal membrane that is worn over the erect penis during sexual intercourse. It serves as a contraceptive device and as a prophylactic preventing venereal infection. Also called *jacket, pro, rubber, safety, skin.*

**con·dy·lo'ma a·cu·min·a'tum,** *pl.* **con·dy·lo'ma·ta a·cu·min·a'ta** Genital wart, q.v.

**con·fine'ment** The period of parturition or childbirth; the lying-in period, accouchement.

**con·gen'i·tal** Present or existing at birth, generally applied to conditions such as malformations or infections which are acquired during intrauterine development but are not hereditary.

**con·ges'tion** A condition in which there is an excessive amount of blood in an organ or tissue.

**con'ju·gal** Of marriage or the marriage state; connubial.

**con·nu'bi·al** Conjugal.

**con·san·guin'e·ous** Of the same ancestry or descent; related by blood.

**con'sum·mate** To complete a marriage by the act of sexual intercourse.

**con'ti·nence** Abstention from sexual activity; self-restraint. *See* abstinence.

**con·tra·cep'tion** The prevention of conception. Term is now generally used as a synonym for birth control and as such includes any measures used to prevent conception or the implantation and development of the embryo or fetus. *See* birth control.

**con·tra·cep'tive** 1. *adj.* Of or pertaining to contraception. 2. *n.* Any agent or device that prevents conception. *See* cervical cap, condom, diaphragm, foam, jelly, pessary, pill.

**con·tra·cep'tive im'plant** A tiny capsule of a progestin, e.g. levo-norgestrol, that is placed under a woman's skin. Protection from five to seven years is provided.

**con·tra·im·plan·ta'tion** The prevention of implantation, as through the action of an intrauterine device (IUD).

**con·tra·sex'ism** The compulsion to have the body transformed surgically and hormonically to that of the opposite sex. *See* transsexual.

**con·tra·sex'u·al** Of, pertaining to, or characteristic of the opposite sex.

**con·trec·ta'tion** Sexual foreplay.

**cooch, cootch** 1. An Oriental (or pseudo-Oriental) dance performed by a woman, characterized by sinuous movements and twisting and shaking pelvic motion. Also called *hootchy-cootchy.* 2. *Slang.* The female crotch.

**cool-out** *Slang.* To have sexual intercourse.

**cooze** *Slang.* 1. The female as a sexual object. 2. The vagina. 3. Sexual intercourse.

**cop a cher'ry** *Slang.* To deflower a female.

**cop a feel** *Slang.* To surreptitiously feel a woman's sexual parts.

**cop·ro·lag'ni·a** A condition in which sexual excitement results from handling, thinking about, or talking about human feces.

**cop'ro·lin** An odoriferous substance produced within the vagina of certain monkeys that acts as a pheromone inducing males to copulate at the time of ovulation in the female.

**cop·rol'o·gy** Pornography, scatology.

**cop·ro·phil'i·a** 1. An abnormal fondness for or attraction to filth, especially human feces. 2. A condition in which the arousal of erotic feelings and achievement of an orgasm depend upon the smell or taste of feces. 3. A condition in which a male is stimulated sexually by the sight of female buttocks, or a female by the sight of male buttocks.

**cop'u·late** To unite and join together, especially to engage in sexual intercourse.

**cop·u·la'tion** The act of sexual intercourse; coitus.

**co·quette'** A flirt, q.v.

**core-gen′der i·den′ti·ty** The sense of maleness or femaleness that develops in an infant or child, dependent principally upon sex assignment and rearing.

**corn hole** *Slang.* Anal intercourse.

**cor′nu,** *pl.* **cor′nu·a** A horn-shaped process or excrescence.

**cor′nu of the u′ter·us** The lateral extension of the uterine cavity into which the uterine tube enters.

**co·ro′na glan′dis** An elevated ridge that forms the posterior border of the glans penis.

**co·ro′na ra·di·a′ta** A layer of follicle cells that surrounds an ovum following its discharge from a mammalian ovary.

**cor′pus,** *pl.* **cor′po·ra** A body; the main portion of a structure.

**cor′pus al′bi·cans** A small, white body in an ovary that constitutes the remains of the corpus luteum after it has passed its peak of functional activity.

**cor′pus cav·er·no′sum** *pl.* **cor′po·ra cav·er·no′sa** One of two columns of erectile tissue in the penis or clitoris.

**cor′pus lu′te·um** A small, yellow body that develops within a ruptured ovarian follicle following the discharge of an ovum. It is an endocrine structure and is the principal source of progesterone. It also secretes estrogens.

**cor′pus spon·gi·o′sum** A column of erectile tissue within the penis that surrounds the urethra. Its expanded distal end forms a conical cap, the *glans penis.*

**coup′ling** Copulation; sexual intercourse.

**court** To attempt to gain the favor of, or to gain the affection and love of; to woo.

**cour′te·san** A high-class prostitute whose clientele includes men of wealth, of power, or of the upper classes.

**cou·vade′** A condition common among primitive peoples in which the husband exhibits symptoms associated with his wife's pregnancy, such as being nauseated and suffering abdominal pains, symptoms which disappear upon delivery of the infant.

**Co′vent Gar′den la′dies** London prostitutes.

**cov′er** 1. To copulate with a female, applied usually to domestic animals, especially fowls. 2. To sit upon and incubate eggs.

**co′vert** Hidden, concealed, disguised, secret.

**cow** 1. A mature, female, bovine animal of the genus *Bos.* 2. The mature female of a number of large mammals, as the elephant, whale, moose. 3. *Slang.* (a) An obese, slovenly woman; (b) a woman with large breasts; (c) a woman with a large number of children and frequently pregnant.

**Cow′per's glands** The bulbourethral glands, q.v.

**crabs** 1. Pubic lice. *See* louse. 2. *Slang.* Syphilis.

**cream** *Slang.* Semen.

**cre·mas′ter** A muscle whose fibers cover the spermatic cord and the testis. Upon contraction, it draws the testis upward.

**cre·mas′ter re′flex** The elevation or retraction of the testis on the same side when the inner surface of the upper portion of the thigh is gently rubbed.

**crib** A brothel, especially a cheap one, usually consisting of a room (pad) with a single bed.

**crime a·gainst′ na′ture** An old expression used to include "unnatural sex acts," i.e., any acts not for the purpose of procreation, such as oral-genital contact, anal intercourse, bestiality, or any of the acts generally included under sodomy laws.

**cross′breed** 1. *v.* To produce a hybrid by mating individuals of different breeds, varieties, or races; to hybridize; to interbreed. 2. *n.* A hybrid produced by crossbreeding.

**cross dress′ing** Wearing clothes of the opposite sex; transvestism.

**crotch** The angle formed by the junction of the inner sides of the thighs and the trunk.

**crou·pade′** The full rear-entry position in coitus.

**crud** *Slang.* 1. Dried semen, especially that which sticks to the body, clothes, or bedclothes following sexual intercourse. 2. Any venereal disease.

**cruise** To move about from place to place, especially in a car, looking for a sexual partner.

**crush** 1. An intense and usually short-lived infatuation, especially of a girl for a boy. 2. The object of such an infatuation.

**cry·o·bank'ing** The storage of spermatozoa or human tissues at an extremely low temperature. *See* sperm bank.

**cry·o·sur'ger·y** The destruction of tissue by application of extreme cold.

**cryp·tor'chism, cryp·tor'chid·ism** A condition in which the testes during development fail to descend into the scrotum and remain within the abdominal cavity or inguinal canal. It results in sterility as the spermatozoa fail to develop because of the higher temperature of the body cavity. Hormone production by the testes is not affected.

**CSC** Collagen sponge contraceptive.

**cuck'old** The husband of a woman who has committed adultery.

**cud'dle** To hold closely in an affectionate manner; to fondle; to hug tenderly.

**cuis·sade'** A half-rear entry in coitus with the female half-turned on her side.

**cul-de-sac** A blind sac, especially the rectouterine pouch, an extension of the peritoneal cavity located between the posterior surface of the uterus and the anterior surface of the rectum. Also called *pouch of Douglas.*

**cul'do·scope** An optical instrument for performing culdoscopy.

**cul·dos'co·py** Visual examination of female pelvic organs by use of a culdoscope, which is introduced into the cul-de-sac through the posterior fornix of the vagina.

**cun·ni·lin'gus** Oral stimulation of the vulva, especially the clitoris; lambitus. Also *cunnilinctus.*

**cun'nus** The vulva.

**cunt** 1. *Vulgar.* The external female genitalia; the vulva or pudendum. 2. *Slang.* A woman as a sex object. 3. *Obscene.* Coitus.

**cu·ret', cu·rette'** A surgical instrument, shaped like a spoon, used for scraping the interior of a cavity, especially the uterus.

**cu·ret·tage'** The scraping of the interior of a cavity, such as that of the uterus, by use of a curet. *See* dilatation and curettage, vacuum curettage.

**cu·rette'ment** Curettage.

**curse** *Slang.* The menstrual period; menstruation.

**cur·va'ceous** Possessing a shapely and voluptuous figure (said especially of women).

**CVA** Cerebrovascular accident.

**cy'cle** *See* menstrual cycle, estrous cycle.

**cyst** A closed sac or cavity that possesses a distinct wall or membrane, especially one that contains fluid or semisolid material. Most cysts are pathological structures that develop within a cavity of the body or in the substance of an organ, such as the ovary.

**cys·ti'tis** Inflammation of the bladder characterized by frequent desire to urinate and a burning sensation upon urination.

**cys'to·cele** A protrusion of the bladder into the vagina.

**cy'to·gen'ic** Producing cells, as a *cytogenic* gland, the testis or ovary.

**cy·to·log'ic** Of or pertaining to cells.

**cy·to·meg'a·lo·vi'rus** One of a group of herpes viruses that infect humans. It is the cause of cytomegalic inclusion disease.

**dai'sy** *Slang.* 1. A beautiful, attractive girl. 2. A male homosexual, especially one who assumes the female role. 3. Any male sexual pervert.

**dai′sy chain** *Slang.* A group of persons engaged in simultaneous sexual activity, especially a group of three or more men in a circle engaged in buggery.

**dal′ly** To play amorously.

**dam′aged goods** *Slang.* A girl who has lost her virginity.

**dame** 1. A married woman. 2. A woman of rank or authority. 3. *Slang.* A girl or young woman, especially a troublesome one.

**dam′sel** A girl or young woman; a maiden.

**Dan′a·zol** A progestin-like, synthetic steroid used experimentally in the treatment of endometriosis and cystic mastitis. It also acts as a contraceptive, inhibiting ovulation by preventing the release of FSH and LH by the pituitary gland.

**D and C** Dilatation and curettage, q.v.

**D and E** Dilatation and evacuation, q.v.

**D and S** Dominance and submission, q.v.

**dan′dy** A man who is extremely fastidious and excessively concerned about his dress and appearance.

**dang** *Slang.* The penis.

**dan′gle queen** *Slang.* A female exhibitionist.

**dar′tos** A thin layer of smooth muscle located beneath the subcutaneous layer of the scrotum.

**date** A social engagement with a person of the opposite sex, or the person with whom the appointment is made.

**daugh′ter** A female child or female adopted child.

**daugh′ter-in-law** The wife of one's son.

**de·bauch′** To corrupt morally; to seduce.

**deb·au·chee′** A person who engages excessively in sensual pleasures; a libertine. *See* rake, roué.

**de·bauch′er·y** Extreme or excessive indulgence in sensual pleasures; intemperance; dissipation.

**de·but′** The formal presentation of a girl or young woman to society.

**deb′u·tante** A young woman making her formal entry into society.

**de·ca·pac·i·ta′tion fac′tor** *Abbr.* DF. A substance present in the male reproductive tract and in seminal plasma that reverses the process of capacitation. It forms a surface layer about the head of each sperm, preventing it from fertilizing an ovum. It is inactivated by acrosomal enzymes.

**de·cid′u·a** The endometrium or mucous lining of the uterus that is shed during menstruation or following parturition.

**de·coy′** A police officer, male or female, usually dressed in casual attire, who loiters in a public rest room or on the street for the purpose of enticing a homosexual or a prostitute to solicit for the commission of a lewd or lascivious act prohibited by law. *See* entrapment.

**de·flo·ra′tion** The act of deflowering; the rupturing of the hymen of a virgin. It occurs naturally at the first sexual intercourse; artificially it may be induced intentionally by the insertion of an object such as a tampon into the vagina or by surgical intervention.

**de·flow′er** To rupture the hymen of a virgin by sexual intercourse.

**de·liv′er·y** Parturition or childbirth; the expulsion of or extraction of a fetus and its membranes from the uterus.

**dem′i·monde** A class of women of questionable respectability who grant sexual favors to men of wealth, power, or influence.

**dem′i·vierge** A promiscuous girl or woman who engages in passionate sexual behavior such as necking and petting but refrains from sexual intercourse.

**de·nude′** To deprive of a covering; to make bare or nude.

**de·ox·y·ri·bo·nu'cle·ic ac'id** One of a number of nucleic acids which, upon hydrolysis, yield deoxyribose. It is present in the nuclei of all living cells and consists of a double helix formed of two long chains of alternating phosphate and deoxyribose units united by purine and pyrimidine bases. It is a constituent of genes and involved in the transmission of genetic information.

**dep·i·la'tion** The removal of hair from a part of the body.

**De'po-Pro·ve'ra** Trade name for medroxy-progesterone acetate (MPA), an injectable, long-term contraceptive, effective when administered once every three months. Commonly called the "shot."

**dep·ri·va'tion ho·mo·sex·u·al'i·ty** Homosexuality resulting when individuals are deprived of the opportunity for heterosexual contacts, as in the harems of Oriental potentates, or among men and women in armies or in prisons.

**der·ri·ere'** The buttocks or rump; the behind.

**DES** Diethylstilbestrol.

**de·ser'tion** The willful abandonment of one's spouse or children or both.

**de·sex'** To castrate or spay.

**de·sex'u·al·ize** To deprive an individual of his or her sexuality. This may be accomplished physically by castration or inactivation of the sex glands or socially and psychologically by education and training.

**de·tu·mes'cence** The restoration of a swollen organ to natural size, especially the subsidence of the erect penis following ejaculation.

**de'vi·ant** 1. *adj.* Deviating from an accepted norm, especially with reference to behavior. 2. *n.* A person who differs markedly in behavior, as a *sexual deviant.*

**de'vi·ate** 1. *v.* To diverge from or move away from an established way or a prescribed mode of behavior; to depart from behavioral norms of a particular society. 2. *n.* A person who engages in unacceptable sexual behavior.

**de·vi·a'tion** The act or process of deviating; a departure from a standard or norm.

**DF** Decapacitation factor, q.v.

**dho'bie itch** *See* jock itch.

**di'a·phragm** 1. In *anatomy,* a musculomembranous sheet that (a) separates the thoracic and abdominal cavities; (b) forms the floor of the pelvis. 2. A cup-shaped contraceptive device that fits over the cervix of the uterus.

**dick** *Slang.* The penis.

**di·eth·yl·stil·bes'trol** *Abbr.* DES. A nonsteroidal, synthetic estrogen used in the treatment of menstrual disorders. It is also the principal constituent in the morning-after pill, a post-coital contraceptive effective when taken within 72 hours after an unprotected coitus. However, if DES fails to intercept a pregnancy or if DES is taken by a pregnant woman, the likelihood of the development of cervical and vaginal abnormalities including cancer in a developing female fetus is high. *See* morning-after pill.

**dil·a·ta'tion** The process of dilating; making wider and larger.

**dil·a·ta'tion and cu·ret·tage'** *Abbr.* D and C. A method of inducing an abortion by dilatation of the cervical canal of the uterus and insertion of a curette by which the embryo is dislodged. The procedure is also employed in securing a sample of the uterine endometrium for diagnostic purposes.

**dil·a·ta'tion and e·vac·u·a'tion** *Abbr.* D and E. Dilatation of the cervix of the uterus and the discharge of the products of conception.

**di'late** To make wider and larger; to expand or enlarge, as an opening.

**di·la'tion** The act of dilating or stretching; dilatation.

**di·la'tor** An instrument or device used in enlarging an opening or passageway.

**dil'do,** *pl.* **dil'does** An object that serves as a substitute penis for insertion into the vagina, commonly used as an aid for masturbation in the female. It may be made of wood, rubber, plastic, horn, ivory, glass, wax, or other materials. Some dildoes are simple in design; others are complex and may be modified to discharge a warm fluid simulating ejaculation. Some contain vibrators to enhance their effect. Dildoes are also used by male homosexuals.

**di·oe'cious, di·e'cious** Unisexual; possessing sexual organs of one sex only. *Compare* monoecious.

**di·phal'lus** A double or bifid penis or clitoris, which may be partial or complete.

**dip'loid** Possessing a double complement of chromosomes—in humans, 23 pairs—as in the fertilized egg and all cells resulting from its division or subdivision. *Compare* haploid.

**dir'ty** In a sexual sense, characterized by obscenity and vulgarity, as *dirty* jokes or *dirty* pictures, the latter showing nudity or sexual activity; pornographic.

**dis'ci·pline** Mutual beating, as by spanking or using switches, to increase erotic sensation. *See* masochism.

**dis·ha·bille'** A state of undress or of being casually or partially dressed.

**dis'so·lute** Morally corrupt; licentious; sexually unrestrained.

**dis·so·lu'tion** The termination of a legal bond such as marriage; loss of restraint, especially in sexual matters.

**dis'tal** Away from the point of attachment; away from a central point or plane.

**dis·til·la'ti·a** A clear, viscid fluid that appears at the tip of the penis when the male is sexually aroused. It is produced by the bulbourethral glands and functions in the alkalization of the urethra and lubrication of the penis.

**di·vorce'** The legal dissolution of a marriage.

**di·vor·cé'** A divorced man.

**di·vor·cée'** A divorced woman.

**di·zy·got'ic** Originating from two fertilized eggs, as in *dizygotic* twins.

**DNA** Deoxyribonucleic acid, q.v.

**dol'ce vi'ta** Italian for "sweet life," a life of pleasure, especially one free from discipline and moral restraint.

**dol'ly mop** *Slang.* 1. A strumpet. 2. An amateur prostitute.

**dom'i·nance and sub·mis'sion** *Abbr.* D and S. Behavioral patterns characteristic of human sexual relations in which the male is considered dominant because of his aggressiveness and superior coital position and the female is considered submissive because of her passivity and inferior coital position.

**Don Juan** A dissolute man obsessed with the seduction of women; a libertine.

**dor'sal** Pertaining to the upper portion or surface of an animal body or one of its parts; in man, pertaining to the back or dorsum; posterior. Opposite of ventral, q.v.

**dose** *Slang.* A venereal infection such as gonorrhea or syphilis.

**dote** 1. To lavish or bestow excessive love or affection on. 2. To be foolish or weak-minded, especially as a result of senility.

**dou'ble stan'dard** This refers to the double standard of sexuality, which establishes different standards of behavior for women and men. Premarital and extramarital sex have generally been condoned in men but condemned in women. The liberation movement for women seeks to do away with the double standard.

**douche** A jet or stream of water directed against the body or into one of its cavities, especially one directed into the vagina for hygienic, contraceptive, or medicinal purposes.

**dow'er** Dowry, q.v.

**Down's syn'drome** Mongolism, q.v.

**dow'ry** Money, goods, or property that a woman brings to her husband at marriage.

**drab** *Slang.* 1. A prostitute. 2. A slattern.

**drag** *Slang.* 1. Female clothing worn by a male, especially a homosexual or transvestite. 2. A homosexual party where the participants wear clothes of the opposite sex. 3. A woman accompanying her escort. 4. Something or someone who is obnoxiously boring.

**drag queen** *Slang.* A male homosexual who dresses in female clothes.

**drag show** *Slang.* A show in which the performers are female impersonators.

**drip** *Slang.* Gonorrhea.

**drug** 1. A substance used in the diagnosis, treatment, and prevention of disease. 2. A chemical substance that has mind-altering properties, affecting mood, perception, judgment, or consciousness; one that stimulates or depresses mental activity or has hallucinogenic properties. Among these drugs are alcohol, nicotine, marijuana, sedatives (barbiturates, tranquilizers), stimulants (caffeine, cocaine, amphetamines), psychedelics (LSD, DOM, mescaline), and narcotics (heroin, morphine, opium).

**drug ad·dic'tion** The compulsive and excessive use of and craving for a drug, often accompanied by physical dependence or addiction, which results from continued usage. When the stimulant is withdrawn suddenly, a syndrome with identifiable symptoms usually appears. The addict's life centers on acquiring the drug at whatever cost and maintaining a constant supply.

**dry birth** One in which the amnion ruptures and the amniotic fluid is discharged prematurely. Also called *dry labor.*

**dry dream** In women who are nonorgasmic when awake, the occurrence of an orgasm during sleep.

**dry hump** *Slang.* 1. *v.* To engage in imitation intercourse with all clothes in place. 2. *n.* A dry fuck.

**dry or'gasm** One in which there is no ejaculation of semen, as occurs in retrograde ejaculation or following the removal of the prostate.

**duc'tus def'er·ens** The vas deferens, q.v.

**duenna** An older woman who serves as a companion or chaperone of a young woman, especially in Spain.

**dy'ad** A couple or a pair, as a pair of chromosomes or a married couple (a *marital dyad*).

**dyke** *Slang.* A female homosexual; a lesbian. *See* bull dyke.

**dys·func'tion** Any impairment or disorder in the functioning of a tissue, organ, or organ system.

**dys·gen'e·sis** Impairment or loss of reproductive function.

**dys·men·or·rhe'a** Painful or difficult menstruation. It may be due to a hormone imbalance, an undeveloped uterus, inflammation of the uterus, a uterine or ovarian cyst or tumor, an occluded cervix, emotional factors, or other causes.

**dys·pa·reu'ni·a** Painful sexual intercourse. In women it may be due to a resistant hymen, an inflamed, excessively dry, or spastic vagina, hormonal disturbances, a displaced uterus, lesions of the genital organs, emotional factors, and other causes.

**dys·pla'si·a** Abnormal growth or development.

**dys·sper'mi·a** 1. Any disturbance in the development of normal spermatozoa or in their deposition within the vagina. 2. The occurrence of pain or discomfort during an ejaculation.

**dys·to'ci·a** Difficult labor or childbirth.

**dys'tro·phy** Atrophy or degeneration of an organ or tissue due to imperfect nutrition or abnormal metabolism.

**eas'y lay** *Slang.* A girl or woman who readily acquiesces in sexual intercourse. Also called *easy make.*

**eat** *Slang.* To use the mouth in oral-genital activities; to engage in fellatio or cunnilingus.

**ec·bol'ic** Inducing an abortion or accelerating childbirth.

**ec·lamp'si·a** Toxemia of pregnancy, a pathological condition of unknown origin characterized by elevated blood pressure, albuminuria, excessive weight gain, edema, convulsions, and coma. Prior to convulsions and coma, it is called *preeclampsia.*

**e·cou'teu·rism** Sexual excitement resulting from hearing sounds associated with sexual activities.

**ec'to·derm** In an embryo, the outermost of the three germ layers. It gives rise to the epidermis and its derivatives (hair, nails), the nervous system and sense organs, and the linings of the nasal and oral cavities and the anal canal.

**ec·top'ic** Occurring in an abnormal position.

**ec·top'ic preg'nan·cy** The development of an embryo in a location other than within the body of the uterus, such as in the uterine tube, the ovary, or the body cavity; extrauterine gestation.

**EDD** Expected date of delivery.

**e·de'ma** The excessive accumulation of fluid within tissue spaces. It results from disturbances in water and electrolyte balance and occurs commonly during pregnancy.

**ef·face'ment** In parturition, the shortening of the cervical canal of the uterus during delivery so that only a circular orifice exists.

**ef·fem'i·nate** Of a male, possessing the characteristics of a female; lacking in manly traits.

**ef·fem·i·na'tion** The development of a feminine personality (feelings, behavior, speech) in a male; eonism; muliebrity.

**ef·fem'i·nize** To bring about effemination.

**ef'fer·ent** Conveying away from an organ or structure, as the *efferent* ducts of the testis.

**ef·fleu·rage'** 1. A delicate stroking motion in massage. 2. A light stroking motion of the lips alternating with tongue caresses.

**egg** A female reproductive cell produced by the ovary; an ovum or female gamete.

**e·jac'u·late** 1. *v.* To discharge semen. 2. *n.* The fluid discharged during an ejaculation; the seminal fluid or semen, consisting of seminal plasma and spermatozoa. It averages 2 to 5 ml in volume and contains from 100 to 600 million spermatozoa.

**e·jac·u·la'tion** The discharge of seminal fluid or semen.

**e·jac·u·la'ti·o prae'cox** Premature ejaculation; an ejaculation that occurs before or shortly after penetration of the vagina.

**e·jac'u·la·to·ry duct** A short duct formed by the union of the vas deferens and the duct from the seminal vesicle. It is embedded within the tissue of the prostate gland and opens into the urethra.

**e·jac'u·la·to·ry in·com'pe·tence** The inability to discharge semen.

**E·lec'tra com'plex** The pathological emotional attachment of a girl or woman to her father. *Compare* Oedipus complex.

**e·lec·tro·cau'ter·y** The destruction of tissue by the application of intense heat, usually provided by a platinum wire heated white-hot electrically.

**e·lec·tro·co·ag·u·la'tion** Coagulation of tissues by the application of a high-frequency electric current.

**El'lis, Hen'ry Have'lock** (1859–1939) English psychologist and writer who conducted research on the psychology and sociology of sex and was the author of seven volumes between 1897 and 1928. He was a pioneer in the development of open discussion of sexual problems.

**e·lope'** To run off and be secretly married, usually without parental consent.

**e·mas'cu·late** 1. To castrate. 2. To deprive of masculine vigor; to make effeminate.

**e·mas·cu·la'tion** Castration; removal of the testes or penis or both.

**em'bo·lism** The blocking of a blood vessel by an embolus. *See* thromboembolism.

**em'bo·lus** An object such as a blood clot, air bubble, or a mass of tissue debris that lodges in a blood vessel forming an obstruction.

**em·brace'** 1. To clasp and hold firmly with the arms as a sign of love or affection. 2. To grasp tightly with the forelimbs, as in the copulation of amphibians.

**em'bry·o** An organism in the early stages of its development. In humans, the developing organism up to the beginning of the third month of pregnancy.

**em·bry·o·gen'e·sis** The development of an embryo.

**em·bry·on'ic disc** A flattened disc of cells in a developing blastocyst from which the embryo proper develops. Also called *germinal disc.*

**e·mis'sion** The discharge of a fluid such as the seminal fluid, whether voluntary or involuntary; an ejaculation. Sometimes the term is restricted to (a) the movement of seminal fluid into the prostatic urethra, or (b) the involuntary discharge of semen, as in a *nocturnal emission.*

**em·men'a·gogue** An agent that increases the menstrual flow.

**em·men'i·a** The menses. *See* menstruation.

**en·ceinte'** *French.* Pregnant; with child.

**en·dear'** To make dear, highly esteemed, or much beloved.

**en·do·cer'vix** The mucous membrane lining the cervical canal of the uterus.

**en'do·crine gland** A gland of internal secretion; one of a number of ductless glands whose secretions, called *hormones,* are discharged into the blood or lymph and distributed throughout the body. The principal endocrine glands are the pituitary, hypothalamus, thyroid, parathyroid, adrenal, islets of the pancreas, ovaries and testes. Other organs that secrete hormones include the pineal body, thymus gland, stomach, duodenum, kidney, placenta, and skin.

**en'do·derm, en'to·derm** In an embryo, the innermost of the three germ layers. It gives rise to the lining of the alimentary canal and its derivatives (liver, pancreas) and the lining of the respiratory passageways and cavities.

**en·dog'a·my** Marriage within a particular tribe, group, or clan; inbreeding. *Compare* exogamy.

**en·do·me'tri·al cy'cle** The menstrual cycle, q.v.

**en·do·me·tri·o'sis** The presence of endometrial tissue in an abnormal location as (a) in the muscle layer or beneath the outer serous layer of the uterus, (b) in the ovary, or (c) within the abdominal cavity.

**en·do·me·tri'tis** Inflammation of the endometrium.

**en·do·me'tri·um** The mucous membrane that lines the cavity of the uterus; the uterine mucosa. *See* menstrual cycle.

**en'do·scope** An instrument for visual examination of a hollow organ or cavity, used especially in examination of the lining of the uterus.

**end'plea·sure** The feelings of profound satisfaction resulting from the release of tension following orgasm. *Compare* forepleasure.

**en'e·ma** The injection of a liquid into the rectum for diagnostic, nutritive, or therapeutic purposes.

**en·gage'ment** A betrothal.

**en'ter** *Slang.* To have intercourse with a female.

**en·tice'** To lure by arousing hope or desire.

**en'to·derm** Endoderm, q.v.

**en·trap'ment** A procedure in which a nonuniformed police officer (a decoy) entices a homosexual or a prostitute to solicit or perform a lewd or lascivious act, the procedure being observed by a second police officer who makes the arrest and serves as a witness.

**en·u·re'sis** Incontinence; inability to control the passage of urine.

**e'on·ism** The assumption of feminine traits and mannerisms and the wearing of female clothes by a male; effemination. *See* female impersonator.

**ep'i·cene** 1. Possessing characteristics of both sexes, as an *epicene* statue. 2. Effeminate, womanish, nonmasculine. 3. Without sex; neuter.

**ep·i·did'y·mis** An elongated structure lying alongside each testis which consists of a tube about 20 feet long condensed into a compact structure about 2 inches in length. It receives spermatozoa from the efferent ducts of the testis and transmits them to the ductus (vas) deferens. Within the epididymis, spermatozoa undergo maturation resulting in an increased capacity for motility and fertility. It usually requires fourteen days for sperm to traverse the epididymis but this time is variable depending upon frequency of ejaculation.

**e·pis·i·ot'o·my** A surgical incision into the vulvar orifice during childbirth to reduce the possibility of laceration.

**ep·i·spa'di·as** A congenital defect in the male in which the urethra opens on the upper surface of the penis. *Compare* hypospadias.

**ep·i·the'li·um** A type of tissue that covers surfaces, lines tubes and cavities, and forms the secreting portions and ducts of glands. It consists of contiguous cells arranged in one or more layers and possesses little intercellular material. It forms the epidermis of the skin and the linings of the digestive, respiratory, and urogenital organs and passageways. Epithelium lining blood vessels is called *endothelium;* that lining the body cavities, *mesothelium.* It functions in the processes of protection, absorption, and secretion, and, by ciliary action, it aids in the movement of substances through tubes. Its regenerative qualities are high.

**e·rec'tile tis·sue** A spongelike tissue which, when filled with blood, brings about the enlargement and rigidity of a structure, as the penis or clitoris. *See* corpus cavernosum.

**e·rec'tion** The process by which an erectile structure such as the penis or clitoris swells and becomes hard and erect, as usually occurs following sexual stimulation. In the erection of the penis, arteries are reflexly dilated, resulting in an increased flow of blood which fills the spaces of the corpora cavernosa. At the same time, veins are constricted preventing the outflow of blood. The resulting turgidity of the cavernous tissue results in the penis becoming hard and assuming an erect position.

**e·rec'tor set queen** *Slang.* A homosexual construction worker.

**er'e·thism** An excessive degree of irritability or sensitivity in an organ or tissue.

**e·rog'e·nous** Arousing sexual desire, stimulating sexual activity and response.

**e·rog'e·nous zones** Regions of the body which, when stimulated, give rise to or increase sexual desire. In the *female,* these include the external genitalia, inner surfaces of the thighs, the breasts, armpits, neck, ear lobes, lips, and mouth. In the *male,* the external genitalia (penis, especially the glans, and scrotum), anal region, breasts, lips, mouth.

**Er'os** 1. The Greek god of love. 2. Sexual love or desire; libido.

**e·rot'ic** 1. Of or pertaining to libido or sexual love and desire; amatory. 2. Tending to excite sexual desire or directed toward the gratification of sex.

**e·rot'i·ca** Written material or art that deals with sexual love, especially in a sensuous manner.

**e·rot'i·cism** 1. Sexual impulses or sexual drive, especially when persistent. 2. The arousal of sexual feelings or desire through the use of suggestion, symbolism, or allusion as occurs in various art forms such as literature, painting, sculpture, and drama.

**e·rot'i·cize** To make erotic; to invest with sexual appeal.

**er'o·tize** To invest with sexual feelings or erotic meaning.

**er·o·to·gen'ic** Causing or giving rise to erotic feelings.

**er·o·to·ma'ni·a** Excessive sexual desire; lagnosis; nymphomania (q.v.) or satyriasis, q.v.

**e·ro'to·path** One afflicted with erotopathy.

**er·o·top'a·thy** Any abnormal state or condition involving sexual feelings and desire.

**es·thet'ic** See aesthetic.

**es·tra·di'ol** A natural estrogen, considered to be the most active of the three estrogens produced in women.

**es'tri·ol** An estrogen present in the urine of women. It is thought to arise as a degradation product of steroid metabolism in the liver. It is produced in large quantities by the placenta.

**es'tro·gen, oes'tro·gen** A female hormone, a substance, natural or synthetic, that is capable of inducing estrus in lower mammals or changes in the uterine endometrium and the development of female characteristics in human females. The principal estrogens are *estradiol, estrone,* and *estriol,* of which estradiol is the most potent. They are produced principally by ovarian follicles and corpora lutea but are also produced by the adrenal cortex, testes, and placenta. Estrogens undergo degradation in the liver and final products are excreted in the urine.

**es·tro·gen'ic** 1. Of, pertaining to, or caused by an estrogen. 2. Promoting estrus.

**es'trone** A natural estrogen produced principally by the ovaries but also secreted by the adrenal cortex, testes, and placenta. It is less active than estradiol.

**es'trous cy'cle** In female mammals other than primates, a series of correlated phenomena involving the reproductive and endocrine systems that culminate in estrus.

**es'trus** Heat, a period in female mammals, except primates, during which females are sexually active and receptive to the male. It occurs at the time of ovulation.

**e·ti·ol'o·gy** 1. A division of medicine that deals with the causes of disease. 2. All the factors or conditions that contribute to the occurrence of a disease or pathologic condition.

**eu·gen'ics** The science that deals with the improvement of the human race through selective breeding. *Compare* euthenics.

**eu'nuch** A castrated man or a male in which there is complete testicular failure. *See* castrate.

**eu'nuch·oid·ism** A condition in which an individual has the characteristics of a eunuch without being castrated, as in agonadism or hypogonadism.

**eu'phe·mism** The substitution of a mild or inoffensive term for one that is offensive or objectionable, as *to sleep with* for *to have sexual intercourse with.*

**eu·tha·na'sia** An easy and painless death, especially the termination of the life of a person suffering from an incurable disease or intractable pain; mercy killing.

**eu·then'ics** The science that deals with the improvement of the human race through improvement of the environment. *Compare* eugenics.

**e·ver'sion** A turning outward.

**e·vert'** To turn inside out or outward.

**ev·i·ra'tion** 1. Castration, emasculation, demasculinization. 2. The assumption of a feminine role by a male, especially in sexual relations.

**ewe** A female sheep, especially when mature.

**ex·ci'sion** The cutting away or removal of a structure, as a tube or organ.

**ex·cite'ment phase** The first phase of an orgasm. In the *male,* erection of the penis occurs resulting from vasocongestion. The penis increases in length and diameter and partial elevation of the testes occurs. In the *female,* beads of moisture occur on the inner surface of the vagina, the clitoris increases in size, and elevation of the uterus occurs.

**ex·hi·bi'tion·ism** A compulsive form of deviant sexual behavior in which a person, usually a male, obtains sexual satisfaction by exposing his genitals to strangers, often women and children. Exhibitionists rarely attack or molest their victims but obtain their gratification from the reaction of the observer, usually that of fright or disgust. In prison slang, exhibitionists are called "flag wavers." *Compare* voyeurism.

**ex·og'a·my** Marriage outside the tribe or clan; crossbreeding. *Compare* endogamy.

**ex·pect'ing** *Euphemism* for anticipating the birth of a baby.

**ex·tra·cur·ric'u·lar ac·tiv'i·ties** *Slang.* Fornication, extramarital sex, adultery.

**ex·tra·em·bry·on'ic mem'branes** Those that develop outside the embryo proper. Included are the amnion, chorion, yolk sac, and allantois.

**ex·tra·gen'i·tal** Outside of, apart from, or unrelated to the genital organs.

**ex·tra·mar'i·tal** Outside of marriage, especially with reference to sexual relationships; comarital. *See* adultery.

**eye'ful** A sight that is pleasing to the eyes, especially a strikingly beautiful woman.

**fac·ul·ta'tive** Optional; resulting from choice. *Compare* obligate.

**fag, fag'got** *Slang.* A male homosexual, especially an extremely effeminate one.

**fail'ure rate** *Abbr.* FR. A figure used to express the effectiveness of a contraceptive method. It is usually the number of unwanted pregnancies that occur among 100 women who use a particular technique for one year. Also called *pregnancy rate. See* HWY.

**fair'y** *Slang.* A male homosexual, especially one who assumes a female role.

**fair'y hawk** *Slang.* A heterosexual who harasses homosexuals.

**fall'ing in love** Limerence, q.v.

**fal·lo'pi·an tube** The uterine tube or oviduct, a tube about 4 inches in length that extends laterally from each side of the uterus. It transmits ova to the uterus and, within the tube, fertilization usually occurs.

**Fal'ope ring ster·i·li·za'tion** A method of tubal sterilization in which the uterine tube is pulled up to the abdominal wall and, by use of a special laparoscope, a tiny silicone rubber band is pushed down over a segment of the tube causing it to become occluded.

**false preg'nan·cy** Pseudocyesis, q.v.

**fal'sie** *Slang.* A padded brassiere worn by a woman to give the appearance of having larger and more shapely breasts.

**fa·mil'i·al** 1. Of or pertaining to a family. 2. Occurring among members of a family; hereditary, as a *familial* disease.

**fam'i·ly** A social group consisting of parents and their offspring, this group comprising a *nuclear family.* When the household includes two or more generations, it is an *extended family* (grandparents, parents, children, grandchildren, uncles, aunts, cousins). The term is sometimes applied to communal households.

**fam'i·ly jew'els** *Slang.* The testes, so called because they are a man's most prized possessions.

**fam'i·ly way, in a** *Euphemism* for the condition of being pregnant.

**fan'cy man** *Slang.* A prostitute's protector and/or lover or husband.

**fan'cy wo'man (girl or la'dy)** *Slang.* A woman of questionable morals; a mistress; a prostitute.

**fan'ny** *Slang.* 1. The buttocks. 2. The female pudenda.

**fan'ta·sy** A mental picture or image actively created when a person is awake; a daydream.

**fa'ther** 1. A male parent. 2. A man who adopts a child.

**fa'ther-in-law** The father of one's husband or wife.

**fe'ces** The excretory material discharged from the intestine through the anus. It consists of undigested food, secretions of digestive glands, mucus, bacteria, exfoliated cells, lymphocytes, and products of metabolism and bacterial decomposition. Also called *stool*.

**fe'cund** Capable of conceiving and producing offspring.

**fec'un·date** To impregnate; to make fertile; to cause to conceive.

**fe·cun'di·ty** Fruitfulness or fertility; the ability to produce young in great numbers.

**fel·la'ti·o, fel·la'tion** Oral stimulation of the penis; irrumation. *Compare* cunnilingus. *See* oral sex.

**fel'la·tor** One who engages in fellatio, especially a male.

**fel'la·trice** A prostitute who specializes in fellating her customers.

**fel'la·trix** A female who practices fellatio.

**fel'low** 1. A man or a boy. 2. *Informal.* A boyfriend.

**fem, femme** 1. A girl or a woman. 2. *Slang.* (a) A passive lesbian; (b) a feminine-appearing lesbian prostitute; (c) an effeminate homosexual, male or female.

**fe'male** 1. *adj.* Of or pertaining to the sex that produces ova; characteristic of the female sex; feminine. 2. *n.* An individual that produces female gametes (ova) that are fertilized by spermatozoa. *Symbol* ○, ♀. 3. *adj.* Designating a part of a device or apparatus that receives a complementary part.

**fe'male im·per'so·na·tor** A male entertainer who dresses and acts as a female.

**fe'male in·ad'e·qua·cy** The inability of a female to obtain sexual satisfaction during coitus, especially failure to experience an orgasm, a condition commonly referred to as *coldness* or *frigidity*. *See* orgasmic impairment.

**fem'i·nine** 1. Of or belonging to the female sex. 2. Of or pertaining to a woman or a girl; possessing womanly qualities; effeminate.

**fem·i·nin'i·ty** The qualities and characteristics of a woman; femaleness.

**fem'i·nism** 1. The doctrine that women should have political, economic, and social rights equivalent to those of men. 2. Organized activity in support of women's rights, especially the movement to eliminate conditions that restrict the activities of women or discriminate against women; the women's liberation movement.

**fem·in·i·za'tion** 1. The normal development in a female of female secondary sex characteristics. 2. The induction and development of female traits and characteristics in a male.

**femme** *French.* 1. A woman. 2. A wife.

**femme fa·tale'** 1. A seductive woman who entices men into difficult, dangerous, or compromising situations. 2. A woman who possesses an aura of charm and mystery which she uses in seducing men.

**fem'or·al in'ter·course** Intercourse in which the penis is placed between the thighs.

**fern** *Slang.* A passive lesbian. *Compare* butch.

**fern'ing** The formation of a fernlike pattern in dried cervical mucus when viewed at time of ovulation.

**fern test** A test to determine the exact time of ovulation. A small amount of mucus obtained from the opening of the cervix is spread on a glass slide. When examined under a microscope, a specimen obtained at the time of ovulation will show a beautiful branching pattern resembling a fern.

**fer'tile** Fruitful, productive, capable of reproducing. *Compare* infertile, sterile.

**fer'tile per'i·od** The time during the menstrual cycle during which conception is most likely to occur, especially the days immediately preceding and following ovulation, which usually occurs between the 12th and 16th days of the cycle. *Compare* safe period. *See* rhythm method.

**fer·til′i·ty** The state or quality of being fertile; the capacity to produce offspring.

**fer·til′i·ty clin′ic** A clinic that seeks to determine the causes of sterility; one that aids in reproduction.

**fer·til′i·ty drug** A drug that induces ovulation in the female or the development of spermatozoa in the male, as clomiphene citrate.

**fer·til′i·ty in·sur′ance** Semen banking or the storage of frozen semen. It is resorted to by men contemplating vasectomy or possible subjection to radioactivity.

**fer·til′i·ty rate** The number of live births per 1000 women between the ages of 15 and 65 that are recorded annually.

**fer·til·i·za′tion** The union of a spermatozoon and an ovum; conception, impregnation, fecundation. In fertilization, the normal diploid number of chromosomes is restored and the development of the ovum is initiated.

**fer·til′i·zin** A substance in the cortex of certain ova that is essential for normal fertilization.

**fe′tal** Of or pertaining to a fetus.

**fe·ta′tion** The development of a fetus within the uterus; pregnancy.

**fe′ti·cide** The killing of a fetus; an illegal abortion.

**fet′ish** An inanimate object or a part of the body other than the sex organs that is used for the arousal of sexual feelings or sexual gratification. Objects may include feminine articles of clothing (panties, bras, girdles, stockings, garters, shoes) or something belonging to the desired person. Parts of the body that may serve as a fetish include the neck, breasts, buttocks, and hair. The use of a fetish is principally limited to males. Fetishes are employed as masturbatory aids or in sadomasochistic activities because of their association with sexual activity or their symbolism.

**fet′ish·ism** The use or employment of a fetish.

**fet′ish·ist** A person who believes in and makes use of a fetish.

**fe′tus, foe′tus** The unborn young of a viviparous mammal, especially in the later stages of development. In humans, the young from the beginning of the third month of development to birth.

**fi·an·cé′** A man engaged to be married.

**fi·an·cée′** A woman engaged to be married.

**fib·ril·la′tion** A local quivering movement of muscle fibers such as that which occurs in the uterus during the excitement phase of orgasm.

**fi·bro·cys′tic dis·ease′** A benign condition characterized by the development of cysts within the breast. Cysts may be solitary or multiple, large or small, asymptomatic or tender and painful. Cysts are common in menopausal women. Most cysts are benign but women with fibrocystic disease are more likely to develop breast cancer than those who have never had it; hence breast examination is recommended every six months. New cysts seldom develop after the menopause.

**fi′broid** 1. A structure composed principally of fibrous tissue. 2. A benign tumor of the uterus. Also called *uterine myoma, fibromyoma,* or *leiomyoma.* Fibroids are a common cause of menstrual disorders and may cause infertility by interfering with implantation or the early development of the embryo. It is generally thought that they are caused by excess production of estrogen.

**fi·bro·my·o′ma** A myoma in which connective tissue is intermingled with muscle tissue.

**fi·del′i·ty** Constancy and faithfulness to one's mate, especially abstention from extramarital intercourse.

**fil′i·al** Of, pertaining to, or associated with a son or daughter.

**fil′ly** 1. A female colt or young mare. 2. *Slang.* A young girl, especially a high-spirited, vivacious girl.

**fim′bri·a** A fringe, edge, or border, such as that surrounding the distal opening (ostium) of the uterine tube, consisting of a variable number of flattened or fingerlike processes (fimbriae).

**first base, get′ting to** *Slang.* Necking, kissing.

**fish** *Slang.* Among homosexuals, a heterosexual female.

**fish′wife** A coarse, vulgar, abusive woman; a shrew.

**fis′sion** 1. A splitting into parts. 2. In *biology,* a form of asexual reproduction in which the organism divides into two or more parts.

**flag′el·lant** A person who obtains sexual satisfaction by beating or being beaten by another person.

**flag·el·la′tion** The act of beating or whipping.

**flag wav′er** *Slang.* An exhibitionist.

**flame** *Slang.* A sweetheart or beloved person.

**flan·quette′** A half-frontal position in coitus.

**flap′per** Around the 1920′s, a term applied to girls and young women who flouted conventional moral restraints by wearing short, straight dresses with no petticoats and having stockings rolled below the knees and bobbed hair, and who generally showed an obvious interest in sex.

**flash′er** *Slang.* 1. An exhibitionist. 2. A French prostitute who solicits by exposing herself along the Bois de Boulogne, Paris.

**flash′y** Cheap, showy, ostentatious, pretentious.

**flea′bag** *Slang.* 1. A prostitute who caters to skid-row males; a slovenly, old woman. 2. A cheap, disreputable hotel or rooming house; a flophouse.

**flesh** 1. The soft tissue of the body, especially muscle. 2. The body as distinguished from the mind or psyche. 3. Man′s physical or carnal nature as distinguished from his spiritual and moral nature. 4. Sensual appetite or desire.

**flesh′ly** 1. Of or pertaining to the body or flesh. 2. Overweight, corpulent; possessing excess flesh, especially adipose tissue. 3. Of a sensuous nature; carnal, worldly.

**flesh ped′dler** *Slang.* 1. A pimp. 2. A prostitute. 3. The manager of a place of entertainment that features seminude girls and women.

**flesh′pot** A brothel or place of entertainment that features luxurious and unrestrained entertainment, especially catering to the vices and weaknesses of the flesh.

**flirt** 1. *v.* To engage in amorous behavior without serious intentions. 2. *n.* A person who engages in flirting, especially a girl; a coquette.

**flit** *Slang.* A male homosexual, especially one who exhibits female mannerisms.

**floo′zie, floo′zy, floo′sie** 1. An attractive girl or young woman, usually of less than average intelligence, who entices men with her sex appeal; an undisciplined, flirtatious, and promiscuous female. 2. *Slang.* A dissolute, coarse, slovenly woman; a cheap prostitute.

**flow′er** *Slang.* A male homosexual.

**flow of wa′ters** The discharge of the fluid contained within the amniotic sac that occurs shortly before childbirth.

**flute, flut′er** *Slang.* A male homosexual.

**foam, con·tra·cep′tive** A chemical substance, usually containing a spermicidal agent, that forms an effervescent mass within the vagina. It acts as a physical barrier preventing spermatozoa from entering the cervix.

**foe′tus** A fetus, q.v.

**fol′li·cle** A small, spherical mass of cells within the ovary that contains a developing ovum (oocyte). Follicles are of three types: *primary, growing,* and *mature* (graafian). The growth, development, and functioning of ovarian follicles depends upon the action of gonadotrophic hormones FSH and LH from the anterior pituitary mediated by estrogens. Normally a single follicle matures and ruptures each month about midway between menstrual periods. Follicles are the principal source of estrogens.

**fol'li·cle-stim'u·la·ting hor'mone** *Abbr.* FSH. A gonadotrophic hormone secreted by the anterior lobe of the pituitary that stimulates the growth, development, and functioning of ovarian follicles.

**fon'dle** To touch lovingly with affection; to caress with the hands.

**fore'bear** A forefather or ancestor.

**fore'fa·ther** An ancestor.

**fore'play** Activities to induce sexual arousal that usually precede sexual intercourse and tend to lead to orgasm. These include expressions of affection, caresses, kissing, tactile stimulation, and sometimes mild pain stimuli; contrectation.

**fore'plea·sure** Pleasure experienced during foreplay preceding sexual intercourse.

**fore'skin** The prepuce, q.v.

**for·ni·ca'tion** Sexual intercourse between unmarried persons, between two married persons not married to each other, or between a married and an unmarried person. *See* adultery.

**for'nix** A circular fold of the vagina that encircles the cervix of the uterus.

**fos'sa** A depression or pit. *See* navicular fossa.

**fos'ter par'ent** One who provides a child with the care of a parent without being related by blood or law.

**foun'tain sy·ringe'** (or **sy'ringe**) An apparatus that injects a liquid by the action of gravity as that used in cleansing the vagina or giving an enema.

**four·chette'** The frenulum of the labia, a fold of skin that joins the posterior ends of the minor labia.

**fox'y la'dy** *Slang.* An attractive young woman.

**FR** Failure rate.

**fra·ter'nal twins** Twins that develop from two ova, each fertilized by a different sperm; dizygotic twins.

**free love** The acceptance of sexual relationships between individuals other than husband and wife as being correct and proper. Premarital and extramarital sex are also condoned.

**French cul'ture** Oral-genital stimulation; fellatio and cunnilingus.

**French job** *Slang.* Fellatio.

**French kiss** A deep, passionate kiss involving intimate contact of the lips, tongue, and teeth and usually tongue-to-tongue contact; a soul kiss. *See* maraichinage.

**French let'ter** A condom.

**French postcards** Obscene, pornographic pictures.

**French tick'ler** A special type of condom with a rough or irregular surface which is thought to increase vaginal stimulation.

**fren'u·lum** A membranous structure that limits or restricts movement of a part; a frenum.

**fren'u·lum of the clit'o·ris** A membrane that connects the minor labia with the undersurface of the clitoris.

**fren'u·lum of the pre'puce** A fold of skin that connects the lower surface of the glans penis to the prepuce.

**fren'um** A restraining structure. *See* frenulum.

**fresh meat** *Slang.* A sexually inexperienced person.

**Freud, Sig'mund** (1856–1939) An Austrian neurologist, physiologist, and psychiatrist, the founder of psychoanalysis. He taught that repressed and forgotten impressions underlie all abnormal mental states and that their revelation often effects a cure; that infantile mental processes are of primary importance in later development; that dreams are the unconscious representation of repressed desires, especially sexual desires.

**fric′a·trice** 1. A prostitute who specializes in masturbating her customers. 2. A lewd woman. 3. A lesbian.

**frig** *Slang.* 1. To have sexual intercourse with a woman. 2. To cheat or trick a person.

**fri·gid′i·ty** In a woman, lack of sexual feelings or desire; coldness; the absence of passion or sexual responsiveness; inability to experience an orgasm. Also called *orgasmic dysfunction, orgasmic impairment, hyposexuality, sexual anesthesia, coldness.*

**frill** *Slang.* A girl or woman.

**front mar′riage** A legal marriage of convenience.

**frot·tage′** The practice of obtaining sexual pleasure by (a) rubbing a part of the body, especially the genitals, against another person, as in a crowd, or (b) having contact with soft fabrics such as silk or velvet; hyphephilia.

**frot·teur′** A person who engages in frottage.

**fruit** *Slang.* 1. A male homosexual. 2. A sexually promiscuous, immoral person.

**fruit′cake** *Slang.* A homosexual.

**fruit fly** A woman who prefers the company of homosexuals.

**frus·tra′tion** Failure to obtain complete sexual satisfaction; sexual activity that does not result in an orgasm.

**FSH** Follicle-stimulating hormone.

**fuck** *Vulgar.* 1. *v.* To have sexual intercourse with. 2. *n.* The act of, or a partner in, sexual intercourse. 3. *v.* To cheat, deceive, or take advantage of.

**fud** *Slang.* The buttocks or rump.

**fun′dus** The base of a hollow organ or the portion most remote from or opposite to its opening, as the *fundus* of the uterus.

**fun′gus,** *pl.* **fun′gi** Any of a large group of simple plants characterized by lack of chlorophyll. They exist as saprophytes or parasites and many are pathogenic. They include the yeasts, molds, mildews, rusts, smuts, and mushrooms. *See* candidiasis, moniliasis.

**gad′get** *Slang.* The penis.

**gal** *Informal.* Term applied to a girl or young woman, especially a pleasant, lively girl. Also commonly applied to a girlfriend, as my *gal.*

**ga·lac′to·gogue** An agent that increases the flow of breast milk.

**ga·lac·to·poi·e′sis** The production of milk by the mammary glands.

**ga·lac·tor·rhe′a** Excessive flow of milk from the mammary gland.

**gam′ete** A mature, haploid, reproductive or sex cell; an ovum or spermatozoon.

**ga·me·to·gen′e·sis** The origin and development of mature gametes. *See* oogenesis, spermatogenesis.

**gams** *Slang.* A woman's legs.

**gang bang** *Slang.* An affair at which several males, generally adolescents, have intercourse with a consenting female. Also called *gang shay.*

**gang rape** An affair in which a woman is forcibly raped by several males.

**gash** *Slang.* A female as a sexual object; sexual intercourse; the vagina.

**gas′tru·la** A stage that follows the blastula in the embryonic development of many metazoans, consisting of a cuplike structure with a wall of two layers, an outer *ectoderm* and an inner *endoderm.* Its cavity, the *archenteron,* opens to the outside through the *blastopore.*

**gay** 1. *adj.* Joyous, spirited, vivacious. 2. *adj.* Licentious, wanton, immoral. 3. *adj.*

*Informal.* Homosexual; pertaining to or involving homosexual behavior. 4. *n.* A homosexual person.

**gay bar** A bar where male and female homosexuals congregate. Distinctive costumes and coded signals are employed in making contacts.

**gay liberation movement** A national movement to encourage homosexuals and lesbians to "come out of the closet" and demand rights denied because of their sexual preference. It seeks to eliminate discrimination in jobs and housing and to do away with penalties for sexual acts engaged in between consenting adults.

**gay life** Homosexual behavior.

**geld** To castrate.

**geld'ing** A castrated animal, especially a male horse.

**gen'der** A category of individuals, words, or objects classed according to sex, termed masculine, feminine, or neuter *gender.*

**gen'der i·den'ti·ty** The development in a child of a sense of maleness or femaleness depending primarily upon how the child is reared.

**gen'der role** The public manifestation of gender identity; the behavior of an individual by which that person indicates to others the degree to which he or she is a male or female. Also called *sex role.*

**gene** A basic unit of heredity consisting of molecules of DNA. It is a self-duplicating particle capable of mutation. Genes are carried by the chromosomes, where each occupies a fixed locus. Genes present in the egg and the sperm are responsible for the hereditary transmission of structural and physiological traits.

**ge·ne·al'o·gy** The study of the ancestry of a person, race, or group, or a record of such a study.

**gen'er·ate** 1. To produce offspring. 2. To give rise to.

**gen·er·a'tion** 1. The act of producing or bringing into existence; procreation. 2. A group of contemporaneous individuals, as the *first generation.*

**ge·net'ic** Of or pertaining to genetics or heredity. 2. Pertaining to, caused by, or resulting from the action of genes or hereditary determiners.

**ge·net'ics** The science that deals with heredity, especially the transmission of hereditary traits and variation among organisms.

**gen'i·tal** 1. Of or pertaining to reproduction. 2. Of or pertaining to the sex organs.

**gen'i·tal her'pes** Herpes genitalis, q.v.

**gen·i·tal'i·a** The male or female reproductive organs, especially the external sex organs. The *external genitalia* in the *male* include the penis, testes, and scrotum; in the *female,* the labia majora, labia minora, clitoris, vestibule, and mons pubis.

**gen'i·tal my·co·plas'mas** Microorganisms intermediate in size between bacteria and viruses, and possessing characteristics of both. They are thought to be the cause of nongonococcal urethritis in humans.

**gen'i·tal or'gans** The reproductive organs; genitalia.

**gen'i·tals** The genitalia.

**gen'i·tal spot** An area on the nasal mucosa that tends to bleed during menstruation.

**gen'i·tal stage** A stage in psychosexual development between the narcissistic and homosexual stages. It occurs between the ages of three and seven and is marked by an increasing awareness of the genital organs and the pleasure arising from their stimulation.

**gen'i·tal wart** A moist or venereal wart occurring commonly on the moist surface of a genital organ, in the rectum, around the anus, and sometimes on the lips or in the oral cavity. It is caused by a papilloma virus. Also called condyloma acuminatum, *pl.* condylomata acuminata.

**gen·i·to·u'ri·nar·y** Of or pertaining to the reproductive and urinary organs. *See* urogenital.

**gen'o·cide** The planned, systematic extermination of a racial or cultural group.

**gen·o·pho'bi·a** A morbid fear of sex or sexuality; an abnormal fear of sexual intercourse.

**gens** A clan, especially an exogamous, patrilineal clan.

**gent** *Informal.* A gentleman.

**gen'tle·man** A polite, well-educated man of good breeding; one with high standards of correct behavior.

**gen'tle sex** Women in general.

**gen'tle·wo·man** A woman of noble birth or superior social position; a gracious, considerate woman.

**germ cell** A reproductive cell or gamete; an ovum (egg) or spermatozoon (sperm).

**germ lay'er** One of the three primary layers of cells (ectoderm, mesoderm, endoderm) in a developing embryo from which the various tissues and organs develop.

**ger'mi·nal** 1. Of, pertaining to, or having the characteristics of germ or reproductive cells. 2. Of, pertaining to, or related to cells in an early embryo that give rise to reproductive cells, as the *germinal epithelium* from which the gonads develop.

**ger'mi·nal disc** *See* embryonic disc.

**ger·on·tol'o·gy** The science that deals with aging and the problems of elderly persons.

**ge·ron·to·phil'i·a** Love for old people, especially sexual love.

**ge·ron·to·sex·u·al'i·ty** Sexual interest in and a desire for sex with an elderly person.

**ges'ta·gen** A hormone with progestational activity, especially progesterone.

**ges·ta'tion** Pregnancy; the act of carrying young within the uterus.

**ges·ta'tion pe'ri·od** The time from conception to childbirth, in humans generally considered to be 266 days from the last ovulation or 276 to 280 days from the beginning of the last menstrual period.

**get some ac'tion** *Slang.* To have sex relations.

**gig** *Slang.* The female genitals.

**gi-gi, gee-gee** *Slang.* The rectum or vagina.

**gig'o·lo** 1. A man who lives off the earnings of a woman, especially a young man who is supported by an older woman in return for sexual favors and companionship. 2. A professional male dancing partner or escort.

**girl** 1. A young, unmarried female. 2. *Slang.* A male homosexual.

**girl'friend** 1. A female friend. 2. A girl or woman to whom a boy or man is attracted and to whom he shows particular affection; a female sweetheart.

**give head, to** *Slang.* To engage in genital kisses or fellatio.

**gland** An organ or structure that manufactures or elaborates a substance. An *exocrine* gland (e.g. the prostate) possesses a duct through which the secretory product is discharged onto a surface; an *endocrine* gland (e.g. the thyroid) lacks a duct, the product being secreted directly into the bloodstream.

**glans** A conical structure that forms the distal end of the penis (*glans penis*) or clitoris (*glans clitoridis*).

**gleet** Inflammation of the urethra characterized by a mucous discharge containing pus, usually the result of a gonorrheal infection.

**go all the way** *Slang.* To complete the act of sexual intercourse, especially after heavy petting.

**go both ways** *Slang.* To practice both heterosexuality and homosexuality.

**go down on** *Slang.* To perform fellatio or cunnilingus on someone.

**go-go boy** A male who dances partially or completely nude.

**go-go girl** An attractive female, usually scantily clad, often topless with breasts exposed, who dances solo in bars, nightclubs, and discotheques. She may dance completely nude (topless and bottomless). Also called *shindig dancer*.

**go'ing off** *See* orgasm.

**go'ing Ro'man** *Slang.* Swinging; mate swapping.

**go'nad** 1. An organ that produces gametes or reproductive cells; an ovary or testis. 2. An undifferentiated structure in an embryo that develops into an ovary or a testis.

**go·nad·o·tro'phic, go·nad·o·tro'pic** Stimulating the growth, development, and functioning of the gonads (testes or ovaries).

**go·nad·o·tro'phic hor'mones** *See* follicle-stimulating hormone (FSH) and luteinizing hormone (LH).

**go·nad·o·tro'phin, go·nad·o·tro'pin** A gonad-stimulating hormone produced by the anterior lobe of the pituitary gland, such as FSH and LH.

**gon·o·coc'cus** The causative organism of gonorrhea, *Neisseria gonorrhoeae*, a bacterium.

**gon·or·rhe'a** An infectious venereal disease that affects the mucous membranes of the urogenital tract; other organs may also be involved. It is caused by a bacterium, *Neisseria gonorrhoeae*, transmitted sexually in most cases. In the *male*, onset is characterized by an acute, purulent urethritis accompanied by difficult urination. In the *female*, the urethra, cervix, and Skene's or Bartholin's glands usually are involved. Symptoms may be severe but sometimes are absent, especially in the female.

**go off** *Slang.* To have an ejaculation and experience an orgasm.

**gos'sy·pol** A chemical that inhibits sperm production and motility, widely used in China as a male contraceptive. A male Pill.

**go to bed with** *Euphemism.* To have sexual relations with.

**graaf'i·an fol'li·cle** A mature vesicular follicle containing an ovum or oocyte that is discharged at ovulation. It is the principal source of ovarian estrogens. *See* follicle.

**grand'child** A child of one's son or daughter.

**grand'dad** A grandfather.

**grand'daugh·ter** The daughter of one's son or daughter.

**grand'fa·ther** The father of one's mother or father.

**grand'ma** 1. A grandmother. 2. *Slang.* An elderly male homosexual.

**grand'mo·ther** The mother of one's father or mother.

**grand'pa** A grandfather.

**grand'son** The son of one's son or daughter.

**gran'ny, gran'nie** 1. A grandmother. 2. *Informal.* An old woman, especially one that is fussy about trivial details.

**gran·u·lo'ma in·gui·na'le** A mildly contagious venereal disease caused by a bacterium. It is a chronic, progressive disease involving the skin and lymphatic organs. It is common in the tropics.

**grav'id** Pregnant; with child.

**grav'i·da** A pregnant woman.

**Greek cul'ture** *Slang.* Anal intercourse.

**groom** A bridegroom.

**group mar'riage** A type of marriage in which several males and females with their children live together under one roof or in a closely knit community. Also called *tribe, commune* (q.v.), or *intentional community.* Within the group, legal marriages and monogamous sexual relationships may prevail, but in most groups polygamous relationships are common.

**group sex** The participation of several couples in indiscriminate sexual intercourse or other sexual activities. *See* commune, orgy, swinging.

**gry·po'sis** (or **gry·pho'sis**) **pe'nis** Chordee, q.v.

**G-string** A loincloth or breechcloth passed between the legs and supported by a cord passed around the waist. It is worn especially by women entertainers in a striptease.

**gum'ma** A lesion characteristic of tertiary or late syphilis. It is an encapsulated, necrotic mass of tissue of variable size occurring most frequently in the liver, but also occurring in the testes, brain, bone, skin, and other organs.

**gun** *Slang.* The penis.

**guy** *Informal.* A man or fellow.

**GYN** Gynecology.

**gy·nan'dro·morph** A sex mosaic, an individual in which sexual characteristics are intermixed, as in certain insects. Individuals may be typically male in certain parts of the body and typically female in other parts. Individuals may have both male and female gonads and genitalia.

**gy'nan·dry** The development of male characteristics in a female; feminine pseudohermaphroditism. *See* virilism.

**gyn'e·coid** Resembling a female; having female characteristics.

**gy·ne·col'o·gist** A physician who specializes in gynecology. *Compare* andrologist.

**gy·ne·col'o·gy** The branch of medicine that deals with pathologic disorders and hygiene of women, especially diseases involving the reproductive organs. *Compare* andrology. *See* obstetrics.

**gyn·e·co·mas'ti·a** Excessive development of the breasts in a male.

**gyn·e·pho'bi·a** An abnormal fear of females or a morbid aversion to associating with women.

**gyn'oid** Like or resembling a female; gynecoid.

**gyn'o·sperm** An X-bearing, female-producing spermatozoon.

**hair pie** *Slang.* The female genitalia.

**hair-trig'ger trou'ble** *Slang.* Premature ejaculation.

**hand** A pledge to marry.

**hand job** *Slang.* Masturbation.

**hand trou'ble** *Slang.* The propensity of a man to touch or caress a woman or a girl.

**hang-up** An idea, belief, or obsession that interferes with or inhibits usual or normal behavior.

**han'ky-pan'ky** *Informal.* Deceitful, unethical behavior, especially philandering and adultery.

**hap'loid** Possessing a single complement of chromosomes. In humans, 23, as in the egg or sperm. *Compare* diploid.

**hard** *Slang.* Stiff, erect, with reference to the penis; an erect penis.

**hard on** *Slang.* An erection of the penis.

**har'em** 1. All the females of an Oriental household. 2. A group of female mammals served by one male. 3. *Slang.* A group of women associated with one man, as a group of prostitutes managed by a pimp.

**har'lot** A prostitute.

**haul (one's) ash'es** *Slang.* To have sexual intercourse (with one).

**have** *Slang.* To have sexual relations with another person, usually of the opposite sex.

**have an af·fair' with** *Slang.* To have an intimate association with another person, usually involving sexual relations.

**HCG** Human chorionic gonadotrophin, q.v.

**HCS** Human chorionic somatomammotrophin, q.v.

**he** 1. A male who has just been mentioned or referred to. 2. A male person or animal. 3. Any person whose sex is unspecified.

**head** *Slang.* 1. A young woman, especially one who is sexually attractive. 2. An erect penis. 3. The glans of the penis.

**head'lights** *Slang.* Well-developed female breasts.

**heat** The period during which a female mammal becomes sexually excited and will accept a male. It is the time at which ovulation occurs. *Compare* rut. *See* estrus.

**he·don'ic** Of, pertaining to, or characterized by pleasure.

**he'don·ism** A philosophy or way of life that holds that happiness or pleasure is the principal goal or aim of life.

**heif'er** 1. A young cow that has not produced a calf. 2. *Slang.* A pretty girl or young woman.

**help'mate** A spouse; one of a wedded pair.

**he-man** A strong, robust, virile man.

**he·ma·tu'ri·a** The presence of blood in the urine.

**he·mo·sper'mi·a** The presence of blood in semen.

**hen** 1. The female of domestic fowl and gallinaceous birds. 2. *Informal.* A fussy, gossipy old woman.

**hen'pecked** Subjected to nagging, scolding, and fault-finding by a domineering wife.

**he·red'i·tar·y** 1. Passing or capable of being transmitted from parents to offspring. 2. Characteristic of or appearing in successive generations.

**he·red'i·ty** The transmission of traits from parents to offspring through hereditary determiners (genes) present in the chromosomes of the germ cells (ova or spermatozoa). *See* genetics.

**her·maph'ro·dite** 1. An individual who possesses both male and female sex organs or a combined ovotestis; a monoecious individual. 2. According to J. Money, a person in whom at least one of five variables of sex is contradictory to the remainder. These are (a) nuclear sex, (b) gonadal sex, (c) hormonal sex and pubertal virilization or feminization, (d) internal accessory reproductive structures, and (e) external genital morphology.

**her'ni·a** An abnormal protrusion of an organ or structure through the wall of the cavity that normally contains it. The most common form is *inguinal hernia*, in which the intestine protrudes through the inguinal canal. Commonly called *rupture.*

**her'pes gen·i·tal'is** Genital herpes (herpes simplex virus, type 2, or HSV-2), an acute, sexually transmitted viral disorder characterized by the formation of fluid-filled vesicles on surfaces of sex organs. Principal sites in the female are the vagina and cervix; in the male, the glans penis and prepuce. Neighboring areas may also become infected. It is a self-limited disease which disappears with or without treatment in a few days or weeks, but infection lies dormant and tends to reappear at irregular intervals. Stress, either physical or emotional, tends to be a precipitating factor. At present (1987), there is no cure and no immunity develops. In a pregnant female, the disease is a common cause of fetal death or birth defects.

**her'pes la·bi·al'is** Labial herpes (herpes simplex virus, type 1, or HSV-1), an acute, infectious, virus-caused disorder characterized by the development of fluid-filled vesicles commonly called *cold sores* or *fever blisters* on the skin, lips, or mucous membranes, almost always occurring above the waist.

**her'pes·vi'rus hom'i·nis (HVH)** The infecting agent that causes herpes simplex. There are two strains of HVH. Type 1 causes *herpes labialis* (q.v.); type 2 causes *herpes genitalis* (q.v.).

**he·tai'ra, he·tae'ra** 1. A woman who uses her charm and sex appeal to achieve fame, wealth, or social position. *See* demimonde. 2. A high-class prostitute in ancient Greece.

**he'tae·rism** Communal marriage as it existed in certain primitive societies; concubinage.

**het·er·og′a·mous** Producing unlike gametes. *Compare* homogamous.

**het·er·og′a·my** The marriage of unlike individuals; crossbreeding. *Compare* homogamy.

**het·er·o·sex′u·al** 1. *adj.* Pertaining to sexual feelings and activities between persons of opposite sex; attracted to the opposite sex. 2. *n.* A heterosexual person; a straight. *Compare* homosexual.

**het·er·o·sex·u·al′i·ty** Sexual attraction to and sexual behavior directed toward persons of the opposite sex; the straight life. *Compare* homosexuality, bisexuality.

**hick′ey** *Slang.* 1. A pimple. 2. A skin blemish, especially the purplish mark left on the skin resulting from kissing or sucking.

**hir′sute** Hairy; covered with excess hair.

**hir′sut·ism** Excessive hairiness, especially in women.

**HMG** Human menopausal gonadotrophin.

**hold′ing back** A conscious effort to delay an orgasm and, in the male, ejaculation.

**home run, get a** *Slang.* To achieve sexual intercourse.

**home′work** *Slang.* Lovemaking, petting, and sometimes, sexual intercourse.

**Ho′mo** The genus of primates that includes modern man, *Homo sapiens,* and some extinct species.

**ho′mo** 1. Any individual of the genus *Homo.* 2. *Slang.* A homosexual.

**ho·mo·e·rot′ic** Characterized by homoeroticism, homosexual.

**ho·mo·e·rot′i·cism** Sexual attraction of an individual for a person of his or her own sex; homosexuality; lesbianism.

**ho·mog′a·mous** Producing like gametes.

**ho·mog′a·my** 1. The mating of like with like, as in reproduction within an isolated group; inbreeding. 2. The marriage of like individuals, as those of the same race, religion, or social status. *Compare* heterogamy.

**ho·mo·gen′i·tal·ism** Sexual relations between members of the same sex.

**ho·mol′o·gous** Corresponding in fundamental structure, embryonic origin, and development, said of organs such as the penis and clitoris. *Compare* analogous.

**ho′mo·phile** A person who practices homophilism.

**ho·mo·phil′ic** Of or pertaining to homophilism.

**ho·moph′i·lism** A fondness for and an attraction to individuals of one's own sex. The association of homophiles may be for social, intellectual, business, recreational, religious, or other reasons. If the association is for sexual purposes, the condition is called *homosexuality.*

**ho·mo·pho′bi·a** A compulsive, unwarranted fear of and intolerance for homosexuals and homosexuality.

**Ho′mo sa′pi·ens** Modern man, the single extant species of the genus *Homo.*

**ho·mo·sex′u·al** 1. *adj.* Pertaining to sexual behavior or feelings between persons of the same sex. 2. *n.* A homosexual person; a lesbian, a gay person, a uranist. *Compare* heterosexual.

**ho·mo·sex·u·al′i·ty** Sexual attraction to and sexual behavior directed toward a person of the same sex; the gay life, lesbianism; uranism. *Compare* heterosexuality. *See* deprivation homosexuality.

**hon′ey-fuck** *Slang.* 1. To have sexual relations in a romantic manner. 2. To have intercourse with a young girl.

**hon′ey man** *Slang.* A pimp; a kept man.

**hon′ey·moon** A trip or vacation taken by a newly married couple; the period immediately following a marriage.

**hon′ey·moon blad′der** In a female, a bruised bladder resulting from violent sexual activity.

**hon'ey pie** *Slang.* The vagina.

**hooked** *Slang.* 1. Addicted to narcotics. 2. Forced into a marriage. 3. Married.

**hook'er** *Slang.* A prostitute.

**hook shop** *Slang.* A cheap brothel.

**hook'y, hook'ey** *Slang.* Semen.

**hoot'chy-coot'chy** *See* cooch.

**hor'mone** A substance, produced by an endocrine gland or tissue, that is transported by the bloodstream to another organ or tissue in which it evokes a response (activation or inhibition). *See* androgen, estrogen, FSH, gonadotrophin, HCG, HMG, HPL, LH, pheromone, progesterone, releasing factor, testosterone.

**hor'ny** *Slang.* Sexually aroused or excited; sexually minded.

**hot** *Slang.* Passionate, easily sexually aroused; possessing strong sexual feelings or desire.

**hot ba'by** *Slang.* An extremely passionate girl or woman.

**hot flash** A sudden, recurrent feeling of warmness accompanied by sweating that occurs in women during the menopause. It is a vasomotor phenomenon due to the dilatation of capillaries of the skin resulting from the withdrawal of estrogen.

**hot flush** The reddening of the skin resulting from the dilatation of capillaries that usually accompanies a hot flash.

**hot pants** *Slang.* Said to be possessed by a promiscuous woman.

**house of ill fame** or **of ill re·pute'** A brothel.

**hoy'den, hoi'den** A boisterous, high-spirited, impudent girl; a tomboy.

**HPL** Human placental lactogen.

**hu'man cho·ri·on'ic go·nad·o·tro'phin** *Abbr.* HCG. A hormone produced by the placenta that appears in the blood and urine in early pregnancy. Its biological effects, which are similar to those of LH, include stimulation of the development of ovarian follicles, induction of ovulation, and formation of a corpus luteum and its secretion of progesterone. Its presence or absence in urine is the basis of most pregnancy tests.

**hu'man cho·ri·on'ic so·ma·to·mam·mo·tro'phin** *Abbr.* HCS. *See* human placental lactogen.

**hu'man men·o·paus'al go·nad·o·tro'phin** *Abbr.* HMC. A substance that has been found in the urine of postmenopausal women and has been used to improve fertility.

**hu'man pla·cen'tal lac'to·gen** *Abbr.* HPL. A polypeptid hormone produced by the placenta in early pregnancy that promotes bone growth and induces lactation.

**hump** *Slang.* To engage in sexual intercourse.

**hung like a horse** *Slang.* Possessing a very large penis.

**hung up** *Slang.* Having old-fashioned ideas and beliefs; frustrated.

**hus'band** A married man; a woman's spouse.

**hus'sy** 1. A lewd woman; a strumpet or trollop. 2. A saucy, impudent girl.

**hus'tle** *Slang.* To solicit customers for a prostitute or as a prostitute.

**hus'tler** *Slang.* 1. One who solicits customers for a prostitute. 2. A female prostitute or streetwalker. 3. A male homosexual prostitute.

**HVH** Herpesvirus hominis, q.v.

**HWY** *Abbr.* for *hundred women years,* i.e., of exposure in determining the use-effectiveness of a contraceptive method.

**hy·a·lu·ron'i·dase** An enzyme present in semen that depolymerizes hyaluronic acid, a substance of importance in maintaining the integrity of tissues. This enzyme disperses cells of the corona radiata surrounding the ovum, thus facilitating fertilization.

**hy'brid** The offspring of two organisms that belong to two different breeds, varieties, races, or species; the offspring of parents genetically unlike.

**hy′brid·ize** To cause to produce hybrids; to crossbreed.

**hy′dro·cele** The accumulation of fluid in a saclike cavity, especially the tunica vaginalis of the testis.

**hy·dro·sal′pinx** The accumulation of fluid within the uterine tube.

**hy′men** A membrane of collagenous and elastic connective tissue that partially closes the external opening of the vagina in most virgins. In some cases it is *imperforate*, completely closing the vaginal orifice. Commonly called *maidenhead*. See defloration.

**hy·men·ec′to·my** Surgical removal of the hymen.

**hy·per·ca·thex′is** Excessive concentration of libido on a single object, such as a person, thing, or idea.

**hy·per·e·me′sis gra·vi·da′rum** Pernicious vomiting of pregnancy, a condition in which excessive vomiting results in dehydration and acidosis and may threaten life.

**hy·per′ga·my** Marriage to a person of higher social status. *Compare* hypogamy.

**hy·per·go′nad·ism** A condition resulting from abnormally increased functioning of the gonads. It may involve either the interstitial cells or gamete-producing cells or both. Precocious sexual development is a common result.

**hy·per·he·do′ni·a** Excessive pleasure derived from gratification of a desire, especially fulfillment of libidinal drive.

**hy·per·men·or·rhe′a** Menstruation of excessive duration or amount; menorrhagia.

**hy·per·pla′si·a** An increase in the size of an organ or tissue resulting from the multiplication of cells. *Compare* hypertrophy.

**hy·per·sex·u·al′i·ty** Excessive sexual desire. *See* nymphomania, satyriasis.

**hy·per·ten′sion** Abnormal elevation of blood pressure.

**hy·per·ton′ic sa′line** A 20 percent salt solution. *See* saline injection method.

**hy·per·tri·cho′sis** Excess growth of hair; excessive hairiness; hirsutism.

**hy·per′tro·phy** An increase in the size of an organ or tissue resulting from increased functional activity but with no increase in the number of cells. *Compare* hyperplasia.

**hyp·he·do′ni·a** A condition in which acts that normally bring pleasure are no longer enjoyed or bring satisfaction.

**hyph·e·phil′i·a** Sexual pleasure obtained from contact with soft fabrics, such as silk or velvet. *See* frottage.

**hy·po·der′mic** Beneath the skin, as in *hypodermic* syringe, by which a solution can be introduced subcutaneously.

**hy·pog′a·my** Marriage to a person of lower social status. *Compare* hypergamy.

**hy·po·gen′i·tal·ism** A condition in which the genital organs are underdeveloped with resulting incomplete development of secondary sex characteristics; hypogonadism.

**hy·po·gon′ad·ism** Inadequate production of sex hormones by the ovaries or testes, such usually resulting in delayed puberty and incomplete development of secondary sex characteristics; hypogenitalism.

**hy·po·men·or·rhe′a** A condition in which uterine bleeding at menstruation is less than normal and the period of flow of short duration.

**hy·poph′y·sis** The pituitary gland, q.v.

**hy·po·pla′si·a** Defective development or abnormal smallness of an organ or tissue.

**hy·po·sex·u·al′i·ty** Reduced ability or inability to experience an orgasm; a condition in which sex drive is markedly reduced or absent.

**hy·po·spa′di·as** A congenital condition in which the urogenital opening is on the undersurface of the penis. *Compare* epispadias.

**hy·po·thal'a·mus** A portion of the diencephalon of the brain (located beneath the cerebrum) which functions in the integration of various autonomic activities of the body, especially the functioning of the nervous and endocrine systems. It is the source of the hormones *vasopressin* and *oxytocin,* which are stored in and released by the posterior lobe of the pituitary. It produces *releasing factors* (RF) for hormones produced by the anterior lobe of the pituitary, especially FSH and LH. It also contains neural centers which, when stimulated, bring about feelings and reactions associated with various emotions such as pleasure, pain, fright, rage.

**hys·ter·ec'to·my** Partial or total removal of the uterus.

**hys·ter'o·gram** A roentgenogram of the uterine cavity following injection of a special dye solution through the cervix.

**hys·ter'o·scope** An instrument with a powerful light that is inserted into the uterus for viewing the uterine cavity.

**hys·ter·os'co·py** Examination of the interior of the uterus by use of a hysteroscope.

**hys·ter·ot'o·my** Surgical incision into the uterus; a cesarean section.

**ICSH** Interstitial cell-stimulating hormone. *See* luteinizing hormone.

**id** In *psychoanalysis,* Freud's term for one of the three fundamental parts of the psyche. It lies in the unconscious and is considered the repository of basic instincts involved in reproduction and self-preservation.

**i·den'ti·cal twins** Twins that develop from a single fertilized ovum; monozygotic twins.

**Ig** Immunoglobulin, q.v.

**il·le·git'i·mate child** One whose parents were not married at the time of his or her birth or subsequently; a bastard.

**im·mod'est** Lacking in modesty; indecent, assertive, unrestrained.

**im·mu·no·glob'u·lin** *Abbr.* Ig. Any of a number of blood proteins, especially gamma globulins, that possess antibody activity. They provide protection against infectious organisms or foreign bodies that act as antigens.

**im·mu·no·log'ic test** A pregnancy test in which sensitized latex particles or red blood cells agglutinate (clump together) when exposed to blood or urine containing human chorionic gonadotrophin.

**im'plant** An object or a substance that is placed within body tissues.

**im·plan·ta'tion** 1. The embedding of a developing zygote (blastocyst) within the endometrium of the uterus; nidation. 2. The placing within body tissues of an object or substance, such as (a) tissue from another person or from another part of the body, or (b) a therapeutic substance, such as a radioactive agent or a hormone.

**im'po·tence, im'po·ten·cy** Inability of a male to perform the sexual act; inability to achieve and maintain an erection to the state of orgasm. It may be due to structural disorders, disease of the genital organs, endocrine dysfunction, systemic disease, or psychic factors.

**im'po·tent** Lacking in strength or power; lacking in ability to perform the sexual act.

**im·preg'nate** To make pregnant; to cause to conceive.

**im·pure'** Not pure; immoral, obscene, unchaste.

**in·ad'e·qua·cy** *See* sexual inadequacy.

**in'born'** Present at birth; inherited.

**in'bred** Resulting from inbreeding.

**in'breed·ing** The mating of closely related individuals, as occurs (a) in isolated communities, (b) as a result of social or religious customs, or (c) as practiced in the improvement of commercial plants and animals. It results in genetic homogeneity and leads in time to the production of a pure breed.

**in'cest** Sexual relations between closely related persons who by law or custom are prohibited from marrying. It may be between parent and child, brother and sister, uncle and niece, aunt and nephew, or certain other relatives. Incestuous relationships are generally prohibited by law.

**in'cest ta·boo'** One of the oldest taboos of mankind, generally considered to be essential for maintaining the stability of the family by preventing sexual relations between members, which would have a disruptive effect. Incest is universally condemned in all societies, although in certain ancient civilizations, as among the Incas and Egyptians, brother-sister matings in royal families were sometimes permitted or required.

**in·ci'sion** A cut made into bodily tissues, especially for surgical purposes.

**in·com'pe·ten·cy** Inability to perform a natural function.

**in·com'pe·tent cer'vix** A uterine cervix that is unable to bear the weight of a developing fetus. It dilates prematurely and is a common cause of miscarriage after the third month of pregnancy.

**in·con'ti·nence** 1. Inability to control fecal or urinary discharge. 2. Lack of self-control with respect to sexual indulgence.

**in'cu·bate** To warm eggs artificially or by bodily heat in order to promote embryonic development and subsequent hatching of the young.

**in'cu·ba·tor** A cabinet, in which a uniform temperature can be maintained, used for the cultivation of bacteriological media, hatching eggs, or maintaining life in premature infants.

**in'cu·bus** 1. An evil spirit or demon in male form that was thought to visit women during sleep and have sexual intercourse with them. 2. A nightmare, or a person, thing, or condition that oppresses one like a nightmare. *Compare* succubus.

**in·de'cent** Offensive to public morals or good taste; immodest; obscene.

**in'fant** A young child from birth to the age of two years.

**in·fan'ti·cide** The killing of an infant.

**in'fan·tile** Of or pertaining to infants; lacking maturity.

**in'fan·til·ism** A state of arrested development characterized by the persistence of infantile or childish characteristics or traits into adolescence or adult life. *See* sexual infantilism.

**in·fat·u·a'tion** The state of being possessed by unreasoning passion or irrational love or desire; being blindly in love. *See* limerence.

**in·fe'cund** Unfruitful, barren, nonproductive, sterile. *See* infertility.

**in·fer'tile** 1. Incapable of conceiving and producing offspring. 2. Incapable of developing, as an *infertile* egg.

**in·fer·til'i·ty** The quality of being unfruitful, barren, or nonproductive of offspring; sterility; inability to conceive; infecundity. Some causes of infertility are (a) in the *female*, malfunction of the ovaries or uterus, obstructed uterine tube, conditions that adversely affect sperm transport; endocrine disorders, psychic factors; (b) in the *male*, impotence, impaired production of spermatozoa, obstruction of sperm ducts, disorders in the secretory function of the accessory sex glands, psychic factors.

**in·fib·u·la'tion** The act of buckling or fastening together, especially the ancient practice of closing the vaginal opening of young girls by stitching together the labia majora or, in boys, of stitching together the prepuce to prevent sexual intercourse. Also called *pharaonic circumcision*.

**in·fi·del'i·ty** Unfaithfulness; disloyalty; adultery.

**in'gui·nal** Of or pertaining to the groin, the region between the abdomen and the thigh.

**in'gui·nal ca·nal'** A passageway in the body wall in the region of the groin through which the vas deferens and accompanying blood vessels and nerves pass from the testis into the body cavity. In the female, it transmits the round ligament.

**in·her'it** To receive from one's parents genetic factors which determine various physical and physiological characteristics. See gene.

**in·hib'it** To arrest, restrain, slow down, or check an impulse or activity.

**in·hi·bi'tion** The act of inhibiting.

**in·ject'** To force a substance, usually a liquid, into a blood vessel, passageway, cavity, or tissue of the body.

**in·jec'ta·ble** 1. adj. Capable of being injected. 2. n. A substance that is injected, especially a contraceptive that is injected into the bloodstream, as Depo-Provera, a long-acting contraceptive.

**in-law** A relative by marriage, as a mother-in-law.

**in'ner cell mass** A cluster of cells that develops on the inner surface of the wall of a blastocyst from which the embryo proper develops.

**in·sem'i·nate** To introduce semen into the vagina or uterus of a female. See artificial insemination.

**in·sem·i·na'tion** The act of inseminating.

**in·sert'** To place in, into, between, or among other things. In sexual relations, to place the penis in the vagina, anus, or mouth.

**in·ser'tee** A passive male homosexual.

**in·ser'ter** A device for placing an IUD within the uterus; an introducer.

**in·ser'tor** An active male homosexual.

**in·spir'er** In pederasty, the older lover. Compare listener.

**in'stant pe'ri·od** Euphemism for menstrual extraction.

**In'sti·tute for Sex Re'search** An institute connected with Indiana University, devoted to the study of sex. It was founded by Alfred Kinsey and contains one of the largest collections of erotica ever assembled.

**in·ter·cep'tion** The process of preventing implantation of a developing blastocyst as by use of DES. Interception may be accomplished by (a) inhibiting the action of the corpus luteum in the production of progesterone essential for early pregnancy; (b) inducing changes in the endometrium, especially a deficiency in carbonic anhydrase and other enzymes essential for development of the blastocyst; or (c) interfering with the passage of a fertilized ovum through the uterine tube, a tube-locking effect.

**in'ter·course** An active connection or association between two persons. See sexual intercourse, coitus.

**in·ter·cru'ral** Between the legs.

**in·ter·eth'nic** Between different racial, religious, cultural, or national groups.

**in'ter·faith** Between different religious groups.

**in·ter·fem'o·ral** Between the thighs.

**in·ter·mam'ma·ry** Between the breasts.

**in·ter·mar'riage** 1. The marriage of individuals belonging to different racial, cultural, or social groups, as interfaith, interethnic, and interracial marriages. 2. The marriage of closely related persons.

**in·ter·men'stru·al** Occurring between menstrual periods.

**in·ter·ra'cial** Between different races or racial groups.

**in'ter·sex** An intersexual individual.

**in·ter·sex'u·al** 1. adj. Of or pertaining to intersexuality. 2. n. An individual who possesses characteristics of both sexes. This condition may result from mutation, chromosome abnormalities, endocrine imbalance, or environmental factors.

**in·ter·sex·u·al'i·ty** The condition in which characteristics of both sexes are intermingled in an individual. *See* hermaphrodite, pseudohermaphrodite.

**in·ter·sti'tial cells** 1. Cells of Leydig, cells that produce male hormones (androgens), located between the seminiferous tubules of the testes. 2. A group of cells or isolated cells in the ovary which are thought to be derived from atretic follicles. They secrete androgens and estrogens.

**in·ter·sti'tial cell-stim'u·la·ting hor'mone** *Abbr.* ICSH. *See* luteinizing hormone.

**in'ti·ma·cy** 1. The state or condition of being close or intimate. 2. The manifestation of a close, personal, affectionate relationship with a person or a group. 3. A sexually familiar act or liberty.

**in'ti·mate** 1. Having a close physical, emotional, or social relationship with another person. 2. Characterized by privacy, informality, and mutual respect and interest. 3. Involved in or characterized by a sexual relationship or activities.

**in·tra·u'ter·ine** Within the uterus.

**in·tra·u'ter·ine con·tra·cep'tive de·vice'** *Abbr.* IUCD. An intrauterine device, q.v.

**in·tra·u'ter·ine de·vice'** *Abbr.* IUD. A contraceptive device which, when placed within the uterus, prevents implantation of the blastocyst. Types include Lippe's Loop, Safe-T Coil, Majzlin Spring, Ypsilon Y, Margulie's Spiral, Progestasert, Copper-T, Copper-7, and others. Each possesses a *tail* or *strings* which, after insertion, protrude from the cervix into the vagina; the presence of these indicates that the device has not been expelled.

**in·tra·va'gin·al** Within the vagina.

**in·tro·duc'er** An inserter, q.v.

**in·tro'i·tus** An entryway, especially the entrance into the vagina; the vaginal orifice.

**in·tro·mis'sion** The insertion of one part into another, especially the insertion of the penis into the vagina.

**in·tro·mit'tent or'gan** The penis.

**in·ver'sion** 1. A turning around or reversing the directions of. 2. Homosexuality. *See* sexual inversion.

**in'vert** A person given to sexual inversion; a homosexual or a lesbian.

**in vi'tro** In glass, with reference to an activity or a reaction that takes place in a test tube, culture dish, or glass container. *Compare* in vivo.

**in vi'vo** Within a living organism, as opposed to *in vitro*.

**in·vo·lu'tion** 1. A turning inward. 2. The return of an organ or structure to normal following enlargement due to functional demands, as occurs in the uterus after childbirth or the mammary gland following lactation. 3. Retrogressive changes that occur in various organs after middle age or during aging, such as changes in the ovaries after menopause.

**ip'sism** Masturbation.

**ir·ru·ma'tion** Fellatio.

**ISD** Inhibited sexual desire.

**it** *Euphemism.* 1. Sex appeal. 2. Sexual intercourse. 3. Variously used to refer to the penis, vagina, vulva, an orgasm, and menstruation.

**ith·y·phal'lic** 1. With an erect phallus, with reference to graphic representations. 2. Obscene; lascivious.

**IUCD** Intrauterine contraceptive device, q.v.

**IUD** Intrauterine device, q.v.

**jack'et** *Slang.* A condom.

**jack off** *Slang.* To masturbate.

**jade stalk** Chinese synonym for the penis.

**jag house** *Slang.* A brothel for male prostitutes who cater to male homosexuals.

**jazz** *Slang.* To copulate.

**jeal'ous** Intolerant of, or troubled by suspicions or fears of, unfaithfulness or loss of a loved one.

**jeal'ous·y** 1. Resentment of another person's success or good fortune. 2. Mental distress resulting from suspicion of or fear of unfaithfulness or infidelity, especially in matters of love.

**jel'ly, con·tra·cep'tive** A semisolid substance of elastic consistency that serves as a vehicle for a contraceptive agent.

**jel'ly roll** *Slang.* 1. The penis. 2. The vagina.

**jerk** *Slang.* One who masturbates.

**jerk off** *Slang.* To masturbate.

**jew'el box** *Slang.* The scrotum.

**jew'els** *See* family jewels.

**jism** *Slang.* Semen.

**jock** *Slang.* 1. An athletic supporter; a jockstrap. 2. A college athlete.

**jock'er** *Slang.* A male homosexual, especially an aggressive one.

**jock itch** A dermatosis, caused by a fungus, affecting the genital region. Also called *tinea cruris, dhobie itch, jockstrap itch.*

**john** *Slang.* 1. A male customer of a prostitute; a trick. 2. A public toilet, especially one for males. 3. A man who keeps a woman in return for sexual favors. 4. A wealthy, elderly homosexual who keeps a young homosexual as a companion or associate.

**John'son, Vir·gin'ia E.** (1925–    ) Associate and wife of Dr. William H. Masters, with whom she has collaborated in sex research at the Reproductive Biology Research Foundation, St. Louis, Missouri.

**joint** *Slang.* 1. The penis. 2. A marijuana cigarette.

**joy house** *Slang.* A brothel.

**joy stick** *Slang.* The penis.

**jugs** *Slang.* The female breasts.

**jun'gle meat** *Slang.* A black man's penis.

**jus pri'mae noc'tis** *Latin* for "right of the first night." In feudal times, the right of the bridegroom to deflower the bride was often claimed by the lord of the manor or the parish priest.

**ju'ven·ile** *adj.* Pertaining to or characterized by youth or young persons. 2. *adj.*Young, youthful, immature, not fully developed. 3. *n.* A young person or child.

**Ka·ma·su'tra** A Hindu love manual of the eighth century A.D.

**ka·rez'za** Coitus reservatus or prolonged sexual union. Also *carezza.*

**ka'ry·o·type** The number, form, and size of the chromosomes within the nucleus of a cell. The term is applied to a photomicrograph or a diagrammatic representation of the chromosomes.

**ka·ry·o·typ'ing** An analysis of the chromosomes of cells, especially fetal cells obtained by amniocentesis, in order to determine the sex of a fetus or to detect possible genetic disorders.

**ka·ta·sex'u·al** Designating sexual relations with a nonhuman.

**Keg'el's ex'er·cis·es** Exercises for strengthening the muscles that play a role in controlling the passage of feces and urine. They consist of tightening the muscles of the pelvic diaphragm by stopping the flow of urine in midstream. Then similar contractions practiced many times a day will aid in controlling incontinence. A side effect is markedly increased sexual response in females. Also called Aswina-Mudra (perineal exercise). *See* pubococcygeus muscle.

**kept** Maintained as a mistress.

**kep'tie** *Slang.* A woman who is supported by a man in return for sexual favors.

**ke·to·ster'oid** A steroid excreted in urine, derived from the breakdown of adrenocortical or gonadal hormones.

**kin** Collective term for one's relatives; kinfolk.

**kin'dred** A group of related persons.

**kin'folk** Members of a family and related persons; kindred. Also *kinsfolk.*

**king** *Slang.* A bull dyke.

**kin'ky sex** Sex involving extreme, deviate behavior.

**Kin'sey, Al'fred M.** (1894–1956) Originally an entomologist and an authority on gall wasps; in later life a pioneer in the field of sex research. He developed one of the first college courses in sex education and marriage. Following extensive research on human sexual behavior, he established the Institute for Sex Research at Indiana University, Bloomington, Indiana. His associates included Wardell B. Pomeroy, Clyde E. Martin, and Paul H. Gebhard. Two of his books, *Sexual Behavior in the Human Male* (1948) and *Sexual Behavior in the Human Female* (1953), laid the foundation for greater freedom in sexual behavior and a better understanding of sex, which are experienced today.

**kin'ship** Relationship by blood or inheritance; family relationship.

**kins'man** 1. A male blood relative or, in a loose sense, any relative by marriage. 2. A person of the same racial or cultural background.

**kins'wo·man** A female blood relative.

**kiss** To touch with the lips slightly pursed as a sign of greeting, affection, reverence, love, or passion. *See* French kiss.

**kiss à la can'ni·bale** An extremely passionate kiss that leaves a bruise.

**klep·to·lag'ni·a** Sexual excitement that accompanies stealing.

**Kline'fel·ter's syn'drome** A condition characterized by small testes, hyalinization of the seminiferous tubules, and androgen deficiency from impaired functioning of the cells of Leydig. It results in sterility, gynecomastia, aspermia, increased excretion of urinary gonadotrophins, and sometimes mental retardation. It is a genetic disorder resulting from an abnormality in the sex chromosomes, an extra X chromosome being present.

**klis·ma·phil'i·a** Erotic pleasure obtained from taking an enema.

**knight** *Slang.* A male homosexual.

**knocked up** *Slang.* With child; pregnant.

**knock'ers** *Slang.* 1. Female breasts, especially when well developed. 2. The testes.

**know, to** *Euphemism.* To have sexual intercourse, especially in biblical terminology.

**Krafft-Eb'ing, Rich'ard von** (1840–1902) A Viennese psychiatrist noted for his work in forensic psychiatry. Author of *Psychopathia Sexualis.*

**la'bi·um** *pl.* **la'bi·a** The labia (literally "lips") are either of two folds of tissue, the outer *labia majora* and the inner *labia minora,* that border the pudendal cleft, which contains the vaginal and urethral orifices.

**la'bor** Childbirth, parturition.

**lac·ta'tion** 1. The secretion of milk by the mammary gland. 2. The period during which a baby feeds on milk from the breast.

**lac·tif'er·ous** 1. Producing milk, as a *lactiferous* gland, the breast. 2. Conveying milk, as a *lactiferous* duct or sinus.

**lac·to·gen'ic** Inducing the secretion of milk.

**lac·to·gen'ic hor'mone** Prolactin, q.v.

**la'cy** *Slang.* Effeminate. 2. Of or pertaining to homosexuals.

**lad** 1. A boy or youth. 2. *Informal.* A familiar form of address for a man.

**la'dy,** *pl.* **la'dies** 1. A woman, an adult female. 2. A woman of refinement and good breeding. 3. A woman of high social rank.

**la'dy's man, la'dies' man** *Slang.* A charming and courtly man with social graces who pursues women politely and with success.

**la′dy kill′er** *Slang.* A lady's man.

**lag·ne′sis** Erotomania.

**lag·neu·o·ma′ni·a** A compulsive state characterized by absorption with or engaging in lustful actions or indulging in lewd talk.

**lag·no′sis** Excessive sexual desire, especially in a male; satyriasis, nymphomania.

**La Le′che League** An organization that encourages and promotes breast-feeding of babies.

**La·maze′ meth′od** or **tech·nique′** A series of physical and mental procedures (psychoprophylaxis) introduced by Dr. Fernard Lamaze, a French physician, designed to alleviate pain during labor and delivery. The development of conditioned responses and controlled respirations play an important role in the method. *See* natural childbirth.

**lam′bi·tive** Licking with the tongue.

**lam′bi·tus** Cunnilingus.

**la·nu′go** The downlike coat of hair that covers the body of a fetus, prominent between the fifth and seventh months but shed before or shortly after birth.

**lap′a·ro·scope** A narrow, elongated, cylindrical instrument containing a light source which, when inserted through the abdominal wall, enables visualization of the abdominal and pelvic organs; an endoscope.

**lap·a·ros′co·py** Examination of the abdominal cavity by use of a laparoscope inserted through its wall.

**lap·a·rot′o·my** A surgical incision through the abdominal wall.

**la pe·tite′ mort** *French* for little death, q.v.

**las·civ′i·ous** Lewd, lecherous, tending to excite sexual desire, salacious.

**lass** 1. A girl or young woman, especially one who is unmarried. 2. A female lover or sweetheart.

**la′tent** 1. Dormant, quiescent; present but not apparent. 2. Present in a concealed or dormant form but capable of being actively expressed, as *latent* homosexuality.

**lay** *Slang.* 1. *v.* To have sexual intercourse with. 2. *n.* A female as a sex object for coitus.

**lay with** *See* lie with.

**leak′age** *Slang.* Involuntary urination.

**Le·bo·yer′, Fred′er·ick** A French obstetrician who advocates techniques in delivery that lessen the trauma a baby experiences in the transition from life in a dark, warm, sheltered womb to the brightly lighted, noisy environment of a delivery room. Gentle handling of the infant, placing it immediately upon the mother's breast, and delayed cutting of the umbilical cord are advocated. *See* natural childbirth.

**lech′er** A man who engages in excessive sexual activity.

**lech′er·ous** 1. Of or pertaining to lechery. 2. Erotically exciting.

**lech′er·y** Excessive indulgence in sexual activity.

**leer** 1. To glance sideways, especially with lascivious or prurient interest or malicious intent or design. 2. A sly, sinister, immodest, or seductive glance.

**left-hand′ed** *Slang.* Homosexual.

**le·git′i·mate** 1. Lawful; according to law. 2. Born in wedlock or to parents married to each other.

**leg man** *Slang.* A man who especially likes women with attractive legs.

**leg show** A show in which women show their legs and bodies with the special intent of arousing men sexually.

**lei·o·my·o′ma** A benign tumor of the uterus consisting principally of smooth muscle cells. Also called *fibroid.*

**les** *Slang.* A lesbian.

**les′bi·an** A homosexual female; a tribade.

**les'bi·an·ism** Sexual love and affection of two women for each other; homosexuality between two females; tribadism; sapphism.

**le'sion** A structural alteration in an organ or tissue resulting from injury, disease processes, faulty metabolism, or maldevelopment.

**leu·kor·rhe'a** A disorder common in women, characterized by an abnormal, whitish, mucopurulent discharge from the vagina. It usually results from an infection of the vagina or cervix by bacteria, protozoa, or fungi. Commonly called the whites.

**lev'i·rate** A marriage custom that requires the marriage of a man to his brother's widow, especially if the widow is childless. *Compare* sororate.

**lev·on·or·ges'trel** A steroid incorporated into intrauterine devices.

**lewd** Licentious, lustful, indecent, obscene.

**Ley'dig's cells** The interstitial cells of the testes, the principal source of androgens (male hormones).

**lez, lezz** *Slang.* A lesbian or female homosexual.

**LGV** Lymphogranuloma venereum, q.v.

**LH** Luteinizing hormone.

**li'ai·son** A sexual relationship without marriage, especially an adulterous relationship.

**lib'er·tine** A dissolute person; one who has no regard for moral or religious restraints.

**li·bid'i·nous** Of or pertaining to libido or sexual drive; lustful; lascivious.

**li·bi'do** 1. In *psychoanalysis,* Freud's term for psychosexual energy, considered to be the dynamic force in one's personality. 2. Sexual desire or sex drive; passion; the force by which the sexual instinct expresses itself.

**lice** *Pl.* of louse, q.v.

**li·cen'tious** Sexually unrestrained; immoral; lewd.

**lie with** *Euphemism* for "to have sexual intercourse with," especially in biblical literature. *Past tense:* lay with.

**li·ga'tion** The act of tying off or constricting a tube or vessel by use of a ligature.

**lig'a·ture** A cord or thread used in tying off a vessel or closing a duct.

**lig·o·tage'** *French* for bondage, the art of tying up a sex partner.

**lil'y** *Slang.* 1. An effeminate male. 2. A homosexual.

**lim'er·ence** Newly coined term to designate lovesickness, being in love, having fallen in love, or being love-smitten.

**lim'er·ence ob'ject** *Abbr.* LO. The object of a person's extreme love and affection.

**limp wrist** *Slang.* An effeminate man; a homosexual.

**lin'e·age** Ancestry or descent; the line of descendents from a single ancestor.

**lin'gam** A phallus, or a representation of one, which among Hindus is a symbol of the god Siva. Also *linga.*

**lin·ge·rie'** Women's underwear, sleepwear, or various undergarments worn by women.

**lin'gual** Of or pertaining to the tongue.

**lin'guist** *Slang.* One who uses the tongue in oral sex.

**lis'ten·er** In pederasty, the younger beloved. *Compare* inspirer.

**lit'ter** A number of young produced at one birth by a multiparous mammal.

**lit'tle death** *Slang.* Passing out or losing consciousness during an orgasm. Also *la petite mort.*

**Lit're, glands of** Numerous small, branched, mucous glands that open into the cavernous urethra (portion within the penis).

**LO** Limerence object; love object.

**lo'chi·a** The discharge from the vagina and uterus during the first few weeks following childbirth.

**lo·co·mo'tor a·tax'i·a** Tabes dorsalis, q.v.

**look in the mir'ror** *Slang.* Female masturbation.

**loose** Characterized by immorality; lewd; unchaste; sexually promiscuous.

**lor·do′sis** 1. Excessive anterior curvature of the spine in the lumbar region. 2. A position assumed by most female mammals preceding and during mating in which the back is arched and haunches elevated, thus facilitating reception of the male.

**lor·do′sis re′flex** The assumption of a lordosis position by an estrous female elicited by gently stroking the back or by mounting on the part of a male, as in rats.

**lo·tha′ri·o** A charming man who deceives and seduces women; a rake or libertine.

**louse,** *pl.* **lice** A small, wingless insect of the order Phthiraptera (suborder Anoplura) that lives as an ectoparasite on humans and other animals. Human lice include the body louse and head louse, *Pediculus humanus,* and the pubic louse or crab, *Phthirus pubis.*

**love** 1. A profound feeling of concern, tenderness, and affection, usually directed toward one of the opposite sex or members of one's family. 2. Sexual passion or desire for another person.

**love af·fair′** A romantic relationship between two lovers; an amour.

**love blos′som** *Slang.* A hickey.

**love child** An illegitimate child; a bastard.

**love juice** *Slang.* Vaginal secretion.

**love life** The affairs, activities, and relationships involved in the affections between two individuals, especially their sexual activities.

**love lips** *Slang.* The labia.

**love mu′sic** *Slang.* The sounds that accompany an orgasm.

**love nest** A habitation of lovers, especially a place where lovers engaged in illicit love meet or live.

**love po′tion** An aphrodisiac; a philter, q.v.

**lov′er** 1. One (most commonly a male) who loves another. 2. Sexual partner.

**love slave** *Slang.* A woman who will submit to sadistic actions.

**L/S ra′tio** The lecithin to sphingomyelin ratio in the amniotic fluid, a test for assaying the maturity of fetal lungs.

**lu′bri·cant** An agent that reduces friction, especially a substance placed on the penis or within the vagina to facilitate coitus. Natural lubricants include vaginal secretions and the secretions of the bulbourethral glands in the male. Additional lubricants include saliva, surgical jelly, silicone lubricants, and others.

**lu′bri·cous, lu·bri′cious** Lewd, lecherous, salacious; sexually unrestrained.

**lu′es** Syphilis.

**luke** *Slang.* Leukorrhea, q.v.

**lu′men** A cavity or space within a tube, duct, canal, or hollow organ.

**lunch-hour a·bor′tion** *Euphemism* for menstrual extraction.

**lust** Sexual desire or appetite especially when compelling and uncontrolled; licentiousness.

**lu′te·al phase** The progestational or secretory phase of the menstrual cycle that follows ovulation. It is controlled by progesterone secreted by the corpus luteum of the ovary.

**lu′te·in·iz·ing hor′mone** *Abbr.* LH. A gonadotrophic hormone secreted by the anterior lobe of the pituitary. In the *female,* it stimulates the secretion of estrogens by the follicle, induces ovulation, promotes development of the corpus luteum, and, with prolactin, stimulates production of progesterone by the corpus luteum. In the *male,* it stimulates the production of testosterone by the interstitial cells of the testes, hence it is called *interstitial cell-stimulating hormone* (ICSH).

**lu·te·ol′y·sis** The process by which a corpus luteum loses its ability to secrete progesterone, degenerates, and becomes a *corpus albicans.*

**lu·te·o·tro'phic hor'mone** Prolactin, q.v. Also called *luteotrophin.*

**ly'ing in** The confinement of a woman at childbirth.

**lym·pho·gran·u·lo'ma ve·ne're·um** *Abbr.* LGV. An infectious venereal disease caused by a viruslike organism of the *Bedsonia* group.

**ma** *Slang.* An effeminate male.

**ma·chis'mo** A Spanish word implying strength and masculinity, especially that demonstrated by the fathering of many children, whether legitimate or illegitimate.

**ma'cho** Characterized by machismo.

**mac'ro·mere** One of the four large cells at the vegetal pole of an ovum undergoing cleavage.

**mac·ro·pe'nis** A macrophallus, q.v.

**mac'ro·phage** A tissue cell that engulfs solid particles; a phagocyte.

**mac·ro·phal'lus** An exceptionally large penis; a macropenis.

**mac·u·lo·pap'u·lar sex flush** *See* sex flush.

**mad'am** 1. A polite form of address for a woman of rank or position. 2. The female head of a household. 3. A female manager of a brothel.

**Madge** *Slang.* The female genitalia.

**Ma·don'na-pros'ti·tute syn'drome** A "split-image" view of women held by some men in which the wife is viewed as a loving, nonsexual, maternal, all-giving figure but sexually is considered as a prostitute or tramp, a toy or hobby for his pleasure or entertainment.

**mae'nad, me'nad** 1. A woman who participates in orgiastic rites. 2. A frenzied female dancer. 3. A raging, distraught woman.

**maid, maid'en** 1. An unmarried woman or girl. 2. A virgin.

**maid'en·head** 1. The state or quality of being a maiden; virginity. 2. The hymen.

**mai'zie** *Slang.* A gay person.

**make it** or **make out** *Slang.* To succeed or fare well in any activity, especially to succeed in seducing a woman.

**make love** *Slang.* To copulate; to engage in coitus.

**make out** *Slang.* To succeed, especially in seducing a woman.

**make whoo'pee** *Slang.* To engage in sexual intercourse.

**mak'ing love** *Euphemism* for sexual intercourse.

**mal·a·dies' d'amour** *French* for sexually transmitted diseases.

**male** 1. *adj.* Of or pertaining to the sex that produces spermatozoa. 2. *adj.* Characteristic of the male sex; masculine. 3. *n.* An individual that produces male gametes (spermatozoa). Symbol □, ♂. 4. *adj.* Designating a part of a device or apparatus that is placed into a complementary part.

**male chau'vin·ism** Exaggerated attachment and devotion to maleness, accompanied by boastfulness and glorification of the male sex, as well as an attitude of superiority to the female sex.

**ma'ma, mam'ma** 1. *Informal.* A mother. 2. *Slang.* A sexually attractive and alluring woman, sometimes referred to as a *sweet mama* or *red-hot mama.*

**mam'ma** A mammary gland; a breast or udder.

**mam'ma·ry gland** The mamma or breast, a compound, alveolar gland that secretes milk.

**mam·mec'to·my** Surgical excision of the breast; mastectomy.

**mam·mog'ra·phy** X-ray examination of the mammary gland.

**man,** *pl.* **men** 1. An adult, human male. 2. In a collective sense, in the singular, the entire human race, including all men and women.

**man'hood** 1. The state of being a man or an adult male person. 2. Manly qualities characteristic of the adult male.

**Mann Act** The Federal White Slave Act, which defines as a felony the providing of transportation or aiding in transportation of a female across state lines for the purpose of prostitution or for any other immoral purpose.

**man-trap** *Slang.* An especially seductive woman.

**ma·raich'in·age** The use of the tongue in erotic kissing; French kissing.

**mar'bles** *Slang.* The testes.

**Marge** *Slang.* A very feminine, passive dyke.

**mar'i·tal** Of or pertaining to marriage.

**mark** The victim of a con game.

**mar'riage** The state of a man and woman being joined legally as a husband and wife; a wedding. *See* closed marriage, open marriage.

**Mar'y** *Slang.* 1. A passive male homosexual. 2. A lesbian.

**mas'cu·line** Having the qualities and characteristics of a male; manly.

**mas·cu·lin'i·ty** Behavior and characteristics generally associated with maleness and with men. *Compare* femininity.

**mas'cu·lin·ize** To bring about the development of male secondary sexual characteristics in a female; to cause virilization.

**mash** *Slang.* Love or affection for one of the opposite sex.

**mask of preg'nan·cy** Melasma, q.v.

**mas'och·ism** 1. Sexual gratification obtained by submission to physical abuse, pain, brutality, or humiliating treatment. 2. A taking pleasure in physical and mental suffering inflicted on oneself; practicing extreme self-denial and self-punishment. *Compare* sadism.

**mas·sage'** Rubbing, stroking, kneading, or tapping the surface of the body for hygienic, therapeutic, or sensual purposes. It may be accomplished manually or mechanically, as by the use of an electric vibrator or a percussion apparatus.

**mas·sage' par'lor** A commercial establishment where massage is provided, often a front for illegal sexual activities such as prostitution.

**mas·seur'** A man who practices massage professionally.

**mas·seuse'** A woman who practices massage professionally.

**mas·tal'gia** Pain in the breast; mastodynia.

**mas·tec'to·my** Surgical removal of a breast. *Simple mastectomy* is removal of the breast only. *Radical mastectomy* is removal of the entire breast and adjacent tissue including the pectoralis muscles and the lymphatic tissue of the chest wall and axilla.

**Mas'ters, Wil'liam H.** (1915–    ) Well-known physician and sex researcher; founder of the Reproductive Biology Research Foundation, St. Louis, Missouri. Coauthor with Virginia E. Johnson, his wife, of *Human Sexual Response, Human Sexual Inadequacy, Pleasure Bond,* and *Homosexuality in Perspective.* Also author of many research papers.

**mas·ti'tis** Inflammation of the mammary gland.

**mas·to·dy'ni·a** Pain in the breast; mastalgia.

**mas·tur·ba'tion** 1. The obtaining of sexual satisfaction through stimulation of one's own genital organs, especially the induction of an orgasm. This may be accomplished by manual manipulation, bodily contact with an object or another person other than by coitus, or by utilization of a mechanical device such as a vibrator. Also called *autoeroticism, ipsism, onanism, self-abuse, self-gratification.* 2. The stimulation of another person to orgasm by means other than coitus.

**mate** 1. *n.* One of a pair of mated animals. 2. *n.* A spouse. 3. *n.* One of a pair of conjoined twins. 4. *v.* To bring a couple together in marriage. 5. *v.* To bring a couple of animals together for breeding.

**ma·ter'nal** Of or pertaining to the mother; inherited from the mother.

**ma·ter'nal im·pres'sions** Prenatal influences. The erroneous belief has often been held that an expectant mother can "mark" or affect her child physically during intrauterine development by her thoughts or actions, especially that birthmarks or abnormalities such as a missing hand can be attributed to a mental shock or a traumatic experience occurring during pregnancy.

**ma·ter'ni·ty** 1. Motherhood; the state of being a mother. 2. Of or pertaining to the period during which a woman is pregnant or has just given birth to a child, as in *maternity* hospital, *maternity* clothes.

**mate swap'ping** The exchange of wives and husbands between married couples; sexual intercourse by a husband and wife with another married couple; swinging.

**mat'ing** The pairing of two persons in marriage or the pairing of two animals for the production of offspring; breeding.

**ma'tri·arch** A woman who holds a position of authority in a family, clan, or tribe. *Compare* patriarch.

**ma'tri·ar·chy** A social system in which the mother is the head of the family and descent is traced through the female line. *Compare* patriarchy.

**mat·ri·clin'ous** Inherited from the mother or maternal line. *Also* matroclinous. *Compare* patriclinous.

**mat·ri·lin'e·al** Tracing descent through the female line. *Compare* patrilineal.

**mat·ri·lo'cal** Maintaining residence with the bride's family or tribe. *Compare* patrilocal.

**mat'ri·mo·ny** 1. The rite, sacrament, or ceremony of marriage. 2. Marriage or married life; wedlock.

**ma'tron** 1. A married woman, especially one with children. 2. A woman of dignity and social position. 3. A woman in charge of a public institution, such as a hospital, school, or prison.

**mat·u·ra'tion** 1. The process of becoming mature. 2. In gametogenesis, the process by which mature, functional germ cells are developed from undifferentiated cells. Changes that occur include a change in size and form and a reduction in chromosome number from the diploid number to the haploid number.

**ma·ture'** Having attained complete natural growth and development.

**ma·tur'i·ty** The state or quality of being mature.

**meat** *Slang.* 1. The penis. 2. The vagina. 3. A sexy woman. 4. Coitus. 5. A male as a sexual object for homosexual gratification.

**meat block** *Slang.* A particular block where homosexuals meet and make contacts.

**me·a'tus** An opening or passageway.

**me·co'ni·um** The dark, greenish material that collects within the intestine of a fetus and forms the first fecal discharge.

**med·u·lar'in** A hormone factor in an embryo that promotes atrophy of the Mullerian duct.

**mei·o'sis** A type of nuclear division occurring in the maturation of gametes in which the chromosome number is reduced from the diploid number (46 in humans) to the haploid number (23).

**mel·a·no·der'ma** Melasma, q.v.

**mel·as'ma** A condition in which brown or yellowish patches appear on the skin of the face and neck during pregnancy. It may also occur during menopause and sometimes is seen in women taking contraceptives. Also called *chloasma, melanoderma, mask of pregnancy.*

**mel'ons** *Slang.* The female breasts.

**mem'ber** *Slang.* The penis or vagina.

**mem'brane** A thin sheet of tissue that covers a structure, lines a tube or cavity, separates parts, or connects organs.

**men** *Plural* of man.

**mé·nage'** The persons living together and forming a household.

**mé·nage' à trois** A triadic relationship involving a married couple and a lover of one of them occupying the same household.

**me'nar·che** The onset of the menses at puberty; the first menstruation. In the U.S., it now occurs between the ages of 11 and 16 (av. 12.9). In 1900, the average age at which the menarche occurred was 14.2.

**men·o·met·ror·rha'gia** Menorrhagia, q.v.

**men'o·pause** The physiological cessation of menstrual cycles and ovarian function in women, usually occurring between the ages of 40 and 50. Also called *change of life* or *climacteric,* q.v.

**men·or·rha'gia** Excessive flow of menstrual fluid either in quantity or in length of time or both; hypermenorrhea, menometrorrhagia.

**men·or·rhal'gia** Pelvic pain or distress associated with menstruation; menstrual pain or cramps; dysmenorrhea.

**men·or·rhe'a** The normal flow of menstrual fluid.

**men·o·tox'in** A toxic substance presumed to be present in menstrual fluid.

**men·o·tro'pin** Human menopausal gonadotrophins (FSH, LH) obtained from the urine of postmenopausal women, used with HCG to induce ovulation.

**men'ses** The normal discharge through the vagina of a bloody fluid from the uterus that occurs monthly. *See* menstruation.

**men'stru·al dis·or'ders** *See* dysmenorrhea, premenstrual syndrome.

**men'stru·al cy'cle** A periodic, recurring series of changes that occurs in the endometrium of the uterus culminating in menstruation. Cycles vary in length from 23 to 35 days; average length is about 28 days. Phases in the cycle include repair, proliferation, secretion, and menstruation. Although menstruation is the terminal phase, length is calculated from the first day of the cycle. In a 28-day cycle, the *menstrual phase* or *menstruation* during which the endometrium is lost occupies days 1 to 4; *repair phase,* days 5 and 6; *follicular or proliferative phase,* days 7 to 14; and *luteal* or *secretory phase,* days 15 to 28. Ovulation occurs about day 14 of the cycle.

**men'stru·al ex·trac'tion** The removal of the endometrium of the uterus by aspiration, a method of fertility control in which a narrow, flexible curet or canula attached to a suction apparatus is employed. In cases of suspected pregnancy, it is performed within two weeks of a missed menstrual period. Also called *instant period, menstrual induction, menstrual planning, menstrual regulation, miniabortion, minisuction, lunch-hour abortion.*

**men'stru·al flu'id** The fluid discharged during menstruation. It consists of necrotic endometrial tissue, extravasated blood, and the secretions of uterine glands. It averages about 50 ml in amount.

**men'stru·al reg·u·la'tion** 1. The regulation of menstrual cycles through the use of contraceptive pills. 2. Menstrual extraction, q.v.

**men·stru·a'tion** The periodic discharge of a bloody fluid from the uterus through the vagina that occurs from puberty to the menopause except during periods of pregnancy. The length of menstrual flow ranges from 3 to 6 days (av. 4½ days). *See* menstrual cycle, menstrual fluid.

**mer·e·tri'cious** 1. Pertaining to, resembling, or characteristic of a prostitute. 2. Exhibiting flashy or vulgar features designed to attract or allure. 3. Deceptive, insincere.

**mer'kin** 1. False hair for the female genitalia. 2. An artificial or substitute vagina.

**mes·al·li'ance** Hypogamy, q.v.

**mes'o·derm** In an embryo, the germ layer that lies between the ectoderm and the endoderm. It gives rise to all connective tissue, the skeleton, muscles, and circulatory system, and parts of the urogenital system.

**mes·o·neph'ric duct** One of a pair of embryonic excretory ducts that, in the male, gives rise to the epididymis, ductus deferens, ejaculatory duct, and seminal vesicle; in the female corresponding vestiges persist. In both sexes, it gives rise to the ureter and pelvis of the kidney. Also called *Wolffian duct.*

**mes·o·sal'pinx** The upper portion of the broad ligament, enclosing the uterine tube.

**mes·o·va'ri·um** A double layer of peritoneum by which the ovary is suspended from the posterior layer of the broad ligament.

**me·tas'ta·sis** The transfer of a disease-producing agent, such as a bacterium or a neoplastic cell, from its original locus to another site in the body with resulting development of a similar lesion in its new locale, the transfer usually occurring by way of the bloodstream.

**me·tror·rha'gia** Bleeding from the uterus, especially bleeding occurring at any time other than during the menstrual period.

**me·tror·rhe'a** Any abnormal or pathologic discharge from the uterus.

**mick'ey** *Slang.* The penis, especially among juveniles.

**mi'cro·mere** One of the four small cells at the animal pole of an ovum undergoing cleavage.

**mi·cro·pe'nis** A microphallus, q.v.

**mi·cro·phal'lus** An exceptionally small penis; a micropenis.

**mic·tu·ri'tion** The act of passing urine; urination.

**mid'wife** An experienced woman (occasionally a man) who cares for and assists another woman in the delivery of a child.

**mi·la'dy** 1. A term of respect applied to an English noblewoman. 2. A woman of fashion with expensive tastes.

**milk** The secretion of the mammary gland used for feeding the young. Human milk contains water (87–88 percent), protein (1.0–1.5 percent), fat (3.5–4.0 percent), carbohydrate (lactose, 6.5–7.0 percent), and minerals (0.2 percent), and is alkaline in reaction.

**milk-e·jec'tion re'flex** The transport of milk from secretory alveoli to the nipple, where it is removed by a suckling infant. It is a neurohormonal reflex initiated by suckling. Resulting nerve impulses pass to the brain where *oxytocin,* a neurohormone secreted by the hypothalamus and stored in the posterior pituitary, is released. Oxytocin passes by way of the bloodstream to the mammary gland, where it causes the myoepithelial cells to contract, bringing about milk ejection.

**milk let'down** *See* milk-ejection reflex.

**milt** 1. The reproductive fluid of male fishes, consisting of sperm and seminal fluid. 2. The testes of fishes when filled with this fluid.

**min'gles** A newly coined term for individuals with nonrelatives in a single household.

**min·i-a·bor'tion** Menstrual extraction, q.v.

**min·i·lap·a·rot'o·my** A method of abdominal tubal ligation in which an instrument is inserted through the cervix into the uterus and the uterus then pushed upward so that it lies against the abdominal wall. Through an incision made over the top of the elevated uterus, the uterine tubes are brought into the operative field and then either cut or tied. Also called *minilap.*

**min·i·men·stru·a'tion** The process of reducing the time of normal menstruation from 4 to 5 days to 15 to 20 minutes by removal of the menstrual endothelium by a suction machine. *See* menstrual extraction.

**min'i·pill** A contraceptive pill that contains only a progestogen. It does not prevent ovulation but it brings about a thickening of the cervical mucus and makes the endothelium of the uterus less receptive to implantation. It must be taken every day without interruption.

**min·i·suc'tion** Menstrual extraction, q.v.

**mi'nor** 1. Lesser in amount, size, extent, or importance. 2. *Law.* A person of either sex who is below legal age, usually 21, at which age civil and personal rights may be legally exercised.

**min'o·taur** A Greek mythological character with the head of a bull and the body of a man.

**mint'ie** *Slang.* An aggressive, masculine dyke.

**mis·al·li'ance** An inappropriate or mismatched marriage.

**mis·an·dry** Hatred of men. *Compare* misogyny.

**mis·car'riage** The expulsion of a fetus before the age of viability; an abortion.

**mis·ceg·e·na'tion** Interbreeding or intermarriage between members of different races; the mixing of races.

**mi·sog'a·my** A dislike for or an aversion to marriage.

**mi·sog'y·nist** A woman hater.

**mi·sog'y·ny** Hatred of women. *Compare* misandry.

**miss** An unmarried woman or girl.

**miss a pe'ri·od** To fail to menstruate at the expected time.

**mis'sion·ar·y po·si'tion** The conventional face-to-face coital position with the female on her back and the male lying on top. The term is attributed to Polynesians, who preferred the squatting position.

**mis'ter** *Abbr.* Mr. A title of respect applied to a man and used as a prefix to his name, as Mr. Jones. 2. *Informal.* (a) Used to address a man, as in "Mister, look out." (b) A husband.

**mis'tress** 1. A woman in a position of authority as the head of a household, institution, or establishment. 2. A woman who has a continuing illicit sexual relationship with a man who, in return, usually contributes financially to her support.

**mit'tel·schmerz** Pain at the time of ovulation; midcyclic pain.

**mix·o·sco'pi·a** Sexual gratification obtained from viewing other couples in coitus.

**mix·os'co·pist** A voyeur, q.v.

**mod'es·ty** The state of being reserved and unpretentious in speech, dress, and behavior.

**mo·lest'** 1. To disturb, annoy, or inconvenience. 2. To accost and harass, making indecent sexual advances.

**mol·i'men** Menstrual distress or dysmenorrhea; nervous and circulatory disorders associated with menstruation.

**moll** 1. A prostitute. 2. A gangster's girlfriend.

**mol·lus'cum con·ta·gi·o'sum** A chronic disease of the skin caused by a virus, characterized by the formation of small, spherical, wartlike elevations. The papules contain caseous material with characteristic capsulated bodies (molluscum bodies). Systemic symptoms are lacking. It is often transmitted through sexual contacts.

**mo·nan'dry** Having only one husband at a time. *Compare* polyandry.

**mon'gol·ism** A congenital condition characterized by various physical abnormalities and mental retardation accompanied by development of mongoloid features. It results from trisomy of chromosome 21. Also called *Down's syndrome.*

**mo·nil·i'a·sis** An infection of the skin and mucous membranes caused by a yeast-like fungus, *Candida albicans.* The mouth, rectum, and vagina are common sites of infection. *See* candidiasis.

**mon'key bite** *Slang.* A hickey.

**mo·noe′cious, mo·ne′cious** Possessing both ovaries and testes; hermaphroditic. *Compare* dioecious.

**mo·nog′a·mous** Of or pertaining to monogamy.

**mo·nog′a·my** Being married to one person at a time. *Compare* bigamy.

**mon·o·sex′u·al** Possessing or exhibiting traits of one sex only. *Compare* bisexual.

**mon·o·zy·got′ic** Developing from a single fertilized egg, as in *monozygotic* twins. *Compare* dizygotic.

**mons pu′bis** An elevated cushion of fat covered by skin that lies over the pubic symphysis. After puberty, it is covered by pubic hair. In women it is called the *mons veneris* (Mountain of Venus).

**mon′ster** A markedly deformed and abnormal fetus or infant, the result of faulty development; a teratism.

**moon′ing** *Slang.* Exhibiting the bare buttocks.

**mo·ral′i·ty** 1. Conformity to rules of proper conduct as established by tradition and custom in a certain group or class, especially with respect to right and wrong. *See* New Morality, Old Morality. 2. Virtuous sexual behavior; chastity.

**mor′als** Rules and standards of conduct, often with reference to sexual behavior, adhered to by an individual or group with respect to principles of right and wrong.

**mo′res** The strictly coercive customs of a particular society, class, or group, especially with reference to moral attitudes and principles.

**mor·ga·nat′ic** Pertaining to a form of marriage in which a person of high social position or, usually, noble rank marries a person of lower social status (a commoner) with the understanding that neither the lower-ranking person nor their children make any claim to the rank or possessions of the higher-ranking person.

**morn′ing-af′ter IUD** An intrauterine device placed within the uterus following coitus employed by women who cannot tolerate DES.

**morn′ing-af′ter pill** A pill containing diethylstilbestrol (DES), a nonsteroidal estrogen that is given sometimes as a contraceptive for an unprotected coitus at the time of ovulation. It is effective when taken three to five days after coitus but its side effects (nausea, vomiting, dizziness, diarrhea, and others) are sometimes severe. Also called *abortion pill*.

**morn′ing drop** *Slang.* Gonorrhea.

**morn′ing sick′ness** Nausea and sometimes vomiting that occur for a few hours in the morning. Morning sickness commonly occurs in early pregnancy.

**mor′u·la** An early stage in mammalian development in which the embryo consists of a solid, spherical mass of cells. It precedes the blastocyst.

**moth′er** 1. A female parent. 2. In a loose sense, a mother-in-law, stepmother, or adoptive mother. 3. *Slang.* (a) One who brings out or introduces another to gay life. (b) A madam or brothel keeper. (c) The most attractive of a group of homosexuals. (d) An effeminate man or homosexual, especially an older one.

**mo′ther-in-law** The mother of a wife or husband.

**mo′tile** Capable of spontaneous movement.

**mo·til′i·ty** Movement; activity.

**mount** 1. In mammals (of a male), to assume a position upon a female, insert the penis into the vagina, and make an initial thrust. 2. In birds (of a male), to climb upon the back of a female and bring the two everted cloacal openings into juxtaposition for the transfer of sperm.

**moun′tain oys′ters** *Slang.* The testes of sheep or hogs when used for food.

**mouse** *Slang.* 1. An affectionate term for an attractive and vivacious girl or young woman. 2. A girlfriend, sweetheart, or wife.

**mouth'ing** Applying the mouth to or nibbling the neck or ear.

**mouth mu'sic** *Slang.* Genital kisses; cunnilingus or fellatio.

**MPA** Medroxyprogesterone acetate.

**M/S** Master/slave.

**mu'coid** Resembling or of the consistency of mucus.

**mu·co·pur'u·lent** Consisting of mucus and pus.

**mu·cor·rhe'a** The excessive flow of mucus, especially cervical mucus from the vagina.

**mu·co'sa** A mucous membrane, q.v.; tunica mucosa.

**mu'cous mem'brane** A membrane that lines tubes, ducts, canals, passageways, or cavities that open to the exterior of the body, such as the lining of the alimentary canal, respiratory passageways, or tracts of the genitourinary organs. Typically a mucous membrane consists of four layers: a surface epithelium, a basement membrane, a lamina propria of connective tissue, and a muscularis mucosa. Mucus is produced by goblet cells in the surface epithelium.

**mu'cus** A viscous, slimy substance present on the surface of a mucous membrane, which it moistens, lubricates, and protects. It is a mixture of mucin (a glyco-protein) and water.

**muff** *Slang.* The vulva, especially when much pubic hair is present.

**muff div'ing** *Slang.* Oral stimulation of the vulva; cunnilingus.

**muf'fins** *Slang.* A woman's or a young girl's breasts.

**mu·li·eb'ri·ty** 1. The possession of a womanly nature or qualities; the condition of being a woman. 2. The assumption of female characteristics by a male; effemination; eonism.

**Mul·le'ri·an duct** One of two embryonic ducts that, in humans, give rise to the uterine tubes, the uterus, and a part of the vagina. In the male, only vestiges persist.

**mul·ti·grav'i·da** A pregnant woman who has had two or more previous pregnancies. *See* multipara.

**mul·ti·lat'er·al mar'riage** Group marriage, q.v.

**mul·tip'a·ra** A pregnant woman who has had two or more pregnancies resulting in viable fetuses. Designated *para II, III,* or *IV* depending upon the number of offspring. *See* multigravida.

**my·o'ma** A tumor that is derived from muscle tissue. *See* leiomyoma.

**my·o·mec'to·my** Removal of a fibroid from the uterus.

**my·o·me'tri·um** A layer of smooth muscle in the wall of the uterus. It lies between the outer peritoneum and the inner endometrium.

**my·o·to'ni·a** Tonic spasm of muscle tissue characterized by increased irritability and decreased ability to relax.

**my·so·phil'i·a** Sexual interest in nonfecal dirt.

**na'ked** Without clothing or covering; nude.

**nance** or **nan'cy boy** *British slang.* A male homosexual.

**nap'kin** *See* sanitary napkin.

**nar'cis·sism, nar'cism** Self-love; excessive love and admiration for one's own body.

**nar-nars** *Slang.* The breasts.

**nar·rat'o·phil'i·a** A condition in which a person requires or depends upon reading or listening to erotic material for the arousal of sexual desire or achievement of an orgasm. *See* paraphilia.

**na'tal** Of, pertaining to, or accompanying childbirth.

**nat'ur·al child'birth** The avoidance of the use of anesthetics for the relief of pain during labor and delivery by reduction of fear and resultant muscular tension. It is accomplished principally through education and physiotherapeutic exercises. *See* Lamaze method.

**na'vel** A depression or scar in the midline of the abdomen that marks the site of attachment of the umbilical cord in a fetus. Also called *umbilicus.*

**na·vic'u·lar fos'sa** 1. In the male, the dilated portion of the urethra within the glans penis. 2. In the female, a depressed region in the floor of the vestibule posterior to the vaginal orifice.

**neck'ing** Affectionate kissing and caressing without complete bodily contact. *Compare* petting.

**nec·ro·phil'i·a** 1. An attraction for and often an erotic interest in a corpse. 2. Sexual intercourse with a dead body. Also *necrophilism.*

**nec·ro·sa'dism** Lust killing; thrill killing usually following sexual abuse.

**ne·cro'sis** The death of tissues within an organism, usually the result of pathological conditions.

**ne·crot'ic** Of or pertaining to necrosis.

**nec·ro·sper'mi·a** A condition in which spermatozoa in semen are inactive or motionless; the absence of viable sperm in semen.

**nec·ro·zo·o·sper'mi·a** Necrospermia, q.v.

**Neis·se'ri·a gon·or·rhoe'ae** A species of bacteria, the causative agent of gonorrhea.

**nel'lie** *Slang.* An effeminate male homosexual.

**ne·o·na'tal** Of or pertaining to the newborn.

**ne'o·nate** A newborn child.

**ne·ot'e·ny** The retention of larval or immature characteristics in the adult state; becoming sexually mature in the larval or immature state.

**neph'ew** 1. The son of a person's brother or sister. 2. The son of a brother-in-law or sister-in-law. 3. *Euphemism* applied to the illegitimate son of a clergyman who had declared vows of celibacy.

**ner'vous blad'der** Urge incontinence, q.v.

**nes'tle** To lie close together in intimate contact.

**neu·ro·gen'ic** Of or originating in the nervous system.

**neu·ro·hor'mone** A hormone produced by a neuron. Important neurohormones are vasopressin, oxytocin, and various releasing factors that act on the anterior lobe of the pituitary. All of these are secreted by neurons in the hypothalamus.

**neu·ro·hu'mor** Any of a number of chemical substances released at the axon terminals of nerve cells that constitute agents in the transmission of impulses across a synapse. Examples are acetylcholine, norepinephrine, dopamine, and serotonin.

**neu·ro·hu'mor·al** Of or pertaining to neurohormones or neurohumors.

**neu·ro·hu'mor·al sys'tem** The nervous and endocrine systems, combined.

**neu'ron** A nerve cell, consisting of a cell body and its processes, a single axon and usually several dendrites. Its primary function is the conduction of impulses but it may also serve a secretory function.

**neu·ro·syph'i·lis** Syphilitic involvement of the nervous system. It may be *asymptomatic* and diagnosed only by examination of the cerebrospinal fluid, or *symptomatic*, characterized by neurologic and often psychotic signs and symptoms. *See* syphilis, tabes dorsalis.

**neu'ter** 1. *n.* An animal with imperfectly developed sex organs. 2. *adj.* Neither male nor female. 3. *n.* A castrated animal.

**neu'tral·ize** To render incapable of reproduction; to castrate; to sterilize.

**ne'vus** A birthmark or mole; a circumscribed, pigmented area present at birth.

**new'born** 1. Recently born. 2. A recently born infant, especially one three days or less in age.

**new'ly·wed** A person just recently married.

**New Mo·ral'i·ty** New attitudes with respect to human sexuality that have developed in recent years. These include approval of sexual intercourse between unmarried persons under certain conditions; approval of the use of contraceptives and of abortion under certain conditions; removal of the stigma applied to masturbation and a more tolerant view toward homosexuality; and the elimination of the double standard of sexuality, thus granting to females the same opportunity for sexual satisfaction granted to males. *Compare* Old Morality.

**NGU** Nongonococcal urethritis.

**nibb′ling** Biting gently various parts of the body, a common precopulatory activity.

**ni·da′tion** Implantation, q.v.

**niece** 1. The daughter of a person's brother or sister. 2. The daughter of a brother-in-law or sister-in-law. 3. *Euphemism* applied to the illegitimate daughter of a clergyman who had declared vows of celibacy.

**night′gown** A loose garment worn at night in bed by women and children; a man's nightshirt.

**night′ie** *Informal* for nightgown.

**nip′ple** A conical process on the breast or mammary gland that bears openings of the lactiferous ducts through which milk is discharged.

**noc·tur′nal e·mis′sion** The involuntary discharge of semen at night during sleep; a wet dream.

**non·gon·o·coc′cal u·re·thri′tis** *Abbr.* NGU. Inflammation of the urethra and associated organs (in the *male*, the prostate gland, seminal vesicles, and epididymis; in the *female*, cervix, uterus, and uterine tubes) usually of venereal origin but not caused by a gonococcus. It may be caused by a number of different organisms or it may occur following infections such as syphilis, chancroid, and others.

**non·or·gas′mic** Incapable of experiencing an orgasm, nonorgastic.

**non·or·gas′tic** Nonorgasmic.

**nook′ie, nook′ey** *Slang.* 1. Sexual activity. 2. A woman as a sex object. 3. Sexual intercourse. 4. The vagina.

**nor′mal** Typical, average, regular, natural; not abnormal.

**NSU** Nonspecific urethritis.

**nu′bile** Suitable for marriage, especially with respect to age and physical development. Said of a girl or young woman.

**nu′cle·ar fam′i·ly** *See* family.

**nu′cle·ar sex′ing** A method of diagnosing sex by the microscopic examination of cells obtained from the skin, blood, or oral mucosa, or from a fetus. The cells of a female possess in their nucleus a mass of sex chromatin (Barr body) not present in the cells of males.

**nu′cle·us** A structure present in most animal and plant cells, typically a spheroid mass enclosed in a membrane, which is essential for the life of a cell. It contains the cell's hereditary material and it regulates the cell's metabolic activities, growth, and reproduction.

**nude** Naked, bare, without clothes.

**nu′dism** A cult or social movement that seeks to promote physical and psychological well-being by doing away with the wearing of clothes.

**nul·li·grav′i·da** A woman who has never been pregnant.

**nul·lip′a·ra** A woman who has never given birth to a viable child. Designated *para 0*.

**nup′ti·al** Of or pertaining to (a) a marriage or a wedding ceremony; (b) mating.

**nuts** *Slang.* The testes.

**nuz′zle** 1. To rub with or touch with the nose. 2. To lie close to; to snuggle or cuddle up to.

**nym′phae,** *pl.* of **nympha** The labia minora.

**nymph·et'** A young girl considered sexually desirable; a sexually precocious girl.

**nym'pho** *Slang.* A nymphomaniac.

**nym'pho·lep·sy** Erotic ecstasy.

**nym·pho·ma'nia** Excessive and sometimes uncontrollable sexual desire in a female. *Compare* satyriasis.

**nym·pho·ma'ni·ac** A person afflicted with nymphomania.

**OB-GYN** Obstetrics-Gynecology.

**ob'li·gate** Essential, compelling; not optional (used in a biological sense). *Compare* facultative.

**ob·scene'** Offensive according to accepted standards of decency and morality; indecent; tending to excite lustful feelings and activities; lewd.

**ob·ste·tri'cian** A physician who specializes in obstetrics. *Compare* gynecologist.

**ob·stet'rics** The branch of medicine that deals with the care of women during pregnancy, labor, and the period following childbirth. *Compare* gynecology.

**Oe'di·pus com'plex** The abnormally strong and persistent attachment of a son to his mother, usually involving incestuous desire or behavior. *Compare* Electra complex.

**oes'tro·gen** Estrogen, q.v.

**off-col'or** 1. Risqué, as in *off-color* anecdotes. 2. Of doubtful propriety; socially unacceptable.

**o'gle** To look or stare, especially in an amorous, impertinent, or flirtatious manner.

**Old Joe** *Slang.* Any venereal disease; a dose.

**Old Mo·ral'i·ty** Views on human sexuality and morality commonly held in the West, derived principally from Judeo-Christian teachings, which held that the sex act was for procreation only, that sexual intercourse either before or outside of marriage was a sin, and sex activity other than sexual intercourse constituted a "crime against nature" or was perverted sexual behavior. Supplemental to this was the assumption that woman constituted the weaker sex, was inferior to man, and was expected to serve him and provide sexual satisfaction without experiencing satisfaction to the same degree. *Compare* New Morality.

**ol·i·go·gen'ics** Limitation of the number of offspring; birth control.

**ol·i·go·sper'mi·a** Condition in which there is a reduced number of spermatozoa in semen, less than 40 million per milliliter. *See* azoospermia.

**ol·i·go·zo·o·sper'mi·a** Oligospermia, q.v.

**o'nan·ism** Old term for coitus interruptus or withdrawal, derived from the biblical character Onan, who "when he went in to his brother's wife, he spilled *it* on the ground, lest that he should give seed unto his brother" (Gen. 38:9). Sometimes the term is applied to masturbation.

**o·nan·is'tic** Of or pertaining to onanism.

**o'o·cyte** One of two types of cells, primary and secondary oocytes, which are stages in oogenesis between the oogonium and the mature ovum.

**o·o·gen'e·sis** The processes involved in the origin, growth, and development of a mature ovum. Stages include *oogonium, primary oocyte, secondary oocyte,* and *ootid* or mature *ovum.* Changes occurring in oogenesis include reduction in the number of chromosomes from the diploid number (46) to the haploid number (23) and an increase in size and yolk content of the ovum. *See* maturation. *Compare* spermatogenesis.

**o·o·go'ni·um,** *pl.* **o·o·go'ni·a** The oogonia are primordial germ cells which, in a fetal mammal, proliferate within the cortex of the ovary. Each becomes encased in smaller cells and together they constitute a follicle. After puberty each oogonium is potentially capable of developing into a mature ovum but, of the million or more oogonia present in the two human ovaries at birth, only about four hundred mature.

**o·oph·o·rec'to·my** Surgical removal of an ovary.

**o·oph·o·ri'tis** Inflammation of the ovary; ovaritis.

**o'o·tid** A ripe ovum.

**o'pen mar'riage** A form of marriage based on mutual trust that permits greater freedom, thus allowing for greater personal growth and independence and more flexible roles for the participants. *Compare* closed marriage.

**o'ral** Of or pertaining to the mouth.

**o'ral con·tra·cep'tive** A contraceptive agent that is taken by mouth, such as the Pill.

**o'ral er'o·tism, o'ral e·rot'i·cism** Sexual drive based on repressed conflicts that originated during infantile feeding experiences.

**o'ral-gen'i·tal sex** Sexual relations involving the mouth of one sexual partner and the genital organs of the other. *See* oral sex.

**o'ral sex** 1. Sexual gratification obtained or given through contact of the mouth with the genital organs. *See* cunnilingus, fellatio. 2. Talking about sex.

**o'ral stage** According to Freud, the initial stage in psychosexual development that occurs during the first eighteen months of life, during which the mouth is the chief source of emotional satisfaction, obtained by sucking. Pleasure accompanying the feeding experience becomes incorporated into the personality and conditions attitudes and behavior in later life. *See* anal stage, genital stage.

**or·chi·dom'e·ter** A device used in measuring the size of a testis.

**or·chi·ec'to·my** Surgical removal of one or both testes.

**or·chi·o·pex'y** Surgical fixation within the scrotum of an undescended testis.

**or·chi'tis** Inflammation of the testis, which may result from trauma, neoplasms, or infections (mumps, gonorrhea, syphilis, tuberculosis, or other diseases).

**or'gan** 1. A part of the body that performs one or more specific functions. 2. *Slang.* The penis.

**or'gans of gen·er·a'tion** The reproductive organs.

**or'gasm** The intense excitement, both physical and emotional, that is experienced at the climax of the sexual act (coitus) or following stimulation of the sex organs, as in masturbation. During an orgasm, breathing becomes more rapid, heart rate is increased, blood pressure rises, and blood flow, especially to the pelvic organs, is increased. In the male, ejaculation of semen occurs; in the female, involuntary uterine and vaginal contractions occur. It constitutes the third phase of sexual response. Also called *climax, going off. See* sexual response.

**or·gas'mic** Of or pertaining to an orgasm; orgastic.

**or·gas'mic dys·func'tion** *See* orgasmic impairment.

**or·gas'mic im·pair'ment** Inability of a person to experience an orgasm. Usually said of a female. Also called *orgasmic dysfunction. See* frigidity.

**or·gas'mic phase** The third phase in sexual response, during which an orgasm occurs. *See* orgasm, sexual response.

**or·gas'mic plat'form** The swelling of tissues about the outer third of the vagina which occurs during the plateau phase of sexual response.

**or·gas'tic** Orgasmic, q.v.

**or·gi·as'tic** 1. Of or pertaining to an orgy. 2. Tending to arouse sexual excitement, often of an unrestrained nature.

**or'gy** 1. Wild, unrestrained activity, usually involving sexual excesses. 2. A party at which participants exchange their sexual partners.

**or'i·fice** An opening.

**os** An opening, as the external opening of the cervical canal (*os externum uteri*).

**os'cu·late** To kiss.

**os·cu·la'tion** The act of kissing.

**os·phre·si·o·lag'ni·a** Sexual stimulation produced by certain odors.

**os'ti·um** An opening, such as (a) the opening of the lateral end of the uterine tube; (b) the opening between the two atria of the heart in a fetus.

**out'breed·ing** The mating of unrelated individuals. *Compare* inbreeding.

**o·va'ri·an** Of, pertaining to, or originating in the ovary.

**o·va·ri·ec'to·my** Surgical removal of an ovary.

**o·va·ri·o·tex'y** A contraceptive technique in which an incision is made into the abdominal cavity and a silastic bag placed around each ovary. This procedure prevents a discharged ovum from entering the uterine tube.

**o·va·ri·ot'o·my** Incision into an ovary; removal of an ovary. *See* ovariectomy.

**o·va·ri'tis** Inflammation of an ovary; oophoritis.

**o'va·ry,** *pl.* **o'va·ries** A female organ that produces ova or eggs. It is also the source of hormones (estrogens, progesterone, androgens, and relaxin). Each human ovary is about 1½ inches long, ½ inch wide, and ½ inch thick. Structurally the ovary consists of a central *medulla* surrounded by a thick *cortex,* which contains many follicles in various stages of development. Each follicle contains a developing oocyte from which a mature ovum develops. Once each month about the middle of the menstrual cycle, a follicle ruptures, releasing a nearly mature ovum. In the ruptured follicle, a small, yellow body, a *corpus luteum,* develops. Follicles and corpora lutea are the source of ovarian hormones.

**o·ver·sexed'** Possessing excessive sexual interest and desire. *See* nymphomania; satyriasis.

**o·vert'** Open, exposed; not concealed or hidden. *Opposite* of covert.

**o'vi·duct** A duct that in female animals conveys eggs from the ovary to the outside or to an organ such as the uterus that communicates with the exterior. In humans, the *fallopian* or *uterine tube.*

**o·vi·pos'it** To lay or deposit eggs.

**o·vo·tes'tis** A reproductive organ that produces both eggs and spermatozoa, present in some hermaphroditic animals and some pseudohermaphrodites.

**o'vu·late** To produce and discharge an egg or eggs.

**o·vu·la'tion** In humans, the discharge of an ovum or oocyte from a vesicular follicle of the ovary, which normally occurs about 14 days prior to the onset of menstruation. *See* thermal shift.

**O·vu·lom'e·ter** Trade name for an experimental device that measures changes in the electrostatic current or voltage in the body. Men are positive throughout the month. Women are usually negative but become positive 3 to 6 days prior to ovulation, reverting to negative 2 to 4 days after ovulation.

**o'vum,** *pl.* **o'va** A female gamete or reproductive cell; an egg. A cell that, when fertilized, is capable of developing into an individual of the same species. A human ovum, when discharged from the ovary, has a diameter of 100 to 150 microns (about 1/200 of an inch). It is surrounded by a clear membrane, the *zona pellucida,* and a layer of follicle cells, the *corona radiata.*

**O·vu·ti'mer** Trade name for a device, a viscometer, that determines the viscosity of the cervical mucus. During the fertile period, the mucus is thin and watery and of low viscosity; at the end of the fertile period, the mucus becomes thick and viscous.

**ox·y·to'cic drug** A drug that stimulates uterine contractions. Among such drugs are oxytocin, some prostaglandins, certain ergot derivatives, sparteine sulfate, norepinephrine, and acetylcholine.

**ox·y·to'cin** A hormone produced by the hypothalamus of the brain but stored and released by the posterior lobe of the pituitary. It is a powerful stimulant of uterine contractions and it brings about milk ejection by the mammary gland.

**pad** *Slang.* 1. One's home, apartment, room, or bed. 2. A drug addict's den. 3. A brothel run by an independent prostitute; a crib.

**pair** 1. An engaged or married couple. 2. Two mated animals.

**pair-bond** The development of a male-female companionship and sexual relationship, generally for life.

**pam·pin'i·form plex'us** A network of veins in the spermatic cord of the male and the broad ligament of the female. They drain the testes and ovaries, transporting blood to the inferior vena cava by way of the spermatic and ovarian veins. *See* varicocele.

**pan'cak·er** *Slang.* A role-switching lesbian.

**pan'der, pan'der·er** 1. A pimp or procurer. 2. An intermediate agent in amorous or sexual intrigues.

**pan·hys·ter·ec'to·my** Total removal of the uterus.

**pan·sex'u·al** Totally or entirely sexual.

**pan·sex'u·al·ism** The view that all conduct, experiences, and desires are derived from or related to the sexual instinct.

**pan'sy** *Slang.* 1. A male homosexual, especially one who plays the passive, female role. 2. An effeminate male.

**pan·tag'a·my** Communal marriage in which every man is the husband of every woman and vice versa.

**Pap** *Abbr.* for Papanicolaou, q.v.

**pa'pa** 1. A father, term used especially by children. 2. *Slang.* A husband, a lover.

**Pa·pa·ni'co·laou, George N.** (1883–1962) An anatomist and cytologist who specialized in the study of exfoliated cells as a means of detection of cancer, especially uterine cancer.

**Pap smear** or **test** A preparation made by taking tissue from the surface of the cervix of the uterus and spreading it onto a slide, which is then examined microscopically, a procedure of especial value in the detection of a malignant growth.

**par·a·bi·ot'ic twins** Conjoined twins; twins whose bodies are united, as Siamese twins.

**par·a·cer'vi·cal** Around the cervix of the uterus.

**par·a·cer'vi·cal block** Injection of an anesthetic into the lateral fornices of the vagina, a region around the cervix of the uterus.

**par·a·me'tri·al** Near or around the uterus.

**par·a·me·tri'tis** Inflammation of the parametrium.

**par·a·me'tri·um** The connective tissue between the two layers of the broad ligament that supports the uterus.

**par'a·mour** A lover, especially a lover of a married person.

**par·a·phil'i·a** A condition in which a person depends upon unusual or socially unacceptable stimuli in order to obtain sexual satisfaction or to experience an orgasm. *See* coprophilia, exhibitionism, masochism, narratophilia, necrophilia, pederasty, pictophilia, sadism, scopophilia, urophilia, voyeurism. *Compare* perversion.

**par·a·phi·mo'sis** Retraction and constriction of the prepuce behind the corona of the glans penis. Also called *Spanish collar. See* phimosis.

**par·a·u·re'thral glands** Skene's glands, q.v.

**par'ent** A father or mother.

**par'ent·age** Lineage, origin, ancestry, descent, or derivation from parents or ancestors.

**pa·ren'tal** Of, pertaining to, or characteristic of a parent.

**par′i·ty** The condition of a woman with respect to having borne one or more children, designated *para I* (one child), *para IV* (four children), etc. *See* nullipara.

**par·the·no·gen′e·sis** The development of an egg without being fertilized; virgin birth.

**par·the·no·pho′bi·a** An abnormal fear of virgin females.

**par′tial·ism** A form of sexual behavior in which a person shows an excessive interest in or an obsession for only one part of the female body such as the breasts, buttocks, or nape of the neck, that part serving as a fetish and being of especial importance in the arousal of sexual desire.

**part′ner** 1. One of a pair engaged in an activity such as a game, sport, or dancing. 2. A spouse.

**par·tu′ri·ent** 1. Of, pertaining to, or associated with parturition. 2. About to bring forth young; being in labor.

**par·tu′ri·ent ca·nal′** The birth canal, q.v.

**par·tu·ri·fa′cient** An agent that induces or facilitates childbirth or labor. *See* oxytocin.

**par·tu·ri′tion** Childbirth; labor; the bringing forth of a newborn child.

**pas′sion** 1. Extremely strong sexual desire; lust. 2. Amorous feelings or desire; extreme love and affection.

**pas′sion·ate** Having intense sexual desire or feelings; amorous, lustful.

**pas′sion ring** A rubber ring with soft, pliable tendrils that slips over the penis, designed to intensify vaginal sensations during intercourse.

**pas′sive part′ner** In heterosexual or homosexual intercourse, the receptive or "feminine" partner who receives the penis. *Compare* active partner.

**pat′ent** Open, expanded; not closed.

**pa·ter′nal** Of or pertaining to the father; inherited from the father.

**pa·ter′ni·ty** The state or condition of being a father; fatherhood.

**pa·ter′ni·ty suit** A legal action taken by a woman, usually in the case of an illegitimate child, to have the court declare a specific man to be the father of a child. By means of blood and other tests, the possibility of paternity can sometimes be established with a fair degree of accuracy, or it can be determined that the man in question cannot be the father.

**pa·ter′ni·ty test** A test to determine if a particular man could be the father of a certain child. By comparison of the blood group of the mother, child, and suspected man, it can be determined that the man could not be the father. However, the test cannot prove that a man is actually the father of a child, only that he is biologically capable of being its father.

**path′o·gen** Any disease-producing organism.

**path·o·gen′ic** Inducing or causing a disease.

**path·o·log′ic, path·o·log′i·cal** Of, pertaining to, caused by, or resulting from disease or disease processes.

**pa·thol′o·gy** The study of disease and disease processes, their causes, nature, course of development, and consequences.

**pa′tri·arch** The paternal leader or ruler of a family, tribe, or clan.

**pa′tri·ar·chy** A form of social organization in which the father is regarded as the supreme authority, and descent and succession are traced through the male line. *Compare* matriarchy.

**pat·ri·cli′nous** Inherited from the father or paternal line. Also *patroclinous*.

**pat·ri·lin′e·al** Tracing descent through the male line. *Compare* matrilineal.

**pat·ri·lo′cal** Maintaining residence with the husband's family or tribe. *Compare* matrilocal.

**pat′ri·mo·ny** 1. An inheritance from one's father or male ancestors. 2. Any trait or characteristic that is inherited; heritage.

**pat'u·lous** Expanded, open, spread widely apart.

**PC** Pubococcygeus muscle, q.v.

**peck'er** *Slang.* The penis.

**peck'er glove** *Slang.* A condom.

**ped'er·ast** A man who engages in pederasty.

**ped'er·as·ty** 1. Anal intercourse between two males, especially an older man and a boy. See sodomy. 2. In ancient Greek city-states, an educational love between and older man (*inspirer*) and a male adolescent (*listener*).

**ped·er·o'sis** An erotic interest in and sexual abuse of children.

**pe·dic·u·lo'sis** A condition resulting from infestation of the skin or scalp by lice, characterized by intense itching. See louse.

**pe'do·phile** A person who directs his sexual interest and desire towards children; a child molester.

**pe·do·phil'i·a** Sexual feelings or passion for children; fondness for children.

**pee** 1. *v.* To urinate 2. *n.* urine.

**Peep'ing Tom** A voyeur, q.v.

**peg-house** A homosexual brothel.

**pel'vic cav'i·ty** The cavity within the bony pelvis consisting of two portions, a true and false pelvis.

**pel'vic gir'dle** The bony pelvis. See pelvis.

**pel'vis** One of certain basinlike structures or cavities, including (a) the *bony pelvis,* a ring of bones consisting of the two hip or innominate bones on the sides, the pubic bones in front, and the sacrum and coccyx behind. It contains the lower abdominal visceral organs, supports the spine, and provides attachment for the muscles of the abdomen and muscles and bones of the lower limbs; (b) a funnel-shaped cavity within the kidney which forms the expanded upper end of the ureter.

**pen·e·tra'tion** The insertion of the penis into the vagina.

**pe'nile** Of, pertaining to, or involving the penis.

**pe'nile aid** A device that enhances sexual pleasure in the male, as a penis extender or penis erector.

**pen·i·lin'gus** Fellatio, q.v.

**pe'nis** The male organ of copulation, which also functions in urination. It is a cylindrical organ consisting of three elongated masses of cavernous erectile tissue comprising two dorsolateral *corpora cavernosa penis* and a ventromedian *corpus cavernosum urethrae* (*corpus spongiosum*), which encloses the penile portion of the urethra. The latter is expanded at its distal end to form the *glans penis,* in the center of which is the urinary *meatus* (opening). The skin of the penis, which is loose and hairless, extends over the glans, forming the *prepuce* or *foreskin.* The penis averages two to four inches in length when flaccid, six to eight inches in length when erect; however, its size is extremely variable. It is homologous to the clitoris of the female. See phallus, priapus, captive penis.

**pe'nis bone** A bony structure in the penis of certain animals, such as apes and dogs; a baculum.

**pe'nis en'vy** An excessive desire or wish for a penis commonly manifested by female children who believe that they have lost a penis or that one has failed to develop.

**pe'nis e·rec'tor** A device of rubber or plastic material made to fit underneath a flaccid penis to provide support during intercourse.

**pe'nis ex·tend'er** A penile aid consisting of a penis-shaped piece of rubber or plastic material which is attached to a male's penis to increase vaginal stimulation.

**pe·o·til·lo·ma'ni·a** A nervous habit involving compulsive handling of the penis; also called *pseudomasturbation.*

**per·form'** *Euphemism.* To accomplish sexual intercourse successfully.

**per·i·men·o·paus'al pe'ri·od** The period in a female characterized by ovarian senescence, accompanied by a slow decline in estrogen production; the climacteric, q.v.

**per·i·me'tri·um** The outermost, serous covering of the uterus.

**per·i·na'tal** Pertaining to or occurring in the period preceding or following childbirth.

**per·i·ne'al ex'er·cise** *See* Kegel's exercises.

**per·i·ne·om'e·ter** A device placed within the vagina for measuring the strength of contraction of vaginal and perivaginal muscles.

**per·i·ne'um** 1. The region of the body comprising the pelvic outlet. 2. In a restricted sense, the region between the vaginal orifice and the anus in a female; in a male, the region between the scrotum and the anus.

**pe'ri·od** The menstrual period or menses. *See* menstruation.

**per·i·to·ne'um** The thin serous membrane that lines the abdominal cavity and covers the surfaces of the organs contained within it.

**per·ver'sion** A sexual practice that deviates from normal and generally accepted behavior. *See* paraphilia, sexual deviation.

**per'vert** One who practices a sexual perversion.

**pes'sa·ry** 1. A device placed within the vagina for various purposes, such as support of the uterus. 2. A suppository used for contraception.

**pe'ter** *Slang.* The penis, especially among juveniles.

**pet'ting** 1. Passionately kissing and intimately caressing one of the opposite sex to effect erotic arousal. 2. Induction of an orgasm without penetration of the vagina by the penis. *Compare* necking.

**Pey·ro·nie's' dis·ease'** A rare condition in which scar tissue develops in the shaft of the penis, resulting in a marked deformity. Also called *bent-nail syndrome.*

**phag'o·cyte** A cell that engulfs and digests foreign particles, such as microorganisms or tissue debris, for example a white blood cell or a macrophage.

**phal'lic** Of or pertaining to the penis.

**phal'li·cism, phal'lism** Adoration for or worship of the phallus or penis or reverence for its creative power in nature.

**phal'lic wor'ship** Nature worship in which special adoration is given to the organs of generation, especially the phallus or penis; phallicism.

**phal'lus** 1. The penis. 2. The embryonic structure that gives rise to the penis or clitoris. 3. An image or representation of the male copulatory organ or penis. *See* lingam.

**phan'tasms** *See* sexual imagery.

**phan'tom preg'nan·cy** False pregnancy. *See* pseudocyesis.

**phar·a·on'ic cir·cum·ci'sion** Infibulation, q.v.

**pher'o·mone** A substance secreted by an organism that is discharged externally and acts on other individuals of the same group, initiating certain behavioral, physiological, or developmental responses; an ectohormone.

**phi·lan'der** To make love in a casual and frivolous manner; to flirt.

**phi·lan'der·er** One that plays at love and courtship in a frivolous, lighthearted manner; one who courts a woman without the intention of marrying her.

**phil'ter** 1. A love potion. 2. A magic drug or potion that will induce a person to fall in love with another person. 3. A drug that stimulates sexual desire. *See* aphrodisiac.

**phi·mo'sis** A condition in which the prepuce cannot be retracted over the glans penis due to a constriction of its orifice.

**pho'bi·a** A persistent, unwarranted, or abnormal fear of an object, condition, situation, or person. *See* androphobia, genophobia, gynephobia, parthenophobia.

**pho·co·me'li·a** A development anomaly characterized by an imperfectly developed body bearing incomplete arms or legs or both. Limbs are often reduced to small, flipperlike appendages. *See* thalidomide.

**pick-up** 1. The act of inviting a person into a car for a free ride, or a person who accepts such a ride. 2. A person, usually a female, who accepts a social invitation to accompany a strange man.

**pic·to·phil'i·a** A condition in which a person requires or depends on seeing erotic pictures for the arousal of sexual feelings or the achievement of an orgasm. *See* paraphilia.

**PIF** Prolactin inhibitory factor.

**pill** 1. A small, globular mass of medicine, usually coated, to be chewed or swallowed whole. 2. An oral contraceptive, usually called "the Pill." There are many brand names but all contain one or two synthetic hormones, an estrogen, a progestogen, or both. Birth-control pills prevent conception by inhibition of releasing factors from the hypothalamus that are necessary for the release of FSH and LH from the anterior pituitary. As a result, when the pills are taken regularly, ovarian follicles do not develop and ovulation does not occur. In addition, the uterine endometrium is changed so that implantation is less likely to occur, cervical mucus is made more viscous, which acts to prevent the entrance of sperm into the uterus, and capacitation of spermatozoa is inhibited. Pills are of two types: *combination pills,* which contain both estrogen and a progestogen; and *sequential pills,* in which the first 15 pills contain only estrogen, the remaining 5 or 6 estrogen and a progestogen. Pills can be obtained by prescription only and should be taken under a physician's direction, as undesirable side effects sometimes occur. *See* morning-after pill. 3. *Slang.* An obnoxious, disagreeable, generally disliked person.

**pimp** 1. A procurer or pander, especially one who solicits customers for a prostitute for compensation provided by the prostitute or patron. 2. A man who cohabits with a prostitute and solicits for her, living off her earnings. *See* cadet.

**pin'ga** The penis (Spanish).

**ping-pong in·fec'tion** An infection that is transmitted back and forth between sexual partners, as trichomoniasis.

**piss** *Vulgar.* 1. *n.* Urine 2. *v.* To urinate.

**pi·tu'i·tar·y gland** The hypophysis cerebri, an endocrine gland located at the base of the brain. It consists of three lobes: (a) the *anterior lobe,* which secretes the gonadotrophic hormones FSH, LH (ICSH), and prolactin and a number of other hormones that regulate various bodily activities; (b) the *intermediate lobe,* which secretes a melanophore-stimulating hormone (MSH) of unknown function in humans; and (c) the *posterior or neural lobe,* which stores and releases oxytocin and vasopressin (antidiuretic hormone or ADH), hormones secreted by the hypothalamus.

**pla·ce'bo** A substance that has no therapeutic effect which is administered to a patient, who assumes that it is a medicine with healing properties.

**pla·cen'ta** A structure composed of embryonic and maternal tissues attached to the inner surface of the uterus and connected to the fetus by the umbilical cord. Through it, the embryo or fetus receives nutritive materials and oxygen from the mother's blood and discharges wastes and carbon dioxide into the mother's blood. It constitutes the principal portion of the afterbirth. It also secretes hormones and thus functions as an endocrine gland. *See* HCG.

**Planned Par'ent·hood** A worldwide organization for promoting the limitation of population growth. Important subdivisions include Planned Parenthood Federation of America, Inc. (PPFA) and International Planned Parenthood Federation (PPF).

**pla·teau′ phase** The second phase in sexual response. It follows the excitement phase (q.v.) and immediately precedes the orgasm (q.v.).

**pla·ton′ic** Purely spiritual and nonphysical; free from lust and sensual desire, as in a *platonic* relationship between a man and a woman.

**pla·ton′ic love** Nonsexual love; filial love.

**PMS** Premenstrual syndrome.

**pock′et pool** *Slang.* The act of playing with the penis and testes with one's hands in one's pants pockets.

**po′lar bod′y** A minute cell produced in the development of an ovum; a polocyte.

**po′lo·cyte** A polar body, q.v.

**pol·y·an′dry** Marriage of a woman to more than one husband at a time. *Compare* monandry.

**po·lyg′a·my** Marriage of a man or a woman to more than one mate at the same time. *Compare* monogamy.

**pol·y·men·or·rhe′a** Menstruation of an excessive amount or duration or both. Also called *menorrhagia.*

**pol′yp** A protruding growth with a stalk or stem that grows from a mucous membrane, as that lining the nasal cavity or the uterus. Uterine and cervical polyps are common in women over 40 and are a common cause of uterine bleeding.

**pom·poir′** The use of the constrictor action of vaginal and pelvic muscles to induce an orgasm in a male.

**ponce** *Slang.* A British pimp.

**poon′tang** *Slang.* 1. Sexual intercourse, especially with a black or mulatto woman. 2. The vagina of a black or mulatto woman.

**pop** 1. *Informal* for father. 2. *Slang.* Any elderly man.

**porn** *Short* for pornography.

**por′ner·ast** An individual who is only potent with a prostitute, the prostitute serving as a fetish.

**por′no·graph** A pornographic picture or writing.

**por·no·graph′ic** Of or pertaining to pornography.

**por·nog′ra·phy** 1. A description of prostitutes or prostitution. 2. The depiction by writing, art, or photography of obscene or indecent activities with the deliberate intent of arousing erotic sensations, especially if such depiction has little or no artistic merit. 3. The display of obscene or erotic behavior.

**POSSLQ** Person of the opposite sex sharing living quarters.

**post·co′i·tal** After or following sexual intercourse.

**post·con·cep′tion** After or following fertilization of the ovum.

**pos·te′ri·or** Toward the hind end of a quadruped; in humans, toward the back side of the body. Opposite of *anterior.*

**pos·til·li·on·age′** Inserting the finger into or applying pressure in the region of the anus of a sexual partner prior to orgasm.

**post·ma′ri·tal** After or following marriage.

**post·or·gas′mic** After or following the climax in sexual intercourse.

**post·ov′u·la·to·ry** After ovulation or rupture of an ovarian follicle and the liberation of the ovum.

**post·par′tum** Following childbirth or delivery.

**po′ten·cy** The ability of a male to perform the sexual act. *Compare* impotence.

**po′tent** Capable of performing sexually, said of a male.

**pouch of Doug′las** *See* cul-de-sac.

**pox** *Slang.* Syphilis.

**pre·ad·o·les′cent** Before adolescence; prepubescent.

**pre·co′i·tal** Before coitus or sexual intercourse.

**pre·co′cious** Prematurely developed; maturing before the usual age.

**pre·co′cious pu′ber·ty** Puberty occurring before the age of eight.

**pre·co′cious sex·u·al′i·ty** The development of sex organs and sexual traits before the usual age of puberty.

**pre·cop′u·la·to·ry** Preceding coitus.

**pre·ec·lamp′si·a** See eclampsia.

**pree′mie, pre′mie** Informal. A premature infant.

**preg′nan·cy** The state of being pregnant; bearing a developing embryo or fetus within the uterus; gestation. The average length of pregnancy from the time of conception is 266 days. The date of delivery is 280 days (ten lunar months) from the beginning of the last menstrual period.

**preg′nan·cy hor′mone** Progesterone, q.v.

**preg′nan·cy rate** See failure rate.

**preg′nan·cy, signs of** Early presumptive signs include cessation of menstruation, occurrence of morning sickness or nausea, increased size and fullness of the breasts, and an increase in the pigmentation of the areola around the nipple. Probable signs include a change in consistency and size of the uterus, occurrence of uterine contractions, softening of the cervix, increased leukorrhea, enlargement of the abdomen, increased tenderness of the breasts, and a positive pregnancy test. Positive signs include hearing and counting fetal heartbeat, detection of fetal movements, and appearance of fetal skeleton in an X-ray film.

**preg′nan·cy test** A test employed to determine if a woman is pregnant or not. Tests depend principally upon detection of human chorionic gonadotrophin (HCG) in the blood or urine or noting various reactions due to its presence. Tests are of four types: biologic, immunologic (q.v.), hormonal, and radioimmunoassay (q.v.). See also rabbit test.

**preg′nant** Carrying an embryo or fetus within the uterus; with child.

**Prem′a·rin** Trade name for a preparation, consisting of conjugated estrogens, used principally in the treatment of menopausal and postmenopausal disorders. It is also used as a postcoital contraceptive.

**pre·mar′i·tal** Preceding marriage.

**pre·ma·ture′ birth** Labor before full term but after the age of viability.

**pre·ma·ture′ e·jac·u·la′tion** Ejaculation prior to, at the time of, or immediately after intromission.

**pre·men′stru·al syn′drome** A condition preceding menstruation, characterized in mild cases by nervousness, irritability, headaches, excessive appetite, fault-finding, and flare-ups of temper. In severe cases depression, crying spells, and paranoic attitudes may develop. It occurs 7 to 10 days before menstruation and ends a few hours after the onset of menstrual flow. Also called premenstrual tension syndrome.

**pre·na′tal** Occurring or existing before birth.

**pre·na′tal in′flu·ences** See maternal impressions.

**pre·ni·da′tion** Before implantation of the blastocyst in the uterus.

**pre·nup′ti·al** Prior to or preceding marriage.

**pre·pu′ber·al, pre·pu′ber·tal** Occurring before puberty, especially with reference to the period of rapid growth that occurs prior to the maturation of the gonads.

**pre·pu·bes′cent** Prepuberal.

**pre′puce** 1. A circular fold of skin that covers the glans of the penis. The prepuce is retractable and withdrawn over the glans when an erection occurs, thus totally exposing the glans. Also called foreskin. See circumcision, phimosis, redundant prepuce, smegma. 2. A fold of skin that covers the clitoris.

**pres·en·ta′tion** A term designating that part of the fetus that can be felt at the cervix at the beginning of labor or that is presented first at birth, such as a *head presentation* or *breech presentation.*

**PRF** Prolactin releasing factor.

**pri·a′pic** 1. Pertaining to the penis; phallic. 2. Emphasizing the penis. 3. Suggestive of or resembling a penis, with reference to an image. 4. Excessively concerned with masculinity or male sexuality.

**pri′a·pism** 1. Persistent, prolonged, and often painful erection of the penis, usually unaccompanied by sexual desire. 2. Lascivious behavior or display.

**pri·a·pi′tis** Inflammation of the penis.

**pri·a′pus** The penis.

**prick** *Slang.* The penis.

**pri·mi·grav′i·da** A woman who is pregnant for the first time.

**pri′vates** The private parts; the external genitalia.

**pro** *Slang.* 1. A prophylactic; a condom. 2. A prostitute.

**pro·an′dro·gen** A compound that is not androgenic when applied locally but acquires androgenic properties when metabolized within an organism.

**pro·cre·a′tion** The production of offspring; reproduction.

**pro·cure′** To secure girls or women for the purpose of prostitution.

**pro·cur′er** One who secures a female for prostitution; a pander or pimp.

**prof′li·gate** 1. Given to dissipation and licentiousness; shamelessly immoral. 2. Reckless; extravagant.

**pro·gam′ic** Preceding fertilization.

**prog′e·ny** Offspring; children or descendents.

**pro·gen′i·tor** An ancestor or forefather; the first in a line of descent.

**Pro·ges′ta·sert** Trade name for an IUD that contains progesterone, which is released slowly into the uterus for a period of a year or more.

**pro·ges·ta′tion·al** Preceding menstruation or pregnancy.

**pro·ges′ter·one** A steroid hormone produced by the corpus luteum of the ovary. In conjunction with estrogens from the follicle, it induces changes in the uterine endometrium preceding menstruation and pregnancy. During early pregnancy it is also produced by the placenta. It acts to maintain pregnancy and stimulates the development of the mammary glands. It is also produced by the adrenal cortex and the testes. Also called *pregnancy hormone.*

**pro·ges′tin** Progesterone or any of a number of synthetic substances that possess the properties of progesterone, especially one that is active when taken orally.

**pro·ges′to·gen** Any substance that possesses progestational properties.

**prog·no′sis** A forecast or prediction of the probable course or outcome of a disease, illness, or any activity.

**pro·grav′id** Progestational.

**pro·lac′tin** A hormone produced by the anterior pituitary that stimulates the production of milk by the mammary gland; the lactogenic hormone.

**pro·lac′tin in·hib′i·to·ry fac′tor** *Abbr.* PIF. A hormone secreted by the hypothalamus that inhibits the secretion of prolactin by the pituitary gland.

**pro·lac′tin re·leas′ing fac′tor** *Abbr.* PRF. A hormone secreted by the hypothalamus that stimulates the release of prolactin by the pituitary gland.

**pro′lapse** The falling down or sinking of a structure from its normal position, as in *prolapse* of the uterus.

**pro·mis′cu·ous** Engaging in frequent and casual intercourse with numerous partners.

**prop′a·gate** 1. To have young or offspring; to multiply. 2. To cause animals or plants to breed and produce offspring. 3. To transmit hereditary traits or characteristics to or through offspring.

**pro·phy·lac′tic** 1. *adj.* Protecting against disease. 2. *n.* An agent or device used for the prevention of disease, especially a venereal disease, such as a condom. *Abbr.* pro.

**prop·o·si'tion** A questionable or immoral proposal, especially a solicitation for illegal sexual relations.

**pros·ta·glan'din** One of a number of biologically active lipids present in seminal, menstrual, and amniotic fluids and in various mammalian tissues. They stimulate the contraction of smooth muscle, especially in the intestine and uterus. Some act as abortifacients.

**pros'tate gland** An accessory sex gland in the male which surrounds the urethra at the base of the bladder. Its secretions, which pass into the urethra, form the second portion of the semen. Prostatic fluid is slightly acid. It contains citric acid and a number of enzymes, especially acid phosphates and various dehydrogenases, as well as the minerals zinc and magnesium.

**pros·tat'ic** Of or pertaining to the prostate gland.

**pros'ta·tism** A condition resulting from various disorders of the prostate gland, especially the obstruction of the bladder outlet by its enlargement, common in men over 60. It may result from hypertrophy, carcinoma, or fibrosis.

**pros·ta·ti'tis** Inflammation of the prostate gland, often associated with inflammation of the urethra and seminal vesicles.

**pros'ti·tute** 1. A person, usually a female, who performs the sexual act for compensation; a harlot, whore, strumpet. 2. A male who engages in homosexual activity for compensation.

**pros·ti·tu'tion** The act or practice of engaging in sexual intercourse or performing other sexual acts for payment.

**pros'ty, pros'tie** *Slang.* A prostitute.

**Pro·ve'ra ring** A contraceptive device consisting of a plastic ring (silastic) containing Depo-Provera or another progestogen, which is inserted into the vagina. The progestogen is slowly released and passes through the vaginal wall into the bloodstream.

**prude** A person who is excessively concerned with proper, modest, or righteous behavior.

**prud'er·y** Excessive modesty in speech and conduct.

**pru'ri·ent** Readily susceptible to lascivious or lustful thoughts or sexual desires.

**pseu·do·co'i·tus** A sexual relationship in which the penis is in contact with the body of a female but is not inserted into the vagina.

**pseu·do·cop·u·la'tion** Close contact of two individuals of different sexes for the transfer of sperm, as in amplexus in frogs.

**pseu·do·cy·e'sis** False pregnancy, a condition in which symptoms of pregnancy, such as cessation of menstruation, morning sickness, and enlargement of the abdomen, occur, brought on by psychogenic factors.

**pseu·do·her·maph'ro·dite** A false hermaphrodite, an individual with the gonads of one sex but who possesses secondary characteristics and external genitalia that resemble those of the opposite sex.

**pseu'do·male** A lesbian with male characteristics; a dyke.

**pseu·do·mas·tur·ba'tion** Peotillomania, q.v.

**pseu·do·pe'nis** An elongated clitoris possessed by certain South American monkeys and by hyenas.

**pseu·do·preg'nan·cy** Pseudocyesis, q.v.

**psy·cho·gen'ic** Of mental or emotional origin; not due to organic causes.

**psy·cho·pro·phy·lax'is** See Lamaze method.

**psy·cho·sex'u·al** Of or related to psychological or emotional factors involved in sexuality.

**pu·bar'che** The beginning of the growth of pubic hair at puberty.

**pu'ber·al, pu'ber·tal** Of or pertaining to puberty.

**pu'ber·ty** The period, also called *pubescence,* during which an individual becomes capable of reproducing, characterized by marked physical, physiological, and psychological changes. It normally occurs in males between the ages of 12 and 16, in females between 10 and 14. In males, spermatozoa begin to be formed and the first ejaculations occur; in females, the first menstruation or menarche occurs and fertile ova begin to be produced. Secondary sex characteristics in both sexes make their appearance.

**pu'bes** The hair that covers the mons pubis; the pubic region.

**pu·bes'cence** 1. Hairiness, especially the presence of fine, soft hair, *See* lanugo. 2. Puberty, especially its onset marked by the appearance of pubic hair.

**pu·bes'cent** 1. Covered with fine, soft hairs. 2. Attaining puberty or characteristic of the state of puberty.

**pu'bic** Of or pertaining to the pubes or the pubis.

**pu'bic bone** The os pubis, a bone which, on each side, forms the anterior portion of the hipbone. *See* pelvic girdle.

**pu'bic sym'phy·sis** The joint between the two pubic bones that forms a bony prominence beneath the region covered by pubic hair.

**pu'bis** The pubic bone or os pubis.

**pu·bo·coc·cy'ge·us mus'cle** *Abbr.* PC. One of two muscles comprising a part of the levator ani muscle that forms the floor of the pelvis. The two PC muscles that extend from the two pubic bones posteriorly to the coccyx enclose the urethra, vagina, and rectum. Contraction of these two muscles constricts these tubes. Their action on the vagina enhances sexual pleasure and facilitates orgasmic response. *See* Kegel's exercises.

**pud, pudd** *Slang.* The penis.

**pud'ding** *Slang.* The penis, especially when used in masturbation.

**pu·den'dal cleft** The region between the labia minora which contains the openings of the urethra and vagina.

**pu·den'dum,** *pl.* **pu·den'da** The external genitalia of a female; the vulva.

**pu'er·ile** Pertaining to childhood or children; childish.

**pu·er'pe·ra** A woman who has just given birth to a baby.

**pu·er'per·al** Of or pertaining to childbirth or parturition.

**pu·er'per·al fe'ver** Childbed fever or puerperal sepsis, an infection of the uterine endometrium accompanied by septicemia, which sometimes follows parturition.

**pu·er·pe'ri·um** 1. The condition of a woman in labor or one who has just given birth to a baby. 2. The period from the time of delivery to complete involution of the uterus, usually lasting from 4 to 6 weeks.

**pull out** *Slang.* Coitus interruptus; withdrawal.

**pup·pet·o·phil'i·a** The use of a lifelike or life-size doll as an aid in masturbation.

**pu'ru·lent** Consisting of, containing, or involved in the formation of pus.

**pus** The semifluid, yellowish-white, viscous substance, formed in infected tissue, consisting of leucocytes, cellular debris, and serum; a product of inflammation.

**push'o·ver** *Slang.* A female who readily acquiesces to sexual intercourse.

**puss** 1. Affectionate term for a girl or young woman. 2. *Slang.* The face or mouth.

**pus'sy** *Slang.* 1. The female pudenda or external genitalia. 2. A female as a sex object. 3. Coitus.

**put'a** *Slang.* 1. A prostitute. 2. A promiscuous female.

**put out** *Slang.* 1. To grant sexual favors readily. 2. To be promiscuous, especially with reference to a female.

**put'ter** *Slang.* The penis.

**pyg·ma'li·on·ism** Sexual interest directed toward a female statue or toward an object of one's own creation.

**quad·rup'let** One of a group of four children resulting from a single pregnancy.

**queen** *Slang.* An effeminate male homosexual, especially one who assumes the female role.

**queer** 1. *adj.* Odd, degenerate, perverted. 2. *n. Slang.* A homosexual of either sex.

**quick** Pregnant, with fetal movements.

**quick'en·ing** The first fetal movements detected by a pregnant woman, usually occurring during the fourth or fifth month of gestation.

**quick'ie** *Slang.* A short, rapid sexual encounter, involving either sexual intercourse or homosexual activity.

**quim** *Slang.* The vulva or vagina.

**quin·tup'let** One of a group of five children resulting from a single pregnancy.

**quix·ot'ic** Extravagantly romantic; idealistic; impractical.

**rab'bit test** A pregnancy test in which the urine of the person being tested is injected into a female rabbit. If the woman is pregnant, the presence of human chorionic gonadotrophin in the urine will cause the release of ova from the rabbit's ovary.

**rack out** *Slang.* To go to bed.

**ra'di·o·im·mu·no·as'say** *Abbr.* RIA. A laboratory procedure that employs radio-chemical and immunological techniques to determine the presence of minute quantities of a substance such as a hormone in a fluid or tissue. The basic principle involved is a reaction between an antigen and an antibody.

**rag** *Slang.* A sanitary napkin.

**rain'coat** *Slang.* A condom.

**raise the dead** *Slang.* To induce a new erection in a male.

**rake** A licentious or dissolute person; a roué.

**ram** A male sheep or the male of a number of other mammals, such as the goat and antelope.

**rape** 1. Illegal sexual intercourse with a woman without her consent. The act may be effected by force, intimidation, threat of bodily harm, or through deception. A woman who consents through fear or consents though mentally deficient or retarded, or a woman whose consent is meaningless because she is unconscious from sleep, alcohol, or drugs, is considered to be a victim of rape. *See* statutory rape. 2. Sexual intercourse forced upon a man or boy by a woman without his consent. 3. Forcible sexual assault against a person of the same sex, such as homosexual assault that occurs commonly in prisons.

**rape cri'sis cen'ter** A counselling center now available in most cities that provides assistance to victims of rape.

**ra'phe** A seamlike union or a fusion of two halves of a structure, usually marked by a line or ridge, as the ridge along the surface of the scrotum.

**raun'chy** Indecent, lecherous, risqué.

**rav'ish** 1. To commit rape upon a woman; to deflower; to violate. 2. To seize and carry away a woman by force.

**RAW** Ready and willing.

**raw** *Slang.* 1. In the nude, naked. 2. Risqué, obscene.

**ream** *See* rim.

**rear end** *Slang.* The buttocks.

**rear en'try** Coitus in which the female assumes the position characteristically assumed by females of most mammals, the opposite of face-to-face. It may be engaged in with the female in various positions—standing, kneeling, sitting, or lying either face down or on her side. *See* croupade, cuissade.

**re·ca·pac·i·ta'tion** The ability of decapacitated spermatozoa to regain their ability to fertilize an ovum.

**rec'tal** Of or pertaining to the rectum.

**rec'to·cele** Protrusion of the rectum into the vagina.

**rec'tum** The distal portion of the large intestine comprising the portion between the sigmoid colon and the anus. It consists of two portions, the rectum proper and the anal canal.

**red hot** Slang. 1. adj. Sexy, enticing, alluring. 2. n. A stag (man without a female partner).

**red-light dis'trict** A region of a town or city characterized by many houses of prostitution.

**re·dun'dant pre'puce** A condition in which growth of the prepuce is so extensive that it cannot be withdrawn over the glans.

**re'flex** A reflex action, an involuntary, stereotyped, and usually adaptive response to a stimulus. Anatomically, it involves a *receptor,* an afferent or sensory neuron that carries the impulses to the spinal cord or brain, where connections are made with efferent or motor neurons that convey the impulses to an *effector organ* (muscle or gland), which responds.

**re·frac'to·ry pe'ri·od** 1. The period of elapsed time during which a structure (cell, tissue, or organ) will not respond to a second stimulus following the primary stimulus. 2. In a male, the period immediately after an orgasm during which an erection cannot occur and another orgasm be experienced.

**re·lax'in** A hormone secreted by the ovaries, uterus, and placenta during pregnancy which promotes relaxation of the tissues surrounding the birth canal in preparation for parturition.

**re·leas'er** Anything that turns a person on sexually, such as clothes in a female that emphasize the breasts and buttocks.

**re·leas'ing fac'tor** Abbr. RF. One of several humoral factors or hormones produced by the hypothalamus that act on the anterior pituitary, bringing about the secretion and release of certain hormones, especially FSH and LH. Also called *releasing hormone* (RH).

**re·mar'ry** To marry again.

**re·pro·duce'** To bring offspring into existence; to produce a new generation.

**re·pro·duc'tion** The act of reproducing.

**Re·pro·duc'tive Bi·ol'o·gy Re'search Foun·da'tion** A center for sex therapy and the training of sex therapists established in St. Louis, Missouri, in 1970 by William Masters and Virginia Johnson.

**re·pro·duc'tive or'gan** A sex organ; a genital organ.

**re·pro·duc'tive sys'tem** The system of organs that is involved in the production of offspring; the male and female sex organs. *See* sex organ.

**res·o·lu'tion** 1. The subsidence or termination of an unusual or abnormal situation. 2. In *medicine,* the reduction or disappearance of a swelling or inflammation.

**re·so·lu'tion phase** In sexual response, the final and recovery phase during which tissues and organs return to normal. *See* sexual response.

**ret·ro·cop·u·la'tion** 1. The act of copulating from behind; coitus à la vache. 2. Among homosexuals, anal intercourse.

**ret'ro·grade e·jac·u·la'tion** The backward flow of semen into the bladder that may result from weakness of or injury to the sphincter muscle of the bladder or constriction of the urethra.

**ret·ro·ver'sion** The state of being tipped or turned backward.

**RF** Releasing factor.

**RG** Abbr. for "real girl" in contrast to a "homosexual girl."

**RH** Releasing hormone.

**rhy'thm meth'od** A method of contraception based on limiting sexual intercourse to times in the menstrual cycle when no ovum is likely to be present to be fertilized. Also called *safe-period method*. (*See* safe period.)

**rid'gel, ridg'ling** A male animal, especially a colt, with one or both testes undescended.

**rim** *Slang*. To insert something into another person's anus.

**rip** *Slang*. A reckless, dissolute person; a libertine.

**ris·qué'** Suggestive of or daringly close to impropriety or indelicacy; off-color, as a *risqué* joke.

**rod** *Slang*. The penis.

**roent'gen·o·gram** A photograph made by X-rays.

**roll** *Slang*. Coitus for a male.

**roll a trick** *Slang*. To rob a trick, said of a prostitute.

**ro·mance'** 1. A passionate love affair. 2. An intense and usually short-lived attachment.

**Ro'man cul'ture** *Slang*. Orgies.

**ro·man'tic** Marked by or involving passionate love or affection.

**root'y** *Slang*. Horny; sexually aroused.

**rou·é'** A debauchee or rake.

**rough trade** *Slang*. 1. A dangerous homosexual. 2. A coarse or tough homosexual prostitute.

**round-heel** *Slang*. A woman who engages readily in sexual intercourse.

**rub'ber** *Slang*. A condom.

**ru·bel'la** German or three-day measles, a mild disease usually of little consequence but, when occurring in a pregnant woman, a common cause of birth defects.

**rub off** *Slang*. To engage in female masturbation.

**ru'ga,** *pl.* **ru'gae** A fold or ridge, as one of the small transverse folds of the mucous membrane of the vagina.

**rump** The buttocks.

**rup'ture** 1. A breaking or bursting. 2. Common term for a hernia, especially an abdominal hernia.

**rut** The recurring period of intense sexual excitement and activity in the males of various mammals, such as cattle, sheep, deer. *Compare* estrus.

**S-A** Sex appeal.

**sac** *Slang*. The scrotum.

**Sach'er-Ma'soch, Le'o·pold von** (1836–1895) The Austrian novelist who described the behavior later known as *masochism* (q.v.), after his name.

**Sade, Mar·quis' de** (1740–1814) French soldier, author, and sex pervert noted for his cruel, orgiastic sexual practices. Sadism was described by him and receives its name from him.

**sad'ism** 1. Sexual gratification obtained through the infliction of pain on one's sexual object. *Compare* masochism. 2. In a general sense, obtaining satisfaction or pleasure through the infliction of physical or psychic pain (fear, embarrassment, humiliation) on another.

**sad'ist** A person who practices sadism.

**sa·dis'tic** Characterized by sadism; deliberately cruel.

**sa·do·ma'so·chism** *Abbr.* S-M. Simultaneous sadism and masochism.

**safe per'i·od** The period in the menstrual cycle during which conception is least likely to occur. Roughly, it includes four or five days following the cessation of menstruation and approximately seven days preceding the beginning of the next menstrual period. The safe period varies with the length of the menstrual cycle. *See* calendar technique, fertile period, rhythm method.

**safe'ty** *Slang.* A condom.

**sa·la'cious** Tending to excite sexual desire; lustful; lecherous; obscene.

**sa'line in·jec'tion meth'od** A procedure for inducing an abortion in which fluid is withdrawn from the amniotic sac (amniocentesis) and an equal amount of hypertonic saline solution injected, a procedure commonly employed after the 15th week of pregnancy.

**sa·li·ro·ma'ni·a** Abnormal sexual behavior characterized by the desire to make unclean the clothes or body of a woman, a mild form of sadism.

**sa·li'va** The fluid that forms in the mouth from the secretions of the salivary glands.

**sa·li'va meth'od** A method of predicting the time of ovulation based upon the concentration in saliva of an enzyme, alkaline phosphatase, which increases at ovulation. There is also an increase in salivary phosphate and glucose.

**sal·pin·gec'to·my** Surgical removal (excision) of a uterine tube or tubes.

**sal·pin·gi'tis** Inflammation of the uterine tubes.

**sal'pinx** A tube, especially the uterine or fallopian tube.

**salt'pe'ter** Potassium nitrate ($KNO_3$), a substance erroneously thought to reduce sex drive and sexual activity. It is a diuretic, increasing the flow of urine.

**Sam** *Slang.* Masculine attractiveness involving sex appeal and personal magnetism.

**Sand'stone** An estate in Topanga, California, at which open sexuality is practiced. Only couples are admitted. Nudity is common and any kind of sexual expression is permitted. Guests may visit once, or they may become members for an annual fee. The aim of Sandstone is to enable straight adults with sexual hang-ups to reassess their goals and self-image and freely express their sexuality.

**Sang'er, Mar'ga·ret** (1883–1966) American nurse and leader of the birth-control movement. Organized the first Birth Control Conference, held in New York in 1921. Author of many books and pamphlets on birth control.

**san'i·ta·ry nap'kin** A disposable pad of cellulose or other absorbent material worn during menstruation or postpartum to absorb the flow from the vagina.

**sap'phism** Lesbianism.

**Sap'pho** A Greek writer of the sixth century B.C., whose poems extolled the love of women for each other. She lived on the island of Lesbos, from which the term "lesbianism" is derived.

**sar·co'ma** A malignant tumor composed of nonepithelial cells, principally connective-tissue cells.

**sar·to'ri·al sex** The sex one is dressed as.

**sat·ur·na'li·a** An occasion or period characterized by unrestrained or orgiastic behavior, revelry, and licentiousness.

**sat'yr** 1. *Mythology.* A woodland god or demon represented as part man and part goat or horse. 2. A lecherous man; one afflicted with satyriasis.

**sat·y·ri'a·sis** Excess and uninhibited sexual desire in a male; lagnosis. *Compare* nymphomania.

**sav'ing it** *Slang.* 1. Preserving virginity until marriage. 2. Reserving sexual favors for a particular person.

**SBK** Spinnbarkeit.

**sca'bies** The itch, an infectious skin disease caused by a mite, *Sarcoptes scabiei,* characterized by intense itching.

**sca·to·log'i·cal** Of or pertaining to (a) excrement or filth; (b) pictorial or verbal material considered filthy or obscene.

**sca·tol'o·gy** 1. In *biology* or *medicine,* the study of fecal excretory material. 2. Obsession by or extreme interest in filthy or obscene matter, especially obscene literature.

**sco·po·phil'i·a, scop·to·phil'i·a** Obtaining sexual gratification through viewing of sexual acts, genital organs, or erotic pictures. *Compare* voyeurism.

**score** *Slang.* To win the favor of a girl or woman and accomplish sexual intercourse with her.

**screw** *Slang.* 1. To have sexual intercourse with; to engage in coitus. 2. To take unfair advantage of; to trick or deceive.

**scro'tum** The external pouch that contains the testes.

**scum** *Slang.* Semen.

**scum bag** *Slang.* A condom.

**seat** *Slang.* The buttocks.

**sec'on·da·ry sex char·ac·ter·is'tics** External structures other than the genitalia that distinguish a male from a female. These include shape of the body, nature and distribution of hair, presence or absence of mammary glands, size of larynx and consequent pitch of voice. Development of these structures depends primarily upon hormones produced by the ovaries and testes.

**sec'ond base, get'ting to** *Slang.* Intimate petting above the waist.

**se·cre'to·ry phase** The phase in the menstrual cycle that follows ovulation, characterized by the development and activity of uterine glands in the endometrium. It is controlled principally by progesterone secreted by the corpus luteum.

**sec'tion** 1. The act of cutting. 2. A cut surface. 3. A segment or subdivision of an organ or structure.

**sec'un·dines** The afterbirth.

**se·duce'** To persuade a person to have sexual intercourse, especially to entice a female to engage in unlawful intercourse without the use of force.

**seed** Common term applied to the male reproductive cells of various animals. *See* milt, semen, sperm.

**see'ing some'one** *Euphemism* for having sexual relations with.

**seg·men·ta'tion** The formation of many cells, as that resulting from repeated cell division in a fertilized egg. *See* cleavage.

**self-a·buse'** Old term for masturbation.

**self-grat·i·fi·ca'tion** *Euphemism* for masturbation.

**se'men** The seminal fluid containing spermatozoa that is discharged by the male during an ejaculation. It consists of *spermatozoa* produced by the testes and *seminal plasma* produced by the accessory reproductive glands. Sometimes referred to as *seed* or *vital fluid*. *See* ejaculate.

**sem'i·nal** Of, pertaining to, or consisting of semen.

**sem'i·nal duct** One of the ducts, the ductus (vas) deferens or the ejaculatory duct, that convey semen from the epididymis to the urethra.

**sem'i·nal flu'id** The semen, composed of seminal plasma and spermatozoa.

**sem'i·nal plas'ma** The noncellular portion of semen (about 90 percent) consisting principally of the secretions of the accessory reproductive glands (bulbourethral and prostate glands and the seminal vesicle).

**sem'i·nal ves'i·cle** A sacular outpocketing of the ductus (vas) deferens close to its junction with the ejaculatory duct just before its entry into the urethra. Its secretion, which is alkaline, contains fructose and prostaglandins, important constituents of semen.

**sem·i·na'tion** The introduction of semen into the vagina or uterus. Also called *insemination*.

**sem·i·nif'er·ous** Producing or conveying semen.

**sem·i·nif'er·ous tu'bule** One of the many convoluted tubules in a testis in which spermatozoa are formed. Each has a wall consisting of a *basal membrane* and a *tunica propria* and is lined with *germinal epithelium* consisting of *Sertoli cells* and *spermatogenic cells* (spermatogonia, primary and secondary spermatocytes, spermatids, and spermatozoa).

**sen′su·al** 1. Pertaining to or concerned with the gratification of physical appetite, especially sexual appetite. 2. Lewd, unchaste. 3. Carnal, worldly, voluptuous.

**sen′su·al·ist** One who indulges in sensual pleasures.

**sen·su·al′i·ty** Lasciviousness, lewdness; excessive devotion to sensual pleasures.

**sep·a·ra′tion** The cessation by mutual consent of conjugal cohabitation or the termination of conjugal cohabitation by a court decree.

**sep·ti·ce′mi·a** A condition resulting from the presence of pathogenic microorganisms or their toxic products in the bloodstream. Commonly called *blood poisoning*.

**sep′tum** *pl.* **sep′ta** A dividing wall or partition.

**se·que′la** That which follows, especially any abnormal condition which develops after a disease.

**se·quen′tial pills** Birth-control pills in which the first 15 or 16 contain only estrogen, the remaining 5 containing both estrogen and a progestogen. This type of pill has been withdrawn from the market.

**se·ra′glio** 1. A place in a Muslim house where wives and concubines are kept. 2. A harem. 3. A place of sexual license, especially a brothel. 4. A Turkish palace, especially for a sultan.

**se·ro·log′ic** Of or pertaining to serology.

**se·rol′o·gy** The science that deals with blood serum and its constituents. It especially deals with antigen-antibody reactions within the body.

**Ser·to′li cells** Large cells in the germinal epithelium of a testis tubule interspersed among sperm-forming cells. Among their functions are (a) nourishment of developing germ cells; (b) secretion of an androgen-binding protein that takes up testosterone and delivers it to the epididymis; (3) phagocytosis of degenerating germ elements; (d) synthesis and transformation of androgens into estrogens; (e) service as a blood testis barrier.

**ser′vice** 1. *n.* The act of a male mating with a female, especially in the breeding of domestic animals. 2. *v.* Said of a male animal, to mate with a female. Sometimes used as a slang term to apply to human intercourse.

**sev′enth heav′en** *Slang.* Perfect happiness.

**sex** 1. The qualities that distinguish an egg-producing individual (female) from a sperm-producing individual (male). 2. The state or condition of being male or female, including all the anatomical, physiological, and psychological characteristics that identify an individual as being male or female. 3. Sexual intercourse or coitus.

**sex act** Sexual intercourse; coitus; copulation.

**sex a′nal·ism** Anal intercourse or the use of the anus in copulation. *See* sodomy, pederasty.

**sex ap·peal′** *Abbr.* S-A. Physical traits and physical attractiveness that arouse sexual interest in a person of the opposite sex.

**sex as·sign′ment** The designation of a newborn baby as a male or female.

**sex cell** An ovum (egg) or a spermatozoon (sperm); a reproductive cell.

**sex-change op·er·a′tion** An operation or a series of operations that a transsexual undergoes to change his or her sex. In the male it involves the removal of the testes and penis and construction of an artificial vagina; in the female, the ovaries and breasts are removed and efforts made to construct a penis. Hormone therapy is usually resorted to.

**sex char·ac·ter·is′tic** An anatomical structure or a physiological activity that is associated with sex. *Primary* sex characteristics are those that involve the male or female reproductive systems; *secondary* sex characteristics are those that result from environmental action or the action of hormones.

**sex chro'mo·some** One of a pair of chromosomes (XX in the female, XY in the male) that are involved in sex determination.

**sex de·ter·mi·na'tion** The establishment of sex and sexual characteristics in a developing individual, primarily determined at fertilization. Fertilization of an ovum by an X-sperm results in the development of a female; fertilization by a Y-sperm results in the development of a male.

**sex drive** The physiological, psychological, and social factors that impel a person to seek to engage in sexual activities, especially sexual intercourse; libido.

**sexed** 1. Possessing sex or sexual instincts; possessing sex appeal. 2. Identified as a male or female.

**sex'er** The person who identifies the sex of an organism, especially one who determines the sex of newly hatched chicks by examining structures within the anal opening.

**sex flush** The reddening of the skin resulting from vasocongestion that occurs over most of the body following erotic stimulation. Also called *maculopapular sex flush.*

**sex gland** A gonad; a testis or an ovary.

**sex hor'mone** A hormone that affects the development and functioning of the reproductive organs and the development of secondary sexual characteristics. *Androgens (male sex hormones)* are produced principally by the testes; *estrogens* and *progesterone (female sex hormones)* are produced principally by the ovaries. Both male and female sex hormones are also produced in the adrenal cortex. Sex hormones influence behavior and various physiological processes such as general metabolism, bone and skeletal development, salt and water metabolism, and the secretory activity of various glands (thymus, thyroid, mammary).

**sex-in'flu·enced char'ac·ter** In *genetics,* a trait or character that behaves as a dominant in males and a recessive in females, as human baldness. Also called *sex-conditioned character.*

**sex'ing** *See* nuclear sexing.

**sex'ism** The granting of favors or advantages to one sex or the other, as in employment practices; the belief that one sex is superior to the other.

**sex'ist** One who believes in and promotes sexism.

**sex job** *Slang.* 1. A promiscuous female. 2. Sexual activity carried on to exhaustion.

**sex'less** 1. Asexual; lacking distinctive sexual characteristics; neither male nor female; neuter 2. Exhibiting sexual apathy.

**sex-lim'i·ted char'ac·ter** A hereditary nonsexual characteristic that occurs predominantly in one sex, such as baldness in males and a hairless face in females.

**sex-linked char'ac·ter** A hereditary characteristic whose gene lies in a sex chromosome, such as color-blindness and hemophilia in humans. Because males possess only one X chromosome, a sex-linked character appears more frequently in males than in females, and males inherit the character from their mother and not their father.

**sex·ol'o·gist** One versed in the science of sexology.

**sex·ol'o·gy** The science that deals with sex, comprising the anatomy and physiology of the organs of the reproductive systems and all aspects of behavior, normal and abnormal, associated with their functioning.

**sex or'al·ism** Oral sex or the application of the mouth, lips, or tongue to the sex organs of one's partner. *See* fellatio, cunnilingus.

**sex or'gan** A genital organ; any organ of the male or female reproductive system. *Primary sex organs* are the ovaries and testes, which produce the sex cells;

*secondary* or *accessory sex organs* include the copulatory structures, the penis and vagina; ducts that transport the spermatozoa and ova; glands that contribute to the genital products; and an organ, the uterus, in which the young develop.

**sex play** Sexual activity among prepubescent children.

**sexpot** *Slang.* An attractive, promiscuous female.

**sex ra′ti·o** The proportion of males to females in a population, usually expressed as the number of males per one hundred females or as the percentage of males to females. For humans at birth, it is 106/100; early adulthood, 100/100; middle age, 85/100; old age, 50/100.

**sex re·as·sign′ment** The change of a person's sex role or gender to agree with his or her anatomy, as in the case of an erroneous sex assignment at birth or following a sex-change operation.

**sex re′search·er** A sexologist who has conducted extensive investigations in the fields of sex and sexual behavior. Important researchers include Richard von Krafft-Ebing, H. Havelock Ellis, Alfred C. Kinsey, William H. Masters, Viriginia E. Johnson.

**sex re·ver′sal** The conversion of an individual of one sex into one of the opposite sex. *See* transsexual.

**sex role** The outward expression of gender identity. *See also* gender role.

**sex-role in·ver′sion** A phenomenon in which a person's anatomy and his or her sex or gender role are incompatible; transsexualism.

**sex skin** The skin, covering the major labia in a female, that exhibits marked color changes in a sexually excited woman, especially during the plateau phase of an orgasm.

**sex stim′u·lant** An aphrodisiac, q.v.

**sex ther′a·pist** One skilled in the practice of sex therapy. Certified therapists are members of the American Association of Sex Educators and Counsellors.

**sex ther′a·py** Treatment for the correction or solution of various sexual difficulties or sexual dysfunction, such as impotence, premature ejaculation, frigidity (orgasmic dysfunction), painful intercourse, and others. Treatment may be on an individual, co-marital, or group basis.

**sex·tip′a·ra** A woman who has borne six children in six pregnancies.

**sex′u·al** 1. Of, pertaining to, or involving sex, the male or female sexes, or the sex organs. 2. Possessing sex or sex organs. 3. Occurring between or involving the two sexes, as in *sexual* relations. 4. Possessing sex organs and reproducing by methods involving both sexes.

**sex′u·al aim** In heterosexual activities, coitus; in deviant sexual behavior, the substitution of aberrant sexual activities (voyeurism, exhibitionism, sadism, masochism) for coitus.

**sex′u·al an·es·the′sia** *See* frigidity.

**sex′u·al ap′a·thy** Complete or nearly complete lack of interest in sex or sexual activities; lack of libido or sex drive.

**sex′u·al as·sault′** An unlawful, intentional threat, attempt, or action to commit a sexual offense against another person against his or her will or without his or her consent.

**sex′u·al cy′cle** A series of anatomical and physiological changes that are associated with sex. *See* estrous cycle, menstrual cycle.

**sex′u·al de·vi·a′tion** Any sexual practice or behavior that differs from the generally accepted norms of a particular society; sexual variance. *See* perversion, paraphilia.

**sex′u·al de·vice′** Any device or object that is used to reach orgasm or to enhance sexual pleasure, such as an artifical penis (dildo), artificial vagina, vibrator, penile aid, or various masturbatory aids.

**sex′u·al dys·func′tion** A condition in which the ability to perform the sex act is impaired or unsatisfactory responses result during or following the act in one or both parties. *See* premature ejaculation, impotence, ejaculatory incompetence, vaginismus, dyspareunia, frigidity, sexual apathy.

**sex′u·al ha·rass′ment** The act of troubling or annoying someone with unwanted or unsought sexual talk, advances, or overt action. Sexual harassment is sometimes exerted by employers against employees, teachers against students, or by persons in authority against those subject to them.

**sex′u·al i·den′ti·ty** The psychic identity of a person with his or her own anatomical and physiological makeup.

**sex′u·al im′age·ry** Mental images, also called phantasms, that serve as erotic stimuli, especially during masturbation or coitus.

**sex′u·al in·ad′e·qua·cy** Inability to engage in coitus or to experience an orgasm. *See* impotence, frigidity, sexual dysfunction.

**sex′u·al in′fan·til·ism** The persistence of infantile or childish sexual traits or characteristics into adult life. Infantile sexuality differs from adult sexuality in that it is closely associated with ingestive and excretory functions and is often directed toward the mother.

**sex′u·al in′ter·course** Sexual contact between individuals of different sexes in which the penis is inserted into the vagina, usually followed by the discharge of semen; coitus, coition, copulation. *Legal terms* include carnal knowledge, sexual union, sexual congress. *Euphemisms* for "to have sexual intercourse" include to sleep with, to go to bed with, to have an affair with, to go into, to go all the way, to see someone, to do it, to know, to lie with, to make love, to be intimate with.

**sex′u·al in·ver′sion** 1. The selection of a person of one's own sex to be the object of sexual interest; homosexuality. 2. The possession of feelings, thoughts, and behavioral characteristics of the opposite sex.

**sex′u·al·ism** Excessive interest in sex and sexuality.

**sex·u·al′i·ty** 1. The condition of being a male or female person and manifesting the attitudes, feelings, and behavior of such an individual *as* a male or female. 2. Placing an emphasis upon sexual matters. 3. Interest and involvement in sexual activities. 4. The state or condition of readiness to engage in sexual activities.

**sex′u·al·ize** To endow with sex or make sexual; to develop sexual characters in a child.

**sex′u·al·ly** Related to or associated with sex.

**sex′u·al·ly trans·mit′ted dis·ease′** *Abbr.* STD. Any disease whose causative organism is *usually* transmitted from one person to another during heterosexual or homosexual intercourse or by intimate contact with the sex organs, mouth, or rectum. Sexually transmitted diseases include not only the common venereal diseases (gonorrhea, syphilis, herpes genitalis, chancroid, granuloma inguinale, lymphogranuloma venereum) but also acquired immune deficiency syndrome (AIDS), nonspecific urethritis (NSU), trichomoniasis, pediculosis (crabs), scabies, genital or venereal warts (condylomata acuminata), hepatitis B infection, molluscum contagiosum, candidiasis, and various oral, rectal, vaginal, and penile infections.

**sex′u·al ob′ject** 1. A person who possesses sexual attraction and arouses sexual desire in another and toward whom sexual activities are directed. 2. An animal or an inanimate thing that arouses sexual desire.

**sex′u·al per·ver′sion** A deviation or variation from what is generally accepted as normal sexual behavior. *See* paraphilia.

**sex′u·al pref′er·ence** The attraction to and selection of a particular sexual object, normally a person of the opposite sex. In deviant behavior the sexual object may be one of the same sex, a child, a close relative, an animal, or an inanimate object.

**sex′u·al re·la′tions** Sexual intercourse; coitus.

**sex′u·al re·pro·duc′tion** Reproduction that involves the union of sex cells (spermatozoa and ova).

**sex′u·al re·sponse′** The changes in the body that result from sexual stimulation, leading to orgasm, followed by the return to the usual unstimulated state. Four phases are included: excitement, plateau, orgasm, and resolution.

**sex′u·al se·lec′tion** A theory proposed by Charles Darwin to account for the existence of many secondary sexual characteristics that could not be accounted for by natural selection. It postulated that males possessing distinctive characteristics such as brilliant colors, conspicuous ornaments, and other attractive features would be more successful in being selected for mating than those lacking those characteristics, hence individuals possessing them would tend to survive and perpetuate them in future generations.

**sex′u·al va′ri·ance** See sexual deviation, sexual perversion, paraphilia.

**sex′y 1.** Excessively concerned with or involving sex, as in a *sexy* book. **2.** Erotically stimulating; suggestive of sex, as in *sexy* pictures. **3.** Sexually attractive and alluring; possessing qualities that tend to arouse sexual interest and desire in one of the opposite sex, as in a *sexy* man or woman.

**shack up with** *Slang.* **1.** To live together in a sexual relationship; cohabit. **2.** To live with one's mistress or paramour.

**shaft** *Slang.* **1.** The penis. **2.** A woman's body. **3.** The vagina. **4.** Female legs, especially when attractive sexually.

**shag** *Slang.* **1.** *v.* To chase after with the intent of coitus. **2.** *n.* A party at which (a) the participants look at and touch the sex organs of one of the opposite sex; (b) several men consort with one or two women.

**Shak′ti.** Also **Sak′ti** In Hinduism, the creative energy of a god such as Siva, personified as his female companion.

**shape′ly** Having a well-proportioned and pleasing shape, with special reference to a woman's body.

**she 1.** A female who has just been mentioned. **2.** A female person or animal. **3.** Anything considered feminine, as a ship.

**sheik** *Slang.* A handsome, romantic male lover; a ladies' man.

**shin′dig dan′cer** See go-go girl.

**shoot a bea′ver** *Slang.* To view the vulva.

**shoot off** *Slang.* To ejaculate.

**shoot one's wad** *Slang.* To ejaculate.

**short eyes** *Slang.* A child molester.

**short hairs** *Slang.* The pubic hairs.

**shot** An injection from a hypodermic needle, especially the injection of Depo-Provera, a long-acting contraceptive.

**shot′gun mar′riage** A marriage in which the father of the bride forces the groom, under threat of bodily harm, to marry his daughter, usually because of pregnancy.

**show** *Slang.* The exposure by a female of the breasts, thighs, or genitalia to males.

**shuck** *Slang.* To undress, especially quickly, in the presence of others.

**Si·a·mese′ twins** Parabiotic twins, q.v.

**sib 1.** A blood relation or kinsman; a sibling. **2.** Relatives or kinfolk considered collectively.

**sib′ling** One of two or more offspring of the same parents; a brother or sister.

**side ef·fect′** An effect of a drug or medicine other than that primarily intended, especially if harmful, disagreeable, or unwanted.

**SIECUS** Sex Information and Education Council of the United States, a nonprofit health organization dedicated to education and the dissemination of information about human sexuality.

**siff** *Slang.* Syphilis.

**sign** An objective manifestation of a disease, pathological disorder, or physiological activity. *Compare* symptom.

**signs of preg'nan·cy** *See* pregnancy, signs of.

**Si·las'tic** Trade name for a silicone material having the properties of rubber. It is biologically inert and tolerated by body fluids and tissues.

**sil'ver ni'trate** AgNO₃, used medicinally as an antiseptic. It is employed as a prophylactic against gonococcal infection of the eyes of a newborn infant.

**sin** A transgression against a religious or moral law, especially when willful and deliberate.

**sin'gles** Term applied to members of a group that includes unmarried individuals, widows, widowers, divorced persons, or any person not living with spouse or family.

**sin'gles bar** A tavern or bar patronized by unaccompanied individuals, frequently with the aim of finding (often sexual) companions.

**sire** 1. *n.* A father or male ancestor. 2. *n.* The male parent of an animal, especially a domestic animal. 3. *v.* To beget or procreate, with reference to domestic animals.

**sis'sy** *Slang.* A cowardly, weak, or effeminate man or boy.

**sis'ter** 1. A female with reference to any other offspring of the same parents. 2. Term commonly applied to a fellow woman, or female friend or companion. 3. *Slang.* An effeminate male homosexual.

**sis'ter-in-law** The sister of one's spouse, the wife of one's brother, or the wife of one's spouse's brother.

**sit·u·a'tion** *Euphemism* for an unwanted pregnancy, especially when the baby is to be adopted.

**six·ty-eight'** *Slang.* Humorous variation on *sixty-nine* (q.v.), in which the second fellatio of the mutual two is omitted, with the words "I owe you one."

**six·ty-nine'** *Slang.* Simultaneous fellatio and/or cunnilingus between individuals of the same or opposite sexes, the term being derived from the position of the participants; also *soixante-neuf* (69).

**Skene's glands** The paraurethral glands, two small glands that open within the urethral orifice in a female. They are homologous to the prostate gland in the male.

**skin** *Slang.* A condom.

**skin-flick** *Slang.* A pornographic movie.

**skin'ny dip** To swim in the nude.

**skiv'vies** Men's underwear consisting of a cotton T-shirt and shorts.

**slat'tern** An untidy, slovenly woman; a slut.

**slave** *Slang.* In S-M, a masochistic partner.

**sleep with** *Euphemism* for "to have sexual intercourse with."

**slut** 1. A slovenly, unclean woman. 2. An immoral woman; a prostitute.

**S-M** Simultaneous sadism and masochism.

**smack** *Slang.* To kiss with a loud sound.

**smeg'ma** 1. A foul-smelling, cheese-like substance, a product of Tyson's glands, that accumulates between the prepuce and the glans penis in uncircumcised males. 2. A similar substance that may accumulate beneath the prepuce of the clitoris in a female.

**smooch** *Slang.* To kiss, pet, neck, or caress.

**smut** Indecent or obscene stories, pictures, or photographs, or any material that violates commonly accepted standards of morality; pornographic material.

**smut ped'dler** A person engaged in the distribution and sale of pornographic material.

**snake** *Slang.* 1. A treacherous or deceitful person, especially a male who deceitfully pursues young females. 2. A deceitful or fickle girl or woman.

**snake pit** *Slang.* 1. A woman's vagina. 2. A place for the treatment of mental disorders.

**snatch** *Slang.* 1. A woman's crotch. 2. The vagina.

**snuff film** *Slang.* A film whose major object is the depiction of a murder; frequently a pornographic film culminating in the murder of a woman.

**snug'gle** To lie close to and press closely together.

**soc'ial dis·ease'** A venereal disease.

**sod'om·ite** A person who commits sodomy.

**sod'o·my** 1. Anal intercourse, especially between two males. *See* pederasty. 2. Anal intercourse or oral sex between members of opposite sexes. 3. Bestiality. *See* crime against nature.

**soi·xante-neuf'** *French* for sixty-nine (69), q.v.

**so'ma** All the cells and tissues of the body, except the reproductive or germ cells.

**so·mat'ic** Of or pertaining to the soma or nonreproductive cells.

**son** A male child or a male adopted child.

**son-in-law** The husband of one's daughter.

**son·og'ra·phy** The use of ultrasound (high-frequency sound waves) directed toward an object (body organ or cavity), which echoes back to form a picture. It is used to show the exact position of the fetus and placenta.

**so'ror·ate** A marriage custom in which a widower marries his deceased wife's sister. *Compare* levirate.

**soul kiss** A deep, passionate, open-mouth kiss; a French kiss.

**spade queen** *Slang.* A homosexual who prefers black partners.

**Span'ish col'lar** *See* paraphimosis.

**Span'ish fly** *See* cantharides.

**spay** To remove the ovaries.

**spec'u·lum** An instrument for enlarging an opening of a cavity of the body in order to permit visual examination. It is used especially in examination of the vagina.

**spend a pen'ny** *Slang.* To urinate involuntarily when sexually aroused.

**sperm** *pl.* **sperm** or **sperms** 1. A male gamete or spermatozoon. 2. Spermatozoa considered collectively. 3. The seminal fluid or semen.

**sper·mat'ic** Of or pertaining to sperm or a sperm-producing structure.

**sper·mat'ic cord** A cord that, in mammals, suspends each testis within the scrotum. In humans, it extends from the deep inguinal ring to the testis and contains blood and lymph vessels, nerves, and the vas deferens.

**sper·mat'ic flu·id** The semen.

**sper'ma·tid** A male germ cell immediately before its transformation into a flagellated, mature spermatozoon.

**sper·ma·to·ci'dal** Destructive to and capable of killing spermatozoa.

**sper'ma·to·cide** An agent that kills or destroys spermatozoa.

**sper'ma·to·cyte** One of two generations of cells (primary and secondary spermatocytes) which occur in the development of spermatozoa. *See* spermatogenesis.

**sper·ma·to·gen'e·sis** The process that takes place within a seminiferous tubule by which functional, haploid spermatozoa are formed from diploid spermatogonia. Cells involved include spermatogonia, primary and secondary spermatocytes, spermatids, and spermatozoa. In the process meiosis occurs, in which the number of chromosomes is reduced from the diploid number (46 in humans) to the haploid number (23), and motility is acquired through the development of a tail or flagellum.

**sper·ma·to·gen'ic** Of or pertaining to spermatogenesis.

**sper·ma·to·go'ni·um** A primordial germ cell that gives rise to primary spermatocytes.

*See* spermatogenesis.

**sper·ma·tor·rhe'a** The involuntary discharge of semen in the absence of an orgasm.

**sper·ma·to·zo'on,** *pl.* **sper·ma·to·zo'a** A mature male germ cell capable of fertilizing an ovum. Spermatozoa, which are produced within the seminiferous tubules of the testes, average 55 to 65 microns in length. Each consists of a *head,* a *body* or *middle piece,* and a *tail* or *flagellum.* They swim at a rate of 1 to 3 mm per minute.

**sper·ma·tu'ri·a** The presence of spermatozoa in urine.

**sperm bank** A place where spermatozoa are preserved for long periods of time by subjecting them to an extremely low temperature (-196° C.).

**sperm cell** A spermatozoon.

**sperm duct** Any of the ducts that convey spermatozoa from the testis to the external urethral orifice. These include the ductus epididymis, ductus (vas) deferens, ejaculatory duct, and the penile portion of the urethra. When unqualified, it generally means the vas deferens.

**sper·mi·ci'dal** Destructive to spermatozoa; spermatocidal.

**sper'mi·cide** An agent that kills spermatozoa; a spermatocide.

**sper·mi·o·gen'e·sis** The transformation of spermatids into mature functional spermatozoa.

**sphinc'ter** A muscle surrounding an opening which, upon contraction, closes the opening.

**spinn'bar·keit** *Abbr.* SBK. A term used to designate the elasticity of cervical mucus.

**spi'ral ar'ter·ies** Arteries which supply the functional zone of the uterine endometrium.

**spi'ro·chete** A flexible, spiral-shaped bacterium such as the organism, *Treponema pallidum,* that causes syphilis.

**sponge** A light, fibrous mass of absorbent material inserted into the vagina for contraceptive purposes. It is intended to function as an occlusive device preventing sperm from entering the cervix. A spermicidal agent is sometimes added to the sponge.

**spoon po·si'tion** A rear-entry coital position in which the male and female participants lie on their sides in semifetal position, a position recommended to improve coitus interruptus.

**sport'ing house** *Slang.* A brothel.

**spot'ting** Break-through bleeding that occurs in some women when taking oral contraceptives.

**spouse** One of a married pair; a husband or wife.

**squeeze tech·nique'** A technique for treating premature ejaculation in which, during coitus, when the male senses an impending ejaculation, he withdraws and the female applies pressure on the penis, stopping ejaculation.

**sta'ble** *Slang.* A group of prostitutes managed by a pimp.

**stag** 1. *n.* An adult male deer or the male of various other mammals. 2. *n.* A male animal castrated after reaching sexual maturity. 3. *n.* A man who attends a social affair without a female partner. 4. *adj.* For men only, as a *stag* party.

**stag film** A pornographic film shown at stag affairs.

**stag par'ty** A party attended by men only, especially one given for a prospective bridegroom.

**stal'lion** 1. An uncastrated adult male horse, especially one used for breeding purposes. 2. *Slang.* A male with a large penis.

**sta'tus sex** Among various mammals, a display of sexual activity to indicate dominance or submission.

**stat'u·to·ry rape** Sexual intercourse between a man and a girl below the age of consent, even though consent may be given.

**STD** Sexually transmitted disease.

**ste·a·to·py'gi·a** The excessive accumulation of fat on or about the buttocks.

**ste·no'sis** A narrowing or constriction of a tube, duct, or opening.

**step'child** A child of one's wife or husband by a previous marriage.

**step'daugh·ter** A daughter of one's wife or husband by a previous marriage.

**step'fath·er** The husband of a person's mother by a subsequent marriage.

**step'moth·er** The wife of a person's father by a subsequent marriage.

**step out** *Slang.* To be sexually unfaithful.

**step'par·ent** A stepfather or stepmother.

**step'son** A son of one's husband or wife by a previous marriage.

**ster'ile** 1. Incapable of producing offspring; infertile; not fertile. 2. Aseptic; free of microorganisms.

**ste·ril'i·ty** The state of being sterile; inability to reproduce.

**ster·i·li·za'tion** The process by which an individual is rendered incapable of reproducing. In the *male,* this is accomplished by castration (orchiectomy) or vasectomy; in the *female,* by castration (ovariectomy), removal of the uterus (hysterectomy), removal of uterine tubes (salpingectomy), or tubal ligation. In both sexes, it may result from irradiation or from the effects of certain drugs.

**ster'oid** Any of a group of fat-soluble compounds that include the sterols, bile acids, D vitamins, and gonadal and adrenocortical hormones.

**stew** *Slang.* 1. A brothel. 2. A district housing many brothels.

**stil·bes'trol** *Synonym* for diethylstilbestrol (DES).

**still'birth** 1. The delivery of a dead child. 2. A dead fetus after delivery.

**still'born** Born dead.

**stool** *See* feces.

**straight** *Slang.* 1. Sexually normal; heterosexual, not homosexual. 2. Not a drug addict.

**straight ar'row** *Slang.* A person who is heterosexual.

**straight sex** *Slang.* Male-female sex.

**streak'ing** Running naked through a public place.

**street'walk·er** A female prostitute, especially one who solicits customers in a public place.

**stress in·con'ti·nence** Loss of urine resulting from any activity that increases intra-abdominal pressure, such as laughing, coughing, or straining. It is due to partial incompetence of the urinary sphincter.

**stric'ture** The abnormal narrowing of the lumen of a tube, duct, or cavity.

**strike out** *Slang.* To be unsuccessful in persuading a woman to engage in sexual intercourse.

**string** *See* tail (2).

**strip** To divest a person or oneself of clothing; to make bare or naked.

**strip'per** A woman who engages in striptease; a female exhibitionist.

**strip'tease** A theatrical performance, usually in a burlesque show, in which a woman removes her apparel, a piece at a time, usually to the accompaniment of music.

**strip'teas·er** A stripper; one who engages in striptease.

**strum'pet** A prostitute; a whore.

**stud** 1. (a) One of a group of animals that are kept for breeding purposes, or (b) a place where they are kept. 2. A male animal, especially a stallion, whose services are used for breeding, usually for a fee. 3. *Slang.* (a) A male who provides heterosexual coital service for compensation; (b) a young, carefree male interested especially in coital activities; (c) a butch.

**sub·fer·til'i·ty** Condition in which a couple is theoretically able to conceive but conception does not readily occur; relative sterility.

**sub·in·cis'ion** The making of a permanent opening into the urethra on the underside of the penis, an initiation rite practiced by primitive peoples in Australia and other parts of the world.

**sub·li·ma'tion** The process by which unacceptable instinctual demands are channeled into socially acceptable activities, e.g. intellectual and physical activities serving vas a substitute outlet for sexual activities.

**sub·pre·pu'tial** Beneath the prepuce.

**suc'cu·bus** 1. An evil spirit or demon in female form that was thought to visit men during the night and have sexual intercourse with them. *Compare* incubus. 2. A demon or fiend. 3. A prostitute.

**suck** 1. To draw into the mouth through the action of the lips, cheeks, and tongue. 2. *Slang*. To engage in fellatio or cunnilingus.

**suck'le** To nurse and draw milk from the breast.

**suck'ling** An infant or young animal that nurses from the mammary gland; one that has not been weaned.

**su'gar dad'dy** *Slang*. A sweet papa, q.v.

**suit'or** A man who courts or woos a woman.

**sunk'en trea'sure** *Slang*. One who is a latent homosexual.

**su·per·fec·un·da'tion** A rare condition in which twins have different fathers.

**su·per·fe·ta'tion** The development of a second fetus after one has already started development within the uterus.

**sup·pos'i·to·ry** A medicated solid body, usually conical in shape, prepared for introduction into one of the body orifices, especially the anus or vagina. It is used for therapeutic or contraceptive purposes.

**sur'ro·gate** A substitute.

**sweet'heart** 1. A term of endearment usually applied to one of a pair of lovers. 2. A lovable person; one who loves and is loved, especially one of an engaged couple.

**sweet'ie** *Informal* for sweetheart.

**sweet'ie pie** *Slang*. 1. A sweetheart. 2. An attractive, young girl.

**sweet ma'ma** *Slang*. A female lover; an extremely sensual young woman.

**sweet man** *Slang*. A male lover; an extremely sensual young man.

**sweet pa'pa** *Slang*. A sugar daddy, a male lover well provided with money who is generous to a female lover who provides sex and companionship.

**sweet pea** *Slang*. A lover of either sex.

**swing** *Slang*. 1. To be sexually promiscuous. 2. To swap partners.

**swing'er** A person who engages in swinging.

**swing'ing** Group sex in which two or more couples (married or unmarried) engage in indiscriminate sexual relations with others of the group.

**swish** *Slang*. An effeminate male homosexual.

**swish'y** *Slang*. Highly effeminate, with reference to males.

**switch hit'ting** *Slang*. Engaging in both heterosexual and homosexual activity.

**syb·a·rit'ic** Devoted to pleasure and luxurious and voluptuous living.

**symp'tom** Any subjective evidence of a disease or bodily disorder. *Compare* sign.

**symp'to·ther'mal meth'od** A method of natural family planning based on charting body temperature and observing the pattern of cervical mucus by which fertile and infertile periods during a woman's menstrual cycle can be determined.

**syn'drome** A group of symptoms and signs which, when considered together, characterize a pathological condition.

**syn·er·gam'ous mar'riage** Double bigamy; a type of marriage in which a man with a legal spouse marries (with a ceremony but not legally) a woman with a legal spouse. Sometimes called *trigamy*.

**syn'ga·my** The union of an egg and a spermatozoon; fertilization.

**syn'or·chism** A condition in which the two testes are fused, forming a single mass which may be present in the scrotum or in the abdomen.

**syph, siff** *Slang.* Syphilis.

**syph'i·lis** A chronic, infectious venereal disease caused by a spirochete, *Treponema pallidum,* and occurring in three stages, primary, secondary, and tertiary. It may be of congenital origin. Also called *lues. See* chancre, gumma, tabes dorsalis.

**sy·ringe'** or **sy'ringe** A device by which a fluid can be injected into or withdrawn from a vessel or cavity. *See* hypodermic, fountain syringe.

**ta'bes dor·sal'is** Syphilitic involvement of the spinal cord in which degeneration of the motor and sensory pathways occurs. It occurs in tertiary syphilis and is characterized by muscular incoordination, loss of reflexes, and marked sensory disturbances, including intense pain. Also called *locomotor ataxia.*

**ta·boo'** A prohibition or interdiction of an action or activity. It may be due to social custom, religion, tradition, superstition, or, according to Freud, it may be an unconscious part of an individual's psychic life.

**tack'i·ness** Term applied to the cohesiveness of cervical mucus, especially the force required to pull it apart.

**tac'tile** Pertaining to or involving the sense of touch.

**tad** *Informal.* A small boy.

**tail** 1. The posterior, vibratile portion of a spermatozoon; a flagellum. 2. A stringlike structure, usually of nylon, attached to an IUD. After insertion of an IUD, the tail protrudes from the cervix into the vagina and its presence is an indicator that the IUD is still in place. It also assists in the removal of the IUD when necessary. Also called *string* or *strings.* 3. *Slang.* (a) Coitus, commonly referred to as a *piece of tail.* (b) The vagina or a woman considered as a sexual object. (c) The buttocks or the rear end of a person.

**tam'pon** A cylindrical plug of absorbent material placed within the vagina to absorb menstrual fluid.

**tar'get or'gan** An organ that is affected and acted on directly by a hormone.

**tart** *Slang.* A prostitute; a promiscuous girl or woman.

**tea'room** *Slang.* A place, usually a public rest room, that serves as a locale for quick homosexual encounters.

**tease** 1. *v.* To annoy, playfully mock, make fun of, or torment. 2. *n. Slang.* A girl or woman who invites a man's attentions and favors but has no intention of returning them.

**teas'er** 1. A woman who stimulates a man sexually but avoids or refuses to engage in sexual intercourse. 2. A male animal such as a bull or stallion used for identifying females in heat or for preparing them to accept another male. 3. A cow in heat used to stimulate a bull so that semen can be obtained for artificial insemination.

**teat** An elongated protuberance through which milk is withdrawn from an udder or mammary gland. *Compare* nipple, tit.

**teen, teen'-ag·er** A person between the ages of thirteen and nineteen inclusive; an adolescent.

**te·leg'o·ny** An erroneous belief that mating with one particular male will influence future progeny even though the future progeny may actually be fathered by another male.

**tem'per·a·ture** *See* basal body temperature, thermal shift.

**ter'as,** *pl.* **te'ra·ta** A fetal monster or monstrosity; a grossly deformed fetus.

**ter'a·tism** An anomalous formation or developmental anomaly; a monster.

**te·rat'o·gen** An agent or factor that causes or induces the development of a physical defect in an embryo or fetus. *See* diethylstilbestrol (DES), rubella, thalidomide.

**ter·a·to·gen'ic** Tending to produce abnormalities in development.

**ter·a·tol'o·gy** The study of abnormal development and abnormalities.

**term** 1. A period of time. 2. In *obstetrics,* the period of pregnancy (nine months) terminating at the time of delivery; the expected time of delivery.

**ter'ma·gant** A violent, turbulent, brawling, shrewish woman.

**ter'mone** A sex-determining hormone.

**Tes-tape** A paper used in certain fertility test kits to determine the amount of glucose, a simple sugar, in cervical mucus. The amount of glucose in cervical mucus increases markedly at time of ovulation. When the mucus is applied to the yellow Tes-tape paper, the color changes to a dark blue at the time of ovulation. Prior to or following ovulation, lesser shades of blue and green are noted, indicating a lessened concentration of glucose.

**tes'ti·cle** A testis, q.v.

**tes'tis,** *pl.* **tes'tes** A male reproductive organ or gonad, which, after puberty, produces spermatozoa. Eash testis is composed principally of *seminiferous tubules* lined with germinal epithelium from which spermatozoa arise. Between the tubules are the *interstitial cells (of Leydig),* which are the principal source of androgens (male hormones). The testes develop within the body cavity but before birth normally descend into the scrotum. *See* cryptorchism.

**tes·tos'ter·one** The principal androgen or male hormone, produced principally by the interstitial cells of the testes. It is also produced by the ovaries in a female and the adrenal cortex in both sexes. It is a steroid and responsible for the development of secondary sexual characteristics in the male and for the sexual drive or libido in both sexes.

**test-tube ba'by** *Euphemism* for a baby that develops from an egg that is fertilized externally. An ovum is secured from an ovary, and spermatozoa obtained from the husband or a donor are added to it in a test tube or glass container, a process called *in vitro fertilization.* Following fertilization, the developing zygote or blastocyst is placed in a woman's uterus and, in some cases, development proceeds to full term.

**tha·li'do·mide** A sedative or hypnotic widely used in Europe in the 1960's. When taken during early pregnancy it was found to be the cause of serious developmental anomalies, especially amelia and phocomelia, q.v.

**the·lal'gia** Pain in the nipple.

**the·lar'che** The beginning of the development of the breasts at puberty.

**ther·a·peu'tic** Of or pertaining to the treatment of disease, pathological conditions, or behavioral disorders; possessing healing or curative powers.

**ther·a·peu'tics** 1. The science and art of healing. 2. The branch of medicine that deals with the treatment of disease.

**ther'a·py** The treatment of illness, disease, or disabilities; therapeutics. *See* sex therapy.

**ther'mal shift** The change in basal body temperature (BBT) that accompanies ovulation. A slight drop, followed by a rise of from 0.4° to 0.8° F., is an indication that ovulation has occurred. The elevation of body temperature occurs within 24 hours after ovulation and persists until the next menstrual period.

**thing** *Slang.* The penis.

**third base, get'ting to** *Slang.* Petting below the waist.

**third sex** *Slang.* Homosexuals.

**three-way girl** *Slang.* A prostitute who engages in vaginal, oral, and anal intercourse.

**throm·bo·em'bo·lism** A condition in which a blood vessel becomes occluded by a thrombus originating at another site.

**throm·bo·phle·bi'tis** Inflammation of veins accompanied by the formation of thrombi. *See* thrombus.

**throm·bo′sis** The formation of a thrombus.

**throm′bus,** *pl.* **throm′bi** A blood clot, consisting of platelets, fibrin, and blood corpuscles, that forms within the heart or a blood vessel and obstructs the flow of blood.

**thrush** Oral candidiasis.

**tin′e·a cru′ris** *See* jock itch.

**tit, titty** 1. A teat. 2. *Vulgar.* A female breast.

**tit′il·late** To excite by stroking lightly.

**tit·il·la′tion** The induction of a pleasurable feeling by stroking lightly.

**TKD** Tokodynamometer. *See* tokography.

**TKG** Tokograph. *See* tokography.

**toad** *Slang.* An ugly and disgusting person, especially a coarse, vulgar woman.

**to′cus** Childbirth, labor, parturition.

**toe queen** *Slang.* A homosexual foot fetishist.

**to·kog′ra·phy** The making and interpretation of graphic recordings or *tokographs* (TKG). The TKG is made by an instrument, a *tokodynamometer* (TKD), which records the amplitude, duration, and frequency of uterine muscular contractions during labor.

**to′kus** *Slang.* The rear end or buttocks; the rectum.

**to·ma′to** *Slang.* A sexually attractive girl or young woman.

**tom′boy** An active, lively girl who behaves like a boy.

**tom′cat** *Slang.* 1. *n.* A man who goes whoring; a woman chaser. 2. *v.* To seek a sexual partner or go whoring.

**tong** *Slang.* The penis.

**tool** *Slang.* The penis.

**top′less** Lacking a covering for the breasts.

**top′less bar** A tavern or bar in which the waitresses serve the customers with breasts exposed.

**tot** A small child.

**to′tal sex′u·al out′let** Kinsey's term for the total number of orgasms achieved during an average week through masturbation, sex dreams, petting or necking, coitus, homosexual activities, or animal contacts over a five-year period.

**to′tem** A natural object or a living organism, as a plant or animal, or a representation of such, that serves as an emblem of a family, clan, or tribe and as a reminder of its ancestry, or an object to which the group considers itself related.

**tox·e′mi·a** A condition in which the blood contains poisonous substances.

**tox·e′mi·a of preg′nan·cy** *See* eclampsia.

**trade** *Slang.* 1. An ambisexual male. 2. A homosexual seeking a companion. 3. A man or woman as a commercial sex object. 4. A very masculine hustler.

**train** *Slang.* A series of males who have intercourse with a woman one after the other.

**tramp** *Slang.* A woman of low morals; a promiscuous girl or woman; a prostitute.

**tran·ex·am′ic ac′id** Aminoethylcyclocarboxylic acid (AMCA), an antifibrinolytic substance that prevents the breakdown of blood in the uterine endometrium.

**tran′qui·li·zer** A drug that acts on the central nervous system, having a quieting or calming effect on the emotional state of a person. Tranquilizers reduce mental tension and anxiety without affecting normal mental activity.

**trans·cer′vi·cal** Through the cervix of the uterus.

**trans·sex′u·al** *Abbr.* TS. A person, usually a male, who either sincerely believes that he is, or else wishes to be, a person of the opposite sex, and feels compelled to dress and act as a member of that sex. Transsexuals believe that they are males in female bodies or females in male bodies. The belief is so strong that some transsexuals resort to a sex-change operation to be transformed into individuals of the opposite sex. *See* sex-change operation, sex reassignment.

**trans·sex′u·al·ism** Sex-role inversion, a condition in which an individual's sex role and gender identity are incompatible. A male possesses XY chromosomes and anatomically possesses male sex organs both internally and externally but he rejects his maleness and wishes to live the life of a female, emotionally and sexually. In a similar way, a female transsexual rejects her femaleness and wishes to be a male.

**trans·u·re′thral** Through the urethra.

**trans·vag′i·nal** Through the vagina.

**trans·ves′tism** The act of dressing as, acting as, and sometimes adopting the sex role of, one of the opposite sex; male and female impersonation.

**trans·ves′tite** A person who practices transvestism. *See* female impersonator.

**trau′ma** 1. A wound or injury, especially one produced by a physical or mechanical agent. 2. In *psychiatry*, a profound emotional shock that results in lasting or permanent psychological pathology.

**tra·vail′** The labor and pain of childbirth.

**tread** To copulate with a female; said of a male bird.

**Tre·po·ne′ma pal′li·dum** The organism, a spirochete, that causes syphilis.

**trib′ade** A lesbian, especially one who assumes the role of a male.

**trib′a·dism** Lesbianism, especially the practice of two women who attempt to simulate heterosexual sexual intercourse.

**tri·cho·mo·ni′a·sis** A condition resulting from the presence of a flagellate protozoan, *Trichomonas vaginalis,* in the vagina. It is a common cause of leukorrhea. The infection may also occur in the urogenital passageways of males.

**trick** *Slang.* 1. A male customer of a prostitute. 2. A partner in a homosexual encounter.

**trig′a·my** *See* synergamous marriage.

**tri′mes·ter** A period of three months; one third of the gestation period or pregnancy.

**tri′o·lism** *See* troilism.

**tri′so·my** A condition in which all the chromosomes of an individual are in pairs except one, which is triploid, as in Down's syndrome, in which the chromosome number is $2n + 1$ (47 instead of 46).

**tro′car** A surgical device for puncturing a cavity and removing the fluid contained within it. It consists of a hollow tube or canula, within which is a sharp-pointed perforator.

**troi′lism** Sexual behavior involving three persons (two females and a male or two males and a female), usually combining coitus with oral-genital stimulation. Also *triolism.*

**trol′lop** A slovenly, unkempt woman; an immoral woman; a strumpet.

**troth** A pledge of fidelity or faithfulness; a promise to marry.

**trous′seau** A bride's clothes and accessories assembled for her marriage.

**tryst** A meeting at a certain time and place, especially a secret meeting of lovers.

**TS** Transsexual.

**tu′bal li·ga′tion** The tying off of both uterine tubes, a method of sterilization.

**tu′bal preg′nan·cy** An ectopic pregnancy in which the embryo develops within the uterine tube.

**tube** A hollow cylindrical structure, especially the uterine or fallopian tube; the oviduct.

**tu·bec′to·my** Excision of a tube or a portion of a tube, especially the uterine tube; salpingectomy.

**tube-lock′ing ef·fect** The slowing down of the passage of a fertilized egg through the uterine tube, as results from the action of DES.

**tu·mes′cence** The state or condition of being swollen.

**tu'ni·ca mu·co'sa** A mucous membrane, q.v.

**tu'ni·ca va·gi·nal'is** A two-layered sac that invests the testis and epididymis.

**Tur'ner's syn'drome** General dysgenesis in a female resulting from the absence of a sex chromosome, females being XO. True ovaries are absent and adults are of short stature with infantile genitalia and accessory sex structures.

**turn on** Slang. 1. To stimulate or excite sexually. 2. To be sexually appealing to (someone).

**TV** Abbr. for 1. Trichomonas vaginalis. 2. A transvestite. 3. Television.

**twat** Slang. 1. The vagina. 2. A young female considered as a sexual object.

**tweed** Slang. A girl or young woman.

**twin** One of two offspring brought forth at the same birth. See fraternal twins, identical twins.

**twink, twink'ie** Slang. An effeminate male, especially an effeminate homosexual.

**two-time** To deceive one's sweetheart or lover; to be unfaithful.

**two-way girl** Slang. A prostitute who engages in genital and oral sex.

**tyke** A small child.

**Ty'son's glands** Modified sebaceous glands located on the inner surface of the male prepuce and on the glans penis. They secret smegma.

**ul'tra·sound** Ultrasonic sound; sound waves with a frequency exceeding 20,000 cycles per second and inaudible to the human ear. See sonography.

**um·bil'i·cal cord** A long cylindrical structure connecting the fetus with the placenta and containing two arteries, a single vein, and a mucous tissue called Wharton's jelly. The arteries carry blood from the fetus to the placenta; the vein from the placenta to the fetus.

**um·bil'i·cus** The navel, q.v.

**un·at·tached'** Not engaged or married; uncommitted.

**un·born'** Not yet born or in existence.

**un·chaste'** Not chaste or virtuous; immodest, impure, lascivious.

**un·de·scend'ed tes'tes** Cryptorchism, q.v.

**un·de·vel'oped** Not adequately developed; immature, as small breasts.

**un'din·ism** The condition in which water, urine, or micturition arouses erotic sensations. See urophilia.

**un·dress'** 1. v. To remove the clothing; to strip or disrobe. 2. n. The state of nakedness.

**un·dressed'** Not fully dressed; naked, nude.

**un·e·rot'ic** Not erotic; not directed toward sexual pleasure or gratification; not affected by sexual desire.

**un'ion** The act of uniting or joining together two persons, as in marriage or sexual intercourse.

**u·ni·sex'u·al** 1. Possessing sex organs of one sex only; dioecious. 2. Of, pertaining to, or involving one sex only.

**un·mar'ried** Not legally joined together in marriage.

**un·nat'ur·al sex** An old term applied to any sexual activity that was not for the purpose of procreation. Sex acts considered to be "unnatural" generally included masturbation, homosexual acts, coitus interruptus, various measures of birth control, acts of sodomy, and others. See aberration, crime against nature, paraphilia, perversion.

**un·sex** To deprive of sexual power; to desexualize; to render impotent.

**u'ra·nism** Male homosexuality; urningism.

**u'ra·nist** A male homosexual; an urning.

**u·re'ter** One of two tubes that carry urine from the kidney to the bladder.

**u·re'thra** A tube that leads from the bladder to the outside. In the female, it is about 1½ inches in length and conveys only urine. Its external orifice, located within the vestibule, lies between the clitoris and the vaginal orifice. In the *male,* the urethra is of variable length and consists of three portions: the *prostatic,* within the prostate gland; *membranous,* which passes through the urogenital diaphragm; and *cavernous,* which is contained within the corpus cavernosum of the penis. In the male it conveys both urine and semen.

**u·re·thri'tis** Inflammation of the urethra, commonly the result of gonorrheal infection.

**u·re'thro·cele** 1. Prolapse of the female urethra through the urinary meatus. 2. Protrusion of the wall of the urethra into the vaginal wall.

**urge in·con'ti·nence** A condition in which, when the bladder reaches a certain capacity, leakage of urine occurs.

**u'ri·nar·y sys'tem** The system concerned with the formation and excretion of urine. It consists of the kidneys, ureters, urinary bladder, and urethra.

**u'rine** The clear, amber-colored fluid excreted by the kidneys. It contains water, salts (chlorides, phosphates, sulfates) and various other substances, such as pigments, hormones, and waste products of metabolism.

**u·ri·no·gen'i·tal** Urogenital.

**ur'ning** A male homosexual.

**ur'ning·ism** Male homosexuality.

**u·ro·gen'i·tal** Of or pertaining to the urinary and reproductive systems.

**u·ro·lag'ni·a** Sexual arousal associated with urine and urination. *See* urophilia.

**u·rol'o·gist** A physician skilled in urology.

**u·rol'o·gy** The scientific study of urine and the organs in which it is produced, stored, and excreted. It also includes the diagnosis and treatment of diseases and disorders of the urinary tract in the female and urogenital tract in the male.

**u·ro·phil'i·a** A condition in which the sight and smell of urine or the sight or sound of a person urinating is essential for the arousal of erotic feelings and the induction of an orgasm. Also called *undinism.*

**u'ter·ine** Of or pertaining to the uterus.

**u'ter·ine cramps** Pain associated with menstruation; dysmenorrhea. The primary cause or causes are unknown, but possible causative factors include uterine muscle spasm, forceful contractions, and vascular changes possibly involving ischemia. Prostaglandins, which increase at menstruation, are thought to be a precipitating factor.

**u'ter·ine my·o'ma** *See* fibroid.

**u'ter·ine sound** A graduated device for inserting into the uterus to measure the extent of its cavity.

**u'ter·ine tube** The fallopian tube or oviduct, one of two tubes that extend laterally from the upper portion of the uterus, each opening into the body cavity near the ovary. They average about four inches in length and convey ova discharged from the ovary toward the uterus. If spermatozoa have been received during coitus, fertilization usually occurs within the tube. *See* fimbria, ostium.

**u'ter·us** A pear-shaped female organ located in the pelvic cavity, consisting of a lower *cervix* and an upper *body,* the two separated by a constriction, the *isthmus.* Its walls consists of three layers: the innermost mucosa or *endometrium,* the middle muscle layer or *myometrium,* and the outermost serosa or *perimetrium.* It receives through the uterine tube or oviduct the product of conception (morula or blastocyst) and within it, the embryo or fetus develops. It is the source of menstrual fluid discharged monthly through the vagina. Also called *womb.*

**u., an·te·flex'ion of** One that is bent forward in a permanent position.

**u., pro·lapsed'** One that protrudes into the vagina and sometimes projects outside the vulva.

**u., ret′ro·flexed** One that is bent backward in a permanent position.

**u., ret·ro·ver′sion of** Condition in which the uterus is tipped backward and is lacking its normal curve.

**ux·o′ri·al** Of, pertaining to, or characteristic of a wife.

**ux·o′ri·ous** Excessively devoted to or overly submissive to one's wife.

**vac·u·rette′** A suction canula or curette introduced into the uterus for aspiration of the products of conception.

**vac′u·um as·pi·ra′tion** Vacuum curettage, q.v.

**vac′u·um cu·ret·tage′** Uterine aspiration, a method used for inducing an abortion, in which suction is used for the removal of the products of conception.

**va·gi′na** A tube that extends from the cervix of the uterus to the vulva, its external orifice lying in a space, the *vestibule,* located between the *labia minora.* During coitus it receives the penis of the male, and during parturition it serves as a birth canal. It also functions in the discharge of menstrual fluid. The vagina averages three to four inches in length, and its wall consists of three layers, a *mucous layer* of stratified epithelium, which is devoid of glands, a *muscular layer* of longitudinal and circular fibers, and a highly vascular *fascial coat* of connective tissue.

**va·gi′na bar′rel** The cylindrical vagina.

**vag′i·nal a·tre′si·a** A congenital condition in which the vagina is absent or is a solid structure.

**vag′i·nal hood** A circular fold in the upper vagina into which the cervix containing adenosis appears to merge, seen in females exposed to DES during fetal development.

**vag′i·nal ring** A Silastic ring containing a progestogen that is inserted into the vagina for contraceptive purposes. *See* Provera ring.

**vag·i·nis′mus** Involuntary contraction or spasm of the vaginal muscles or muscles bordering the vaginal orifice which prevents entry of the penis in coitus. If penetration has occurred, in rare cases withdrawal may be impeded. *See* dyspareunia.

**vag·i·ni′tis** Inflammation of the vagina, usually accompanied by a bloody discharge. *See* leukorrhea.

**vag·i·no·plas′ty** The construction or reconstruction surgically of a vagina.

**vamp** *Slang.* 1. *n.* A sexually attractive and seductive woman. 2. *v.* To seduce a man through the use of sex appeal.

**vam′pire** One who preys mercilessly on others, especially an unscrupulous woman who uses her sex appeal to seduce and exploit men.

**va·nil′la bar** A gay bar frequented by effeminate gay males.

**var′i·co·cele** Excessive dilatation of veins within the spermatic cord leading from the testis, a common cause of sterility in males.

**vas** A tube or duct.

**vas′cu·lar·ized** Well supplied with blood vessels.

**vas def′er·ens,** *pl.* **va′sa de·fe·ren′ti·a** The ductus deferens, a duct that conveys sperm from the testis (more specifically, the epididymis) to the ejaculatory duct and thence to the urethra.

**vas·ec′to·my** Ligation or cutting of the vas deferens, a method of sterilization in the male.

**vas oc·clu′sion** Blockage of the vas deferens.

**vas·o·con·ges′tion** *See* congestion.

**vas·o·con·stric′tion** Constriction of blood vessels.

**vas·o·di·la·ta'tion** Dilatation of blood vessels.

**vas·o·li·ga'tion** Surgical ligation of the ductus (vas) deferens.

**vas·o·vas·os'to·my** Surgically rejoining the cut ends of a sectioned vas deferens.

**Vat'i·can rou·lette'** *Slang.* The rhythm or safe-period method of birth control.

**VD** Venereal disease.

**Vel'de, The'o·door Hend'rik van de** (1873–1937) A Dutch gynecologist, author of *Ideal Marriage* and other books dealing with sex in marriage.

**vel'vets** *Slang.* Swish male clothing or drag.

**ve·ne're·al** 1. Of, pertaining to, associated with, or resulting from sexual intercourse or sexual contact. 2. Of or pertaining to sexual desire or intercourse. 3. Tending to excite sexual love or desire; aphrodisiac.

**ve·ne're·al dis·ease'** A sexually transmitted disease (STD), q.v.; one that is usually acquired through sexual intercourse or sexual contact. Common venereal diseases are gonorrhea, syphilis, AIDS, and genital herpes. Other venereal diseases include chancroid (q.v.), granuloma inguinale (q.v.), lymphogranuloma venereum (q.v.), cytomegalovirus (q.v.), and genital mycoplasmas (q.v.).

**ve·ne're·al wart** A genital wart, q.v.

**ven'er·y** Indulgence in sexual pleasures; sexual intercourse.

**ven'tral** Of, pertaining to, toward, or near the under surface of the body of a quadruped; in man and other upright animals, the anterior or front surface of the body. *Opposite* of dorsal.

**ver·ru'ca a·cum·i·na'ta** A moist venereal or genital wart. *See* genital wart.

**ver'sa·tile** *Slang.* Willing to engage in any form of sexual activity.

**ver·u·mon·ta'num** The colliculus seminalis, an elevation on the floor of the prostatic urethra forming a part of the urethral crest. It bears the openings of the prostatic utricle and the ejaculatory ducts, and lateral to it are the openings of the ducts of the prostate gland.

**ves'i·cle** A small bladder or sac that usually contains a fluid.

**ve·sic'u·lar** Pertaining to or consisting of a vesicle or vesicles.

**ve·sic'u·lar fol'li·cle** A graafian follicle; a follicle containing a cavity filled with a fluid. *See* follicle.

**ves'tal** 1. *adj.* Of or pertaining to the Roman goddess Vesta, goddess of the hearth. 2. *n.* A vestal virgin; a virgin woman; a nun.

**ves'tal vir'gin** One of six virgin priestesses who served in the temple of the goddess Vesta in ancient Rome.

**ves·tib'u·lar glands** Compound tubulo-alveolar glands found in the wall of the vestibule of the vagina.

**ves'ti·bule** A cavity or chamber that serves as an approach or entrance to another cavity, as the space in a female between the labia minora into which open the urogenital tracts (urethra and vagina).

**V-girl** *Slang.* A woman suspected of having a venereal disease.

**vi'a·ble** 1. Capable of living outside the uterus, with reference to a fetus. 2. Capable of being fertilized and developing, with reference to an ovum. 3. Capable of fertilizing an ovum, with reference to spermatozoa.

**vice** 1. Immoral, evil, or wicked conduct; depravity; corruption. 2. Sexual immorality, especially prostitution.

**vice squad** A police squad charged with enforcing laws dealing with various forms of vice, especially prostitution.

**Vic·to'ri·an Age** The age of Queen Victoria of England (approximately the latter half of the nineteenth century), characterized by extreme prudishness, modesty, and conventional behavior. It was an age of sexual repression. Sex and sexual energy were to be saved, not wasted. Sexuality was permissible in marriage only and the sex act was for procreation only. Females were not supposed to experience sexual pleasure in the sex act and lost self-esteem if they did. Masturbation was condemned. Sexual activities or matters pertaining to sex or reproduction were not considered permissible topics of conversation.

**vi'o·late** To harm or injure a person; to rape.

**vi·ra'go** A loud-mouthed, ill-tempered, and scolding woman; a shrew.

**vir'gin** 1. A person, especially a female, who has never had sexual intercourse with one of the opposite sex. 2. A pure, chaste, unmarried woman. 3. One who has not had a homosexual or lesbian experience.

**vir'gin·al** Of, pertaining to, or characteristic of virgins; chaste; pure.

**vir'gin birth** Parthenogenesis, q.v.

**vir·gin'i·ty** The state or condition of being a virgin.

**vir'ile** 1. Having masculine strength, vigor, and force. 2. Of or pertaining to male sexual activity; capable of procreation.

**vir'il·ism** The development of male traits in a female; masculinity. It is commonly the result of a tumor of the adrenal cortex that results in excessive production of androgens. It occurs to a mild degree following menopause, as a result of reduced production of female hormones. Also called *gynandry.*

**vi·ril'i·ty** Masculine strength and vigor; masculinity; potency.

**vir'tue** 1. Moral excellence and responsibility; goodness; righteousness. 2. Conformity to generally accepted standards of behavior; abstention from vices. 3. Chastity, especially in a girl or woman; faithfulness.

**vis'cer·al** Of or pertaining to the internal organs.

**vi'tal** Of or pertaining to life; essential for life.

**vi'tal sta·tis'tics** 1. Public records of significant events in the lives of individuals of a community, as births, marriages, deaths, and certain illnesses. 2. *Slang.* A woman's measurements.

**vix'en** An ill-tempered, quarrelsome, and malicious woman; a shrew.

**void** To discard the contents of; to urinate or defecate.

**vo·lup'tu·ary** A person whose life is devoted to pleasure and the gratification of sensual appetites; a sensualist.

**vo·lup'tu·ous** Directed toward, concerned with, or involved in sensuous activities or sensual pleasures; suggesting sensuality.

**vom'it·ing** The forceful ejection of the stomach contents through the mouth; emesis. Pernicious *vomiting* sometimes occurs during pregnancy. *See* hyperemesis gravidarum.

**vo·yeur'** One who engages in voyeurism; a peeping Tom.

**vo·yeur'ism** Deviant behavior in which a person obtains sexual pleasure and gratification from viewing the sexual acts of others or seeing people or pictures of people in the nude. *See* mixoscopia.

**vul'va** The structures surrounding the genital orifice in a female; the external genitalia or pudendum. These include the labia majora, labia minora, clitoris, vestibule, and mons pubis.

**vul·vec'to·my** Excision of the vulva or a portion of it.

**vul·vi'tis** Inflammation of the vulva.

**vul·vo·vag'i·nal glands** Bartholin's glands.

**vul·vo·vag·in·i'tis** Simultaneous inflammation of the vulva and the vagina.

**wack off** *Slang.* To masturbate.

**wang** *Slang.* The penis.

**wan'ton** Immoral, lewd, unchaste.

**wart** A verruca, a growth resulting from hyperplasia of the papillae of the skin, forming a small projection. Some warts are hard and cornified; others soft and moist. *See* genital wart.

**Was'ser·man test** A specific blood test for the diagnosis of syphilis.

**watch queen** *Slang.* 1. A homosexual who serves as a lookout, especially at a "tearoom." 2. A homosexual voyeur.

**wa'ter sports** *Slang.* Urolagnia.

**wa'zoo** *Slang.* A person's buttocks or rear.

**wean** To deprive of a mother's milk; to cause a suckling infant to cease to nurse at the breast.

**wed'lock** The state of being married; matrimony.

**well hung** *Slang.* Possessing a large penis.

**wench** 1. A young girl or woman, especially a peasant girl. 2. A promiscuous woman; a prostitute.

**wench'er** One who consorts with wenches.

**wet deck** *Slang.* A female who has just completed coitus with a male and is willing to accept another male.

**wet dream** A nocturnal emission, a dream accompanied by the ejaculation of semen. Sometimes referred to as *nocturnal pollution.*

**wet nurse** A woman who provides milk from the breast to an infant not her own.

**whack off** *Slang.* To masturbate.

**wham-bang** *Slang.* To have intercourse quickly and unerotically; a casual and quick coitus.

**whites** *Colloquial* term for leukorrhea.

**white slave** A woman who has been seduced and forced into prostitution against her will. *See* Mann Act.

**white slav'er** A person who acts as a procurer in the white slave trade.

**whore** A female prostitute, especially a streetwalker; a harlot.

**whor'ing** Consorting with and engaging in sexual activities with whores.

**wid'ow** A woman whose husband has died and who has not remarried.

**wid'ow·er** A man whose wife has died and who has not remarried.

**wie'nie** *Slang.* The penis, especially when flaccid.

**wife,** *pl.* **wives** 1. A woman married to a man; a female spouse. 2. *Slang.* The more submissive partner in a homosexual relationship.

**wise** *Slang.* Knowledgeable about homosexuality; tolerant of homosexuality.

**witch** 1. One who bewitches, especially a charming and alluring young woman. 2. An unattractive, disagreeable old woman; a hag. 3. *Euphemism* for *bitch.* 4. A woman who takes pleasure in arousing a man sexually but refuses to satisfy him.

**witch's milk** Milk secreted by mammary glands of newborn infants of both sexes for a few days following birth.

**with·draw'al** Coitus interruptus, a method of contraception in which the penis is withdrawn from the vagina before ejaculation.

**with young** Pregnant.

**wolf** *Slang.* 1. A sexually aggressive male who is constantly pursuing women. 2. A seducer or a sexually uninhibited potential seducer. 3. An aggressive male homosexual, especially one engaged in seductive activities such as pederasty.

**wolf'ess** *Slang.* A sexually aggressive female who pursues men; a seductress.

**Wolff'ian duct** The mesonephric duct.

**wo'man** *pl.* **wo'men** 1. An adult female. 2. Women considered collectively; womenkind. 3. *Informal.* A wife, mistress, or paramour.

**wo'man ha'ter** A person, usually a man, who has an extreme dislike for women; a misogynist.

**wo'man·ize** 1. To give feminine characteristics to; to make effeminate. 2. To pursue and court women illicitly.

**wo'man·iz·er** A woman chaser; a philanderer; a libertine.

**womb** The uterus.

**work** *Slang.* To engage in sexual intercourse.

**work'ing girl** *Slang.* A prostitute or call girl.

**world'ly** Of or pertaining to temporal or secular matters; not religious or ecclesiastical; pertaining to matters of the flesh, especially sensual pleasures; pertaining to matters of a practical nature, as opposed to moral or spiritual matters.

**X chro'mo·some** One of a pair of sex-determining chromosomes present in all eggs and in one half of the spermatozoa. In the zygote, when paired with another X chromosome (XX), it brings about the development of female sex organs and characteristics. In the male, it is paired with the Y chromosome (XY).

**X sperm** A spermatozoon that bears an X chromosome; a gynosperm, q.v.

**XX** Sex chromosomes in cells of a human female.

**XXY** The sex chromosomal make-up of an individual in Klinefelter's syndrome, q.v.

**XY** Sex chromosomes in cells of a human male.

**yang** In Chinese philosophy, one of two contrasting cosmic forces, *yang* and *yin.* Yang signifies heaven, yin the earth; yang the sun, yin the moon; yang the male, yin the female. In sexual intercourse, yang and yin come together, creating and maintaining the principle of universal life.

**Y chromosome** One of a pair of sex-determining chromosomes present in one half of spermatozoa produced. It is absent in ova and is practically devoid of genes. When present in a zygote, it is paired with an X chromosome and brings about the development of male sex organs and characteristics.

**yeast** Any of a number of various unicellular fungi that reproduce by budding and are capable of fermenting carbohydrates. Some are pathogenic organisms. *See* candidiasis.

**yeast in·fec'tion** *See* candidiasis.

**yin** *See* yang.

**yolk sac** A saclike structure attached to the digestive cavity of an early human embryo. It is a vestigial structure containing no yolk, but blood cells and primordial germ cells originate from its walls. It is discharged with the placenta as a part of the afterbirth.

**yo'ni** The external female genitalia or vulva considered as a symbol of Shakti in Indian and Tibetan religion. *Compare* lingam.

**young** 1. *adj.* In an early stage of development; not old; youthful, fresh, vigorous. 2. *n.* Offspring or brood.

**Yp'si·lon** A Y-shaped IUD especially designed to adjust to the uterine cavity.

**Y sperm** An androsperm, q.v.

**Ze'ro Pop·u·la'tion Growth** *Abbr.* ZPG. A worldwide organization that seeks to limit population to its present level.

**zo'na pel·lu'ci·da** A thin, noncellular, transparent membrane that surrounds the human egg at ovulation. It persists throughout early cleavage.

**zo·o·e·ras'ti·a, zo·o·e·ras'ty** Bestiality, q.v.

**zo·o·phil'i·a** 1. Extreme fondness for animals. 2. Sexual excitement resulting from real or fancied contact with animals.

**zoo queen** *Slang.* A homosexual who engages in bestiality.

**ZPG** Zero Population Growth.

**zy'gote** A fertilized egg or ovum.